SOLIDWORKS 2022
Intermediate Skills

Expanding on Solids, Surfaces, Multibodies,
Configurations, Drawings, Sheet Metal, and Assemblies

Paul Tran, CSWE, CSWI

SDC Publications
P.O. Box 1334
Mission, KS 66222
913-262-2664
www.SDCpublications.com
Publisher: Stephen Schroff

ISBN-13: 978-1-63057-470-3
ISBN-10: 1-63057-470-8

Printed and bound in the United States of America.

Acknowledgments

Thanks as always to my wife Vivian and my daughter Lani for always being there and providing support and honest feedback on all the chapters in the textbook.

Additionally, thanks to Jennifer Douglas for writing the forewords.

I also have to thank SDC Publications and the staff for their continuing encouragement and support for this edition of **SOLIDWORKS 2022 Intermediate Skills**. Thanks also to Tyler Bryant for putting together such a beautiful cover design.

Finally, I would like to thank you, our readers, for your continued support. It is with your consistent feedback that we were able to create the lessons and exercises in this book with more detailed and useful information.

Foreword

Paul Tran and I met professionally more than 25 years ago.

I have learned through the years that Paul has a unique ability to deeply understand technical concepts, along with a passion to help students learn - I marvel at his ability to combine these talents.

It takes an exceptional person to get students to learn technical skills quickly, especially when these skills are directly tied to career success. Time and time again, when students return to take additional classes, they specifically request Paul for their instructor. That kind of endorsement speaks for itself.

His commitment to help others improve their technical skills runs deep and seems without end as evidenced by this latest book *SOLIDWORKS 2022 Intermediate Skills*. The lessons and exercises in Paul's books are based on real world projects. The lectures and delivery methods are extremely easy to follow. Literally anyone can learn SOLIDWORKS from his books.

Besides over 13,000 engineering professionals that have been trained by Paul, there are around 600 schools (high schools, colleges, universities, and other technical institutions) that have adopted Paul's books to teach on the subject of SOLIDWORKS.

Paul's has 35 years of experience in the fields of mechanical and manufacturing engineering, more than 2/3 of those years were spent in teaching and supporting SOLIDWORKS, make him a trusted partner. But beyond all that, Paul has become our friend.

Many others and I at GoEngineer feel privileged to work with Paul. He has been a mentor for other trainers and aspiring authors, and we could not be more grateful that he is chosen to commit so much of his training time and talent to us - our customers do have the best in the business.

To all of those wanting to take your SOLIDWORKS skills to the next level, you are in good hands!

Jennifer Douglas
Former Vice President Marketing, GoEngineer

Images courtesy of C.A.K.E. Energy Corp. Designed in SOLIDWORKS by Paul Tran.

Author's Note

SOLIDWORKS 2022 Basic Tools, Intermediate Skills, and Advanced Techniques is comprised of lessons and exercises based on the author's extensive knowledge on this software. Paul has more than 32 years of experience in the fields of mechanical and manufacturing engineering; 24 years were in teaching and supporting the SOLIDWORKS software and its add-ins. As an active Senior SOLIDWORKS instructor and design engineer, Paul has worked and consulted with hundreds of reputable companies including IBM, Intel, NASA, US-Navy, Boeing, Disneyland, Medtronic, Terumo, Toyota, Kingston and many more. Today, he has trained more than 10,000 engineering professionals and given guidance to nearly half of the number of Certified SOLIDWORKS Professionals and Certified SOLIDWORKS Experts (CSWP & CSWE) in the state of California.

Every lesson and exercise in this book was created based on real world projects. Each of these projects have been broken down and developed into easy and comprehendible steps for the reader. Learn the fundamentals of SOLIDWORKS at your own pace, as you progress from simple to more complex design challenges. Furthermore, at the end of every chapter, there are self test questionnaires to ensure that the reader has gained sufficient knowledge from each section before moving on to more advanced lessons.

Paul believes that the most effective way to learn the "world's most sophisticated software" is to learn it inside and out, create everything from the beginning, and take it step by step. This is what the **SOLIDWORKS 2022 Basic Tools, Intermediate Skills, and Advanced Techniques** manuals are all about.

About the Training Files

The files for this textbook are available for download on the publisher's website at www.SDCpublications.com/downloads/978-1-63057-470-3. They are organized by the chapter numbers and the file names that are normally mentioned at the beginning of each chapter or exercise. In the Built Parts folder, you will also find copies of the parts, assemblies, and drawings that were created for cross references or reviewing purposes.

It would be best to make a copy of the content to your local hard drive and work from these documents; you can always go back to the original training files location at anytime in the future, if needed.

Who this book is for?

This book is for the mid-level user, who is already familiar with the SOLIDWORKS program. It is also a great resource for the more CAD literate individuals who want to expand their knowledge of the different features that SOLIDWORKS 2022 has to offer.

The organization of the book

The chapters in this book are organized in the logical order in which you would learn the SOLIDWORKS 2022 program. Each chapter will guide you through different tasks, from navigating through the user interface, to exploring the toolbars, from some simple 3D modeling to more complex tasks that are common to all SOLIDWORKS releases. There is also a self-test questionnaire at the end of each chapter to ensure that you have gained sufficient knowledge before moving on to the next chapter.

The conventions in this book

This book uses the following conventions to describe the actions you perform when using the keyboard and mouse to work in SOLIDWORKS 2022:

Click: means to press and release the left mouse button. A click of a mouse button is used to select a command or an item on the screen.

Double Click: means to quickly press and release the left mouse button twice. A double mouse click is used to open a program or to show the dimensions of a feature.

Right Click: means to press and release the right mouse button. A right mouse click is used to display a list of commands, a list of shortcuts that is related to the selected item.

Click and Drag: means to position the mouse cursor over an item on the screen then press and hold down the left mouse button; still holding down the left button, move the mouse to the new destination and release the mouse button. Drag and drop makes it easy to move things around within a SOLIDWORKS document.

Bolded words: indicate the action items that you need to perform.

Italic words: side notes are tips that give you additional information or explain special conditions that may occur during the course of the task.

Numbered Steps: indicate that you should follow these steps in order to successfully perform the task.

Icons: indicate the buttons or commands that you need to press.

SOLIDWORKS 2022

SOLIDWORKS 2022 is a program suite, or a collection of engineering programs that can help you design better products faster. SOLIDWORKS 2022 contains different combinations of programs; some of the programs used in this book may not be available in your suites.

Start and exit SOLIDWORKS

SOLIDWORKS allows you to start its program in several ways. You can either double click on its shortcut icon on the desktop or go to the Start menu and select the following: All Programs / SOLIDWORKS 2022 / SOLIDWORKS or drag a SOLIDWORKS document and drop it on the SOLIDWORKS shortcut icon.

Before exiting SOLIDWORKS, be sure to save any open documents, and then click File / Exit. You can also click the X button on the top right of your screen to exit the program.

Using the Toolbars

You can use toolbars to select commands in SOLIDWORKS rather than using the drop-down menus. Using the toolbars is normally faster. The toolbars come with commonly used commands in SOLIDWORKS, but they can be customized to help you work more efficiently.

To access the toolbars, either right click in an empty spot on the top right of your screen or select View / Toolbars.

To customize the toolbars, select Tools / Customize. When the dialog pops up, click on the Commands tab, select a Category, then drag an icon out of the dialog box and drop it on a toolbar that you want to customize. To remove an icon from a toolbar, drag an icon out of the toolbar and drop it into the dialog box.

Using the task pane

The task pane is normally kept on the right side of your screen. It displays various options like SOLIDWORKS resources, Design Library, File Explorer, Search, View Palette, Appearances and Scenes, Custom Properties, Built-in Libraries, Technical Alerts and News, etc.

The task pane provides quick access to any of the mentioned items by offering the drag and drop function to all of its contents. You can see a large preview of a SOLIDWORKS document before opening it. New documents can be saved in the task pane at anytime, and existing documents can also be edited and re-saved. The task pane can be resized, closed, or moved to different locations on your screen if needed.

TABLE OF CONTENTS

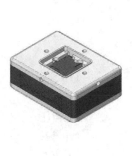

Chapter 17:   **SOLIDWORKS MBD**   **17-1**

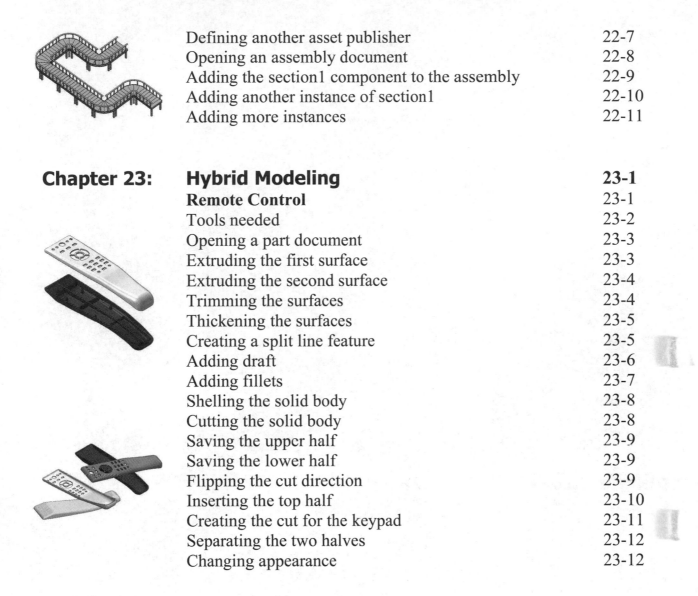
Glossary

Index

SOLIDWORKS 2022 Quick-Guides

INTRODUCTION

SOLIDWORKS User Interface

The SOLIDWORKS 2022 User Interface

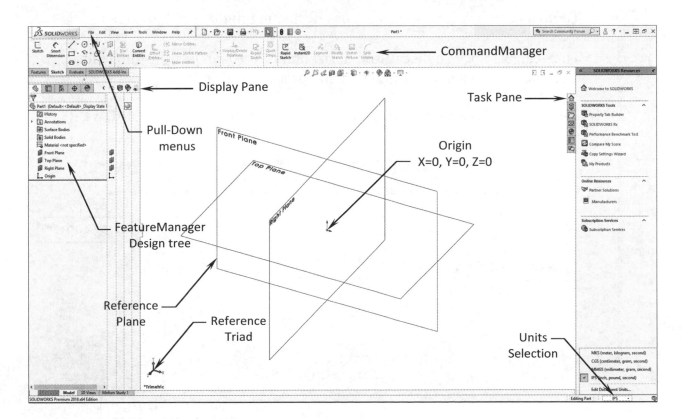

The 3 reference planes:

The Front, Top and the Right plane are 90° apart. They share the same center point called the Origin.

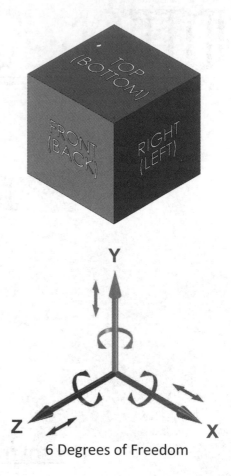

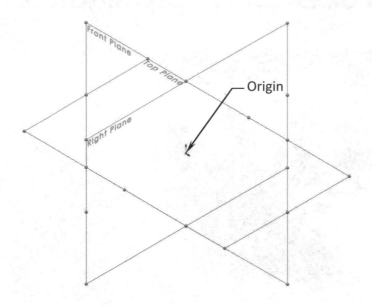

Y

6 Degrees of Freedom

The Toolbars:

Toolbars can be moved, docked, or left floating in the graphics area.

They can also be "shaped" from horizontal to vertical, or from single to multiple rows when dragging on their corners.

The CommandManager is recommended for the newer releases of SOLIDWORKS.

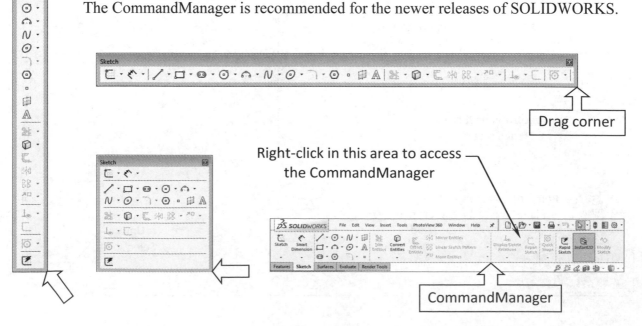

Drag corner

Right-click in this area to access the CommandManager

CommandManager

If the CommandManager is not used, toolbars can be docked or left floating.

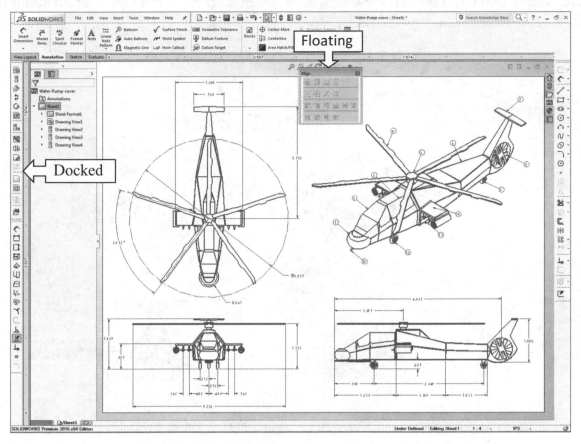

Toolbars can be toggled on or off by activating or de-activating their check boxes:

Select **Tools / Customize / Toolbars** tab.

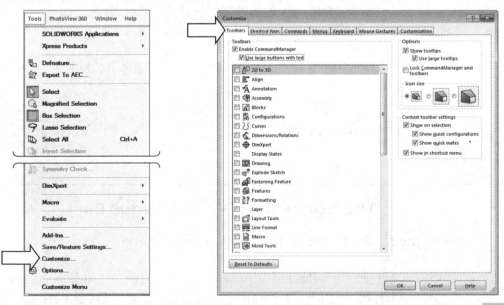

The icons in the toolbars can be enlarged when its check box is selected ☐ Large icons

The View ports: You can view or work with SOLIDWORKS model or an assembly using one, two or four view ports.

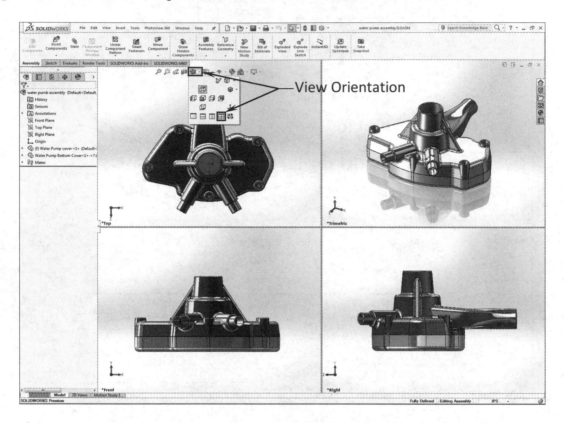

Some of the **System Feedback symbols** (Inference pointers):

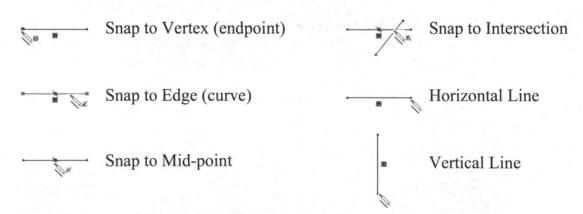

Snap to Vertex (endpoint)   Snap to Intersection

Snap to Edge (curve)   Horizontal Line

Snap to Mid-point   Vertical Line

The Status Bar: (View / Status Bar)

Displays the status of the sketch entity using different colors to indicate:

Green = Selected   **Blue** = Under defined
Black = Fully defined   **Red** = Over defined

2D Sketch examples:

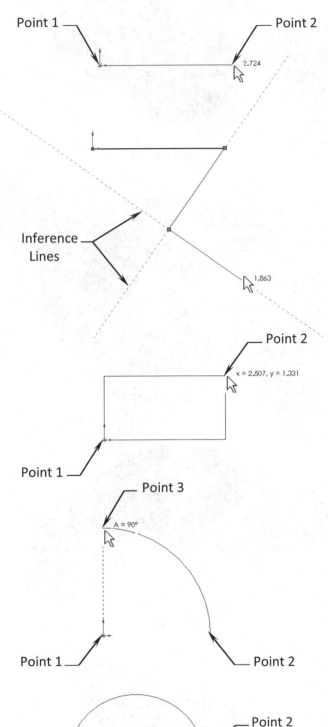

Click-Drag-Release: Single entity.

ck Point 1, hold the mouse button, drag oint 2 and release.)

ck-Release: Continuous multiple entities.

e Inference Lines appear when the sketch ties are Parallel, Perpendicular, or Tangent i each other.)

Click-Drag-Release: Single Rectangle

(Click point 1, hold the mouse button, drag to Point 2 and release.)

Click-Drag-Release: Single Centerpoint Arc

(Click point 1, hold the mouse button and drag to Point 2, release; then drag to Point 3 and release.)

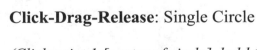

Click-Drag-Release: Single Circle

(Click point 1 [center of circle], hold the mouse button, drag to Point 2 [Radius] and release.)

3D Feature examples:

2D sketch   **Extrude** →   3D feature

2D sketch   **Revolve** →   3D feature

2D sketch   **Sweep** →   3D feature

2D sketch   **Loft** →   3D feature

Box-Select: Use the Select Pointer 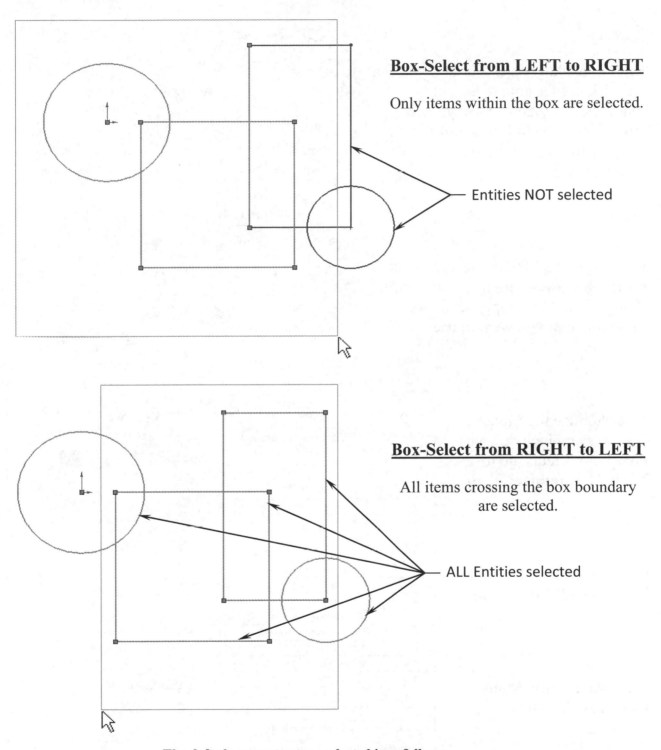 to drag a selection box around items.

Box-Select from LEFT to RIGHT

Only items within the box are selected.

Entities NOT selected

Box-Select from RIGHT to LEFT

All items crossing the box boundary are selected.

ALL Entities selected

The default geometry type selected is as follows:

* Part documents – edges * Assembly documents – components * Drawing documents - sketch entities, dims & annotations. * To select multiple entities, hold down **Ctrl** while selecting after the first selection.

The <u>Mouse Gestures</u> for Parts, Sketches, Assemblies and Drawings

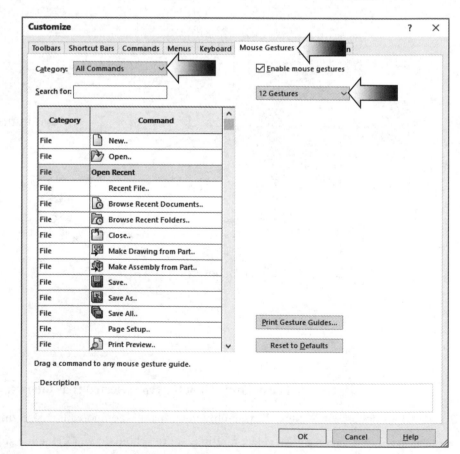

Similar to a keyboard shortcut, you can use a Mouse Gesture to execute a command. A total of 12 keyboard shortcuts can be independently mapped and stored in the Mouse Gesture Guides.

To activate the Mouse Gesture Guide, **right-click-and-drag** to see the current 12 gestures, then simply select the command that you want to use.

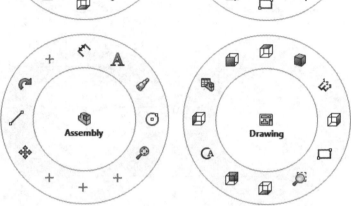

To customize the Mouse Gestures and include your favorite shortcuts, go to **Tools / Customize**.

From the **Mouse Gestures** tab select **All Commands**.

Click the **Enable Mouse Gestures** checkbox.

Select the **12 Gestures** option (arrow).

Customizing Mouse Gestures

To reassign a mouse gesture:

1. With a document open, click **Tools > Customize** and select the **Mouse Gestures** tab. The tab displays a list of tools and macros. If a mouse gesture is currently assigned to a tool, the icon for the gesture appears in the appropriate column for the command.

For example, by default, the right mouse gesture is assigned to the Right tool for parts and assemblies, so the icon for that gesture (🖱→) appears in the Part and Assembly columns for that tool.

To filter the list of tools and macros, use the options at the top of the tab. By default, four mouse gesture directions are visible in the Mouse Gestures tab and available in the mouse gesture guide. Select 8 gestures to view and reassign commands for eight gesture directions.

2. Find the row for the tool or macro you want to assign to a mouse gesture and click in the cell where that row intersects the appropriate column.

For cxample, to assign Make Drawing from Part 🖼 to the Part column, click in the cell where the Make-Drawing-from-Part row and the Part column intersect.

A list of either 4 or 12 gesture directions appears as shown, depending on whether you have the 4 gestures, or 12 gestures option selected.

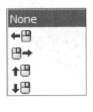
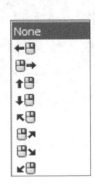

4 gestures   **12 gestures**

Some tools are not applicable to all columns, so the cell is unavailable, and you cannot assign a mouse gesture. For example, you cannot assign a mouse gesture for Make Drawing from Part in the Assembly or Drawing columns.

3. Select the mouse gesture direction you want to assign from the list. The mouse gesture direction is reassigned to that tool and its icon appears in the cell.

4. Click OK.

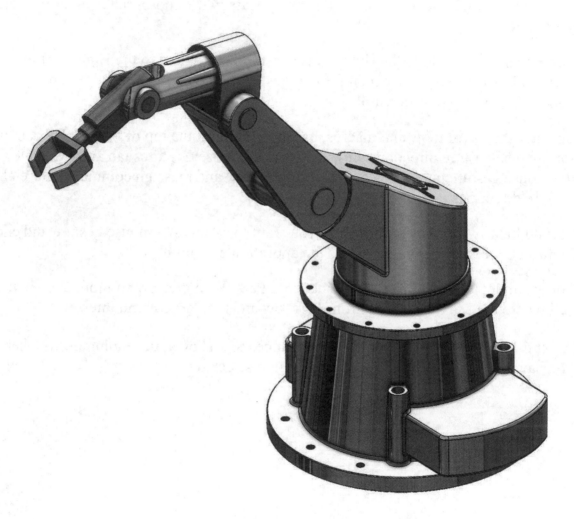

Designed with SOLIDWORKS 2022, SP0

CHAPTER 1

Document Properties

Document Properties - Overview
Setting up the Document Template

In SOLIDWORKS, there are two types of settings which allow the users to customize their own parameters and use them to create models, assemblies, and drawings.

The first type is called System Options. They are settings and parameters that are stored in the registry and are not part of the documents. Therefore, these changes affect all documents, current and future.

The second type is called Document Properties. They are parameters that you can preset and save in the templates; they affect only the **current document.**

SOLIDWORKS provides templates for parts, assemblies, and drawings. You can create custom templates by opening existing templates, setting options, and inserting items, then saving the documents as templates.
Templates files have the following extensions:

    * .prtdot (parts)   * .asmdot (assemblies)   * .drwdot (drawings)

The Document Templates are stored in one of the following directories:
C:\Program Files\SolidWorks Corp\SolidWorks\Data\Templates.
- or -
C:\Program Files\SolidWorks Corp\SolidWorks\Lang\English\Tutorial.

You can specify the document-level overall detailing drafting standards such as dimensions, annotations, view labels, tables available for all document types.

This chapter discusses the parameters that are needed to use throughout the textbook, and they are intended for use with this textbook only; you may need to modify them to ensure full compatibility with your applications.

Document Properties
Setting up the Document Template

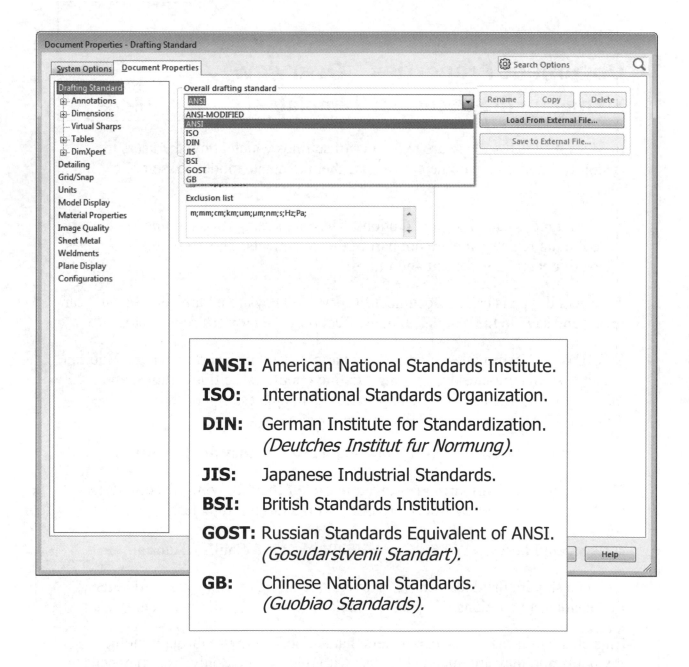

ANSI: American National Standards Institute.

ISO: International Standards Organization.

DIN: German Institute for Standardization. *(Deutches Institut fur Normung).*

JIS: Japanese Industrial Standards.

BSI: British Standards Institution.

GOST: Russian Standards Equivalent of ANSI. *(Gosudarstvenii Standart).*

GB: Chinese National Standards. *(Guobiao Standards).*

The settings shown in this chapter are intended for use with this textbook only. The Standards will be set to ANSI and the Units are Inch, Pound, and Second.

1. Starting a new part document:

Select **File / New**.

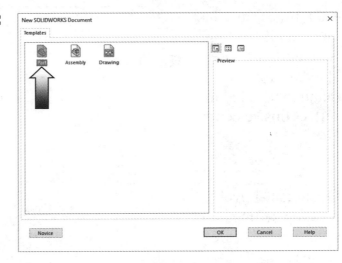

Select the **Part** template from the New SOLIDWORKS Document Dialog box.

The document properties apply only to the current document, and the Document Properties tab is available only when a document is open. New documents get their document settings (such as Units, Image Quality, and so on) from the document properties of the template used to create the document. Use the Document Properties tab when you set up document templates.

Select **Tools / Options**.

Click the **Document Properties** tab (arrow).

Click the **Drafting Standard** section and select the **ANSI** std.

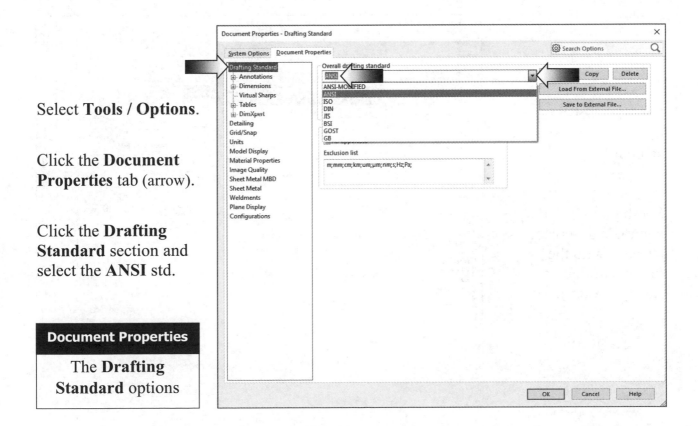

Document Properties
The **Drafting Standard** options

Each time you select an Overall Drafting Standard, the base standard for each detail is updated to reflect this selection.

Document-Level drafting settings for all dimensions are set here.

Document Properties

The **Dimensions** options

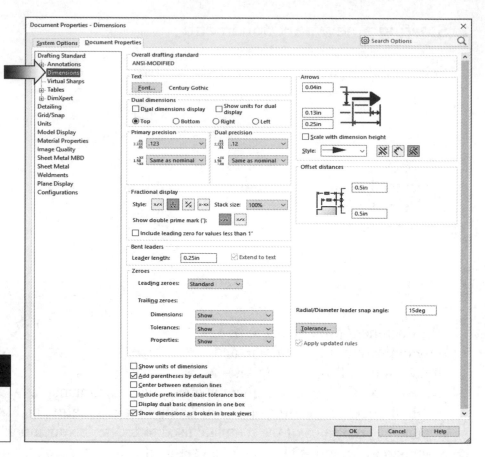

Set all dimension options to ANSI standards.

Set the number of decimals to 3 places (for use with the lessons in this textbook only).

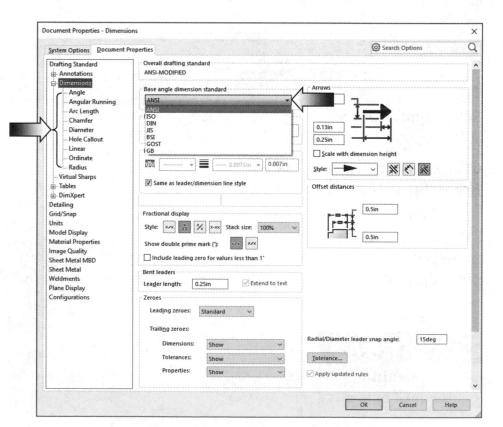

Select the Plus
symbol for
Virtual Sharp
(the intersection
between the two
entities).

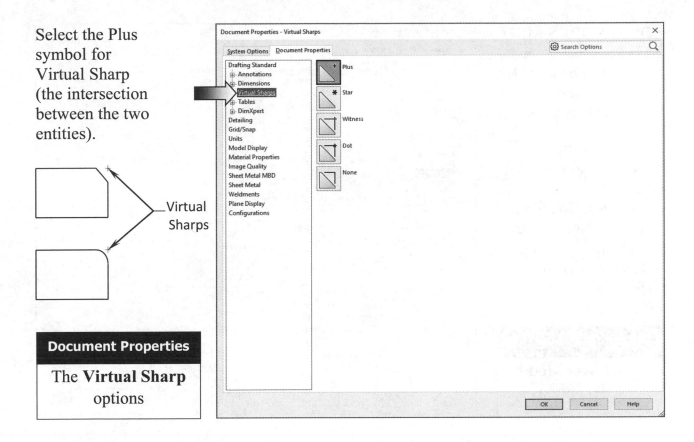

Virtual
Sharps

Document Properties

The **Virtual Sharp**
options

Set the Unit options to
IPS, and the number of
decimals as indicated.

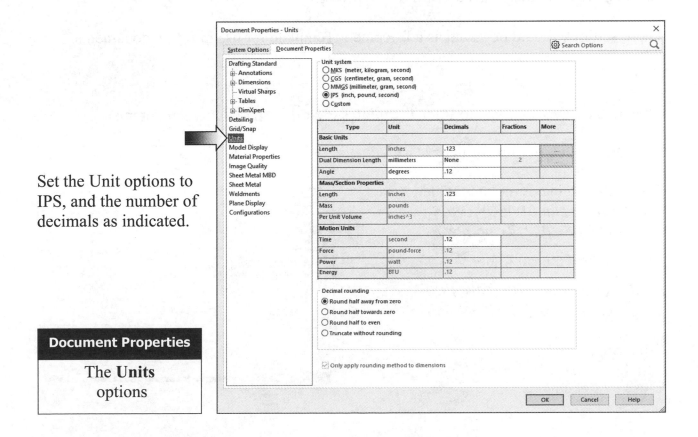

Document Properties

The **Units**
options

SOLIDWORKS MBD supports sheet metal bend notes, bend tables, bend lines, and bounding box lines.

You can specify the color and line type for:
Bend lines, Bounding box lines, Sheet metal sketch entities, Output to DXF and DWG files.

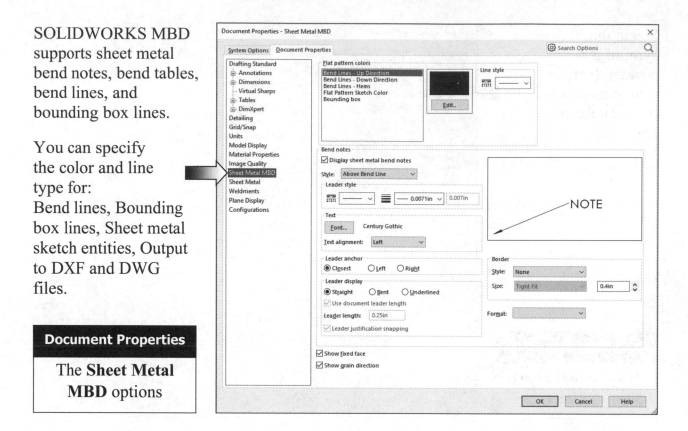

Document Properties

The **Sheet Metal MBD** options

Saving the settings as a Part Template:

These settings should be saved as a Document Template for use in future documents.

Go to **File / Save As.**

Change the Save-As-Type to **Part Templates** (*.prtdot - part document template).

Enter **Part-Inch.prtdot** for the file name.

Save either in the Templates folder
(C:\Program Data\SOLDWORKS\ SOLIDWORKS 2022\ Templates).

– or – in the Tutorial folder

(C:\Program Files\SOLIDWORKS Corp\Version\Lang\English\Tutorial).

Click **Save**.

Close all documents.

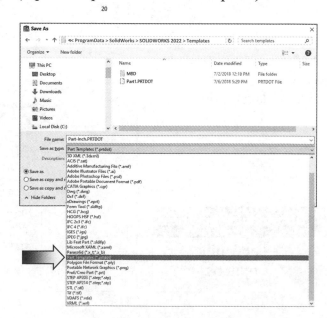

Customizing Keyboard Shortcuts

You can customize keyboard shortcuts to a command from the Keyboard tab of the Customize dialog box.

There are many shortcuts available in SOLIDWORKS but you can assign your own shortcut keys to any of your frequently used commands.

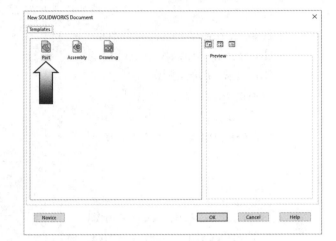

1. Starting a new part document:

Click **File / New / Part**.

Select **Customize** from the **Options** drop down menu (arrow).

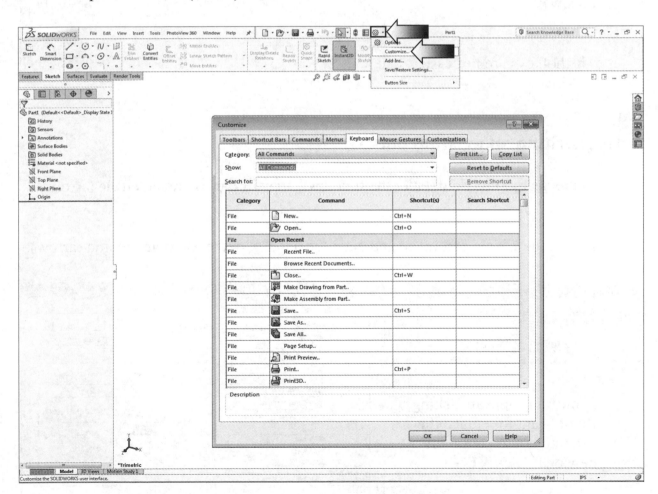

2. Creating a new hotkey:

Select the **Keyboard** tab (arrow).

It is quicker to search for the command than to scroll through a long list of commands. Click in the field Search For, and enter the command name: **Line** (arrow).

Locate the **Line** command and enter the letter **L** under the Shortcut column (arrow).

The hotkeys are not case sensitive.

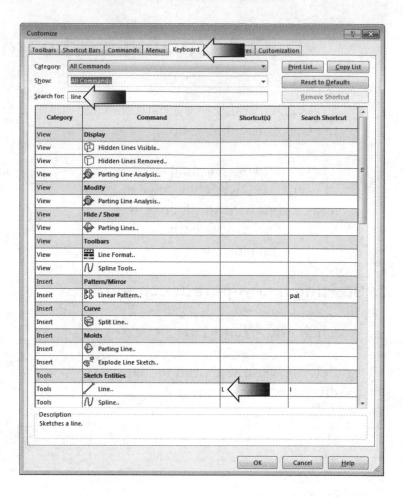

3. Overwriting an existing hotkey:

In the Search For field, enter the first few letters of the command **circle** (arrow).

Locate the **Circle** command and enter the letter **C** in the Shortcut column (arrow).

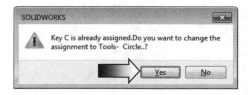

A message appears asking if you want to replace the existing hotkey with your new one; click **YES**.

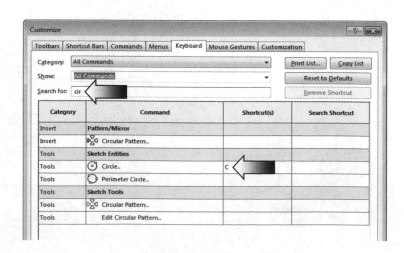

4. Adding other hotkeys:

For practice purposes, we will assign a couple hotkeys to the Rectangle and the Dimension commands.

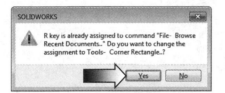

Assign the letter **R** to the **Corner Rectangle** command. If the shortcut shows more than one key (Alt+Shift+R), delete it and press the letter **R** again to reassign it.

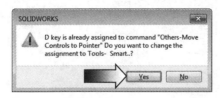

Click **YES** to overwrite the existing hotkey with with your new key.

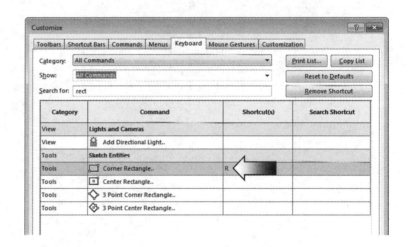

Type the word **smart** (for Smart Dimension) in the Search For field.

Enter the letter **D** as the new hotkey for the Smart Dimension command (arrow).

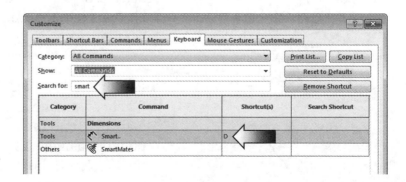

Click **YES** to confirm the assignment of the new key.

5. Testing the new hotkeys:

The four hotkeys are sketch commands: they must be tested in the sketch mode. Select the **Front** plane from the FeatureManager tree and click the **Sketch** icon to open a new sketch (arrow).

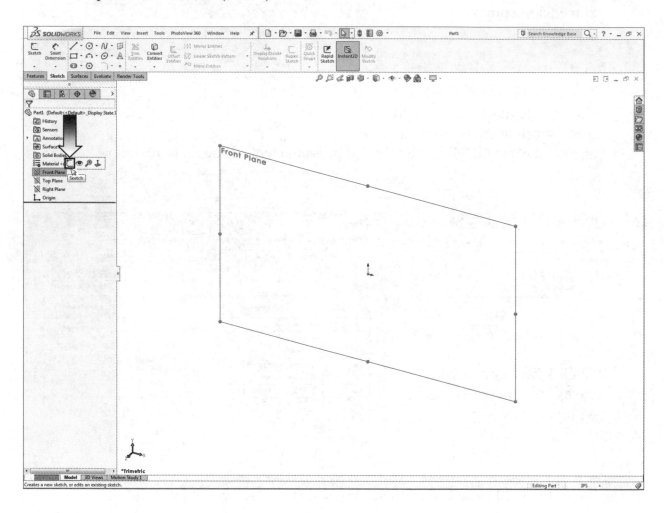

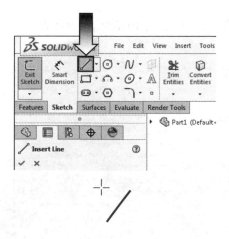

Press the **L** key, the Line command is activated, and the cursor changes into the Line symbol.

Press the **C** key, the Circle command is activated, and the pointer changes into the Circle symbol.

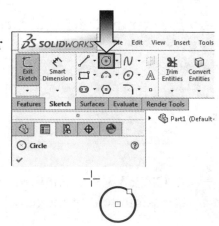

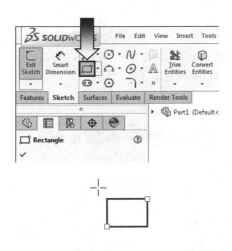

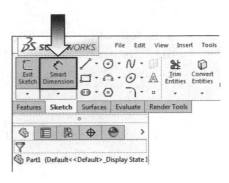

Press the **R** key, the Rectangle command is activated and the cursor changes into the Rectangle symbol.

Press the **D** key, the Dimension command is activated and the cursor changes into the Dimensions symbol.

6. Resetting the hotkeys back to default (optional):

The new hotkeys can be reset back to their default commands or simply erased from the Shortcut column.

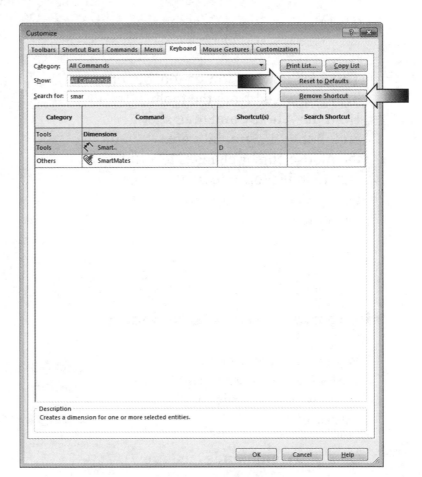

The list of shortcut keys can be printed from the Keyboard tab of the Customize dialog box.

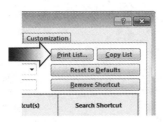

Click **OK** to exit the dialog box.

Customizing Tool Buttons

You can add tool buttons to one or more active toolbars from the Keyboard tab of the Customize dialog box, including the CommandManager, and the Menu Bar toolbar. This includes buttons with flyout controls such as View Orientations.

You can use drag and drop to remove, rearrange, or move tool buttons from one toolbar to another.

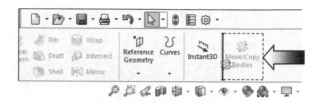

The first two images on the right show an example of a tool button being dragged from the right side and dropped onto the left side of the same toolbar.

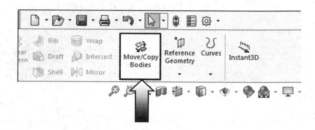

The third image shows a tool button being removed from the toolbars: Hold the **Alt** key and drag a tool button into the graphics area. When the mouse cursor changes to a Red Delete indicator, release the mouse button to drop and remove it from the toolbar.

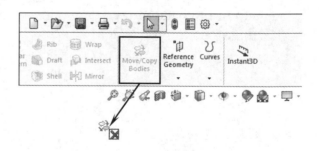

1. Adding tool buttons to toolbars:

With a document open, click **Tools / Customize** or right-click in the window border and select Customize.

Click the **Commands** tab (arrow).

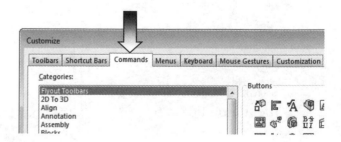

Click the **Sketch tab** on the upper left side of the screen.

Select **Sketch** under the Categories section (arrow).

Locate the **Modify Sketch** command from the Buttons list (arrow).

Drag the **Modify Sketch** command into the Sketch toolbar as shown below. The new command is added to the toolbar.

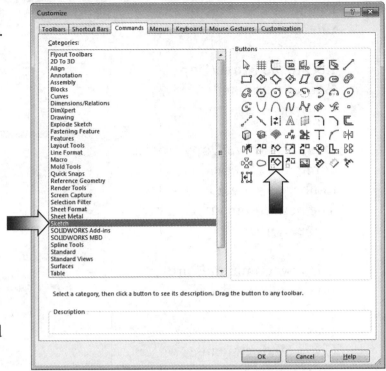

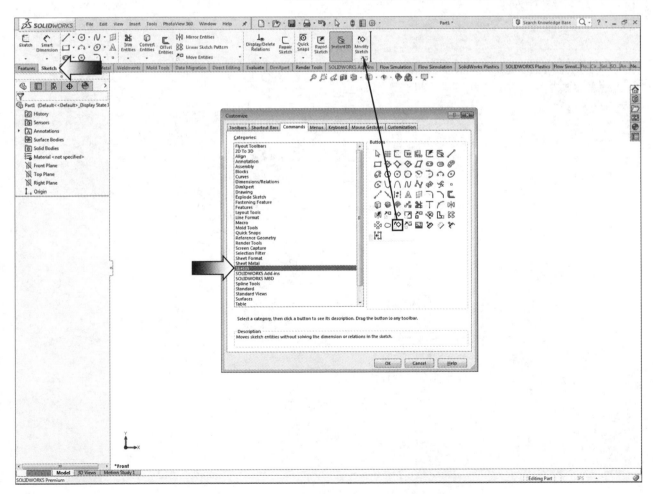

Change to the **Features tab** (arrow).

Locate the **Move/Copy Bodies** command (arrow).

Drag the Move/Copy Bodies command into the Features toolbar as shown below. The new command is added to the toolbar.

The new commands are saved automatically.

Press **OK**.
Close all documents.

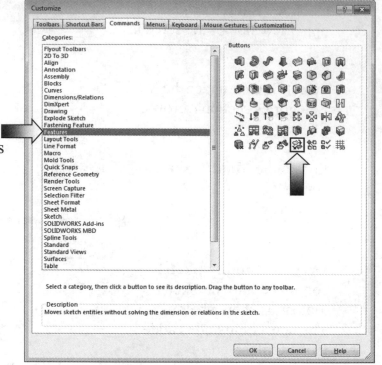

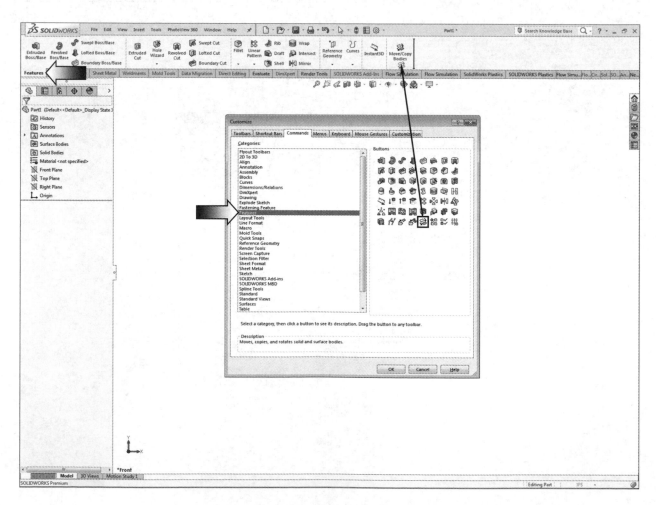

CHAPTER 2

Sketching Skills

Sketching Skills
Handle

Most features in SOLIDWORKS start with a sketch. The sketch is the basis for a 3D model. You can create a sketch on any of the default planes (Front Plane, Top Plane, and Right Plane), or a created plane.

There are two modes for sketching in 2D: click-drag or click-click. The click-drag method will create a single entity each time, but the click-click creates multiple, connecting lines instead.

While sketching the lines you can change from sketching a line into sketching a tangent arc, and vice versa, without selecting the arc tool. Simply press the A key on the keyboard to switch from a line to a tangent arc – OR - start the line with the 1st click, move the pointer outward and back to the starting point, then away again.

Inferencing lines are dotted lines that appear as you sketch, displaying relations between the pointer and existing sketch entities or model geometry. When your pointer approaches highlighted cues such as midpoints, the inferencing lines guide you relative to existing sketch entities.

There are two types of Snaps in the sketch mode:
Sketch Snap and Quick Snap.
Each Sketch Snap allows you to automatically snap to selected entities as you sketch. By default, all Sketch Snaps except Grid are enabled.

Quick Snaps are instantaneous, single operation Sketch Snaps. Sketching any sketch entity (such as a line) from start to finish is a single operation.

Sketching Skills
Handle

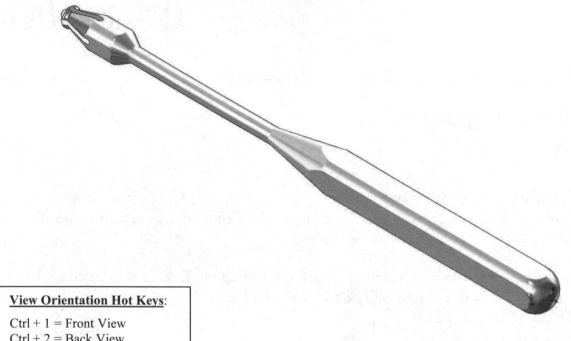

Dimensioning Standards: **ANSI**

Units: **INCHES** – 3 Decimals

Tools Needed:

 Insert Sketch   Line   Circle

 Straight Slot   Centerline   Smart Dimension

 Extruded Boss   Extruded Cut   Chamfer

1. Starting a new part document:

Select **File / New**.

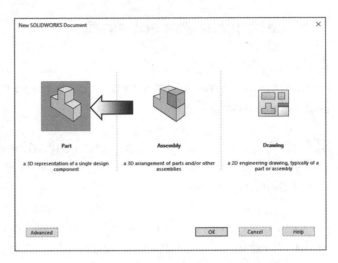

If the Novice dialog box is the default, it is showing three template options: **Part, Assembly**, and **Drawing**.

Select the **Part** template and click **OK**.

2. Changing the System Options:

Click the small arrow 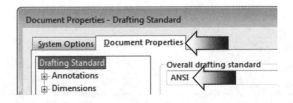 at the bottom right corner of the screen and select **IPS (Inch, Pound, Second)**.

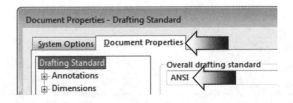

Select **Tools, Options**.

Click the **Sketch** option.

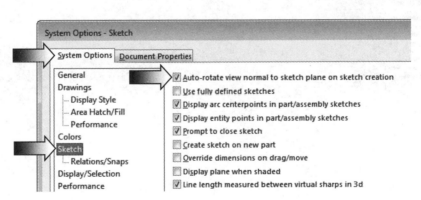

Enable the checkbox: **Auto-Rotate View Normal to Sketch plane on Sketch Creation**.

Switch to the **Document Properties** tab.

Click the **Drafting Standard** option and select the **ANSI** standard from the drop down list.

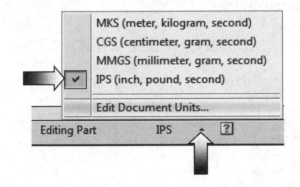

Click **OK**.

3. Creating the Parent sketch:

Sketching in SOLIDWORKS is the basis for creating features. Features are the basis for creating parts, which can be put together into assemblies. Sketch entities can also be added to drawings.

SOLIDWORKS sketch entities can snap to points (endpoint, midpoints, intersections, and so on) of other sketch entities. With Quick Snaps, you can filter the types of sketch snaps that are available.

Inferencing displays relations by means of dotted inferencing lines, pointer display, and highlighted cues such as endpoints and midpoints.

The sketch status appears in the window status bar. Colors indicate the state of individual sketch entities.

The first sketch in a part document is considered the parent sketch.

Select the Front plane and open a **new sketch** (arrow).

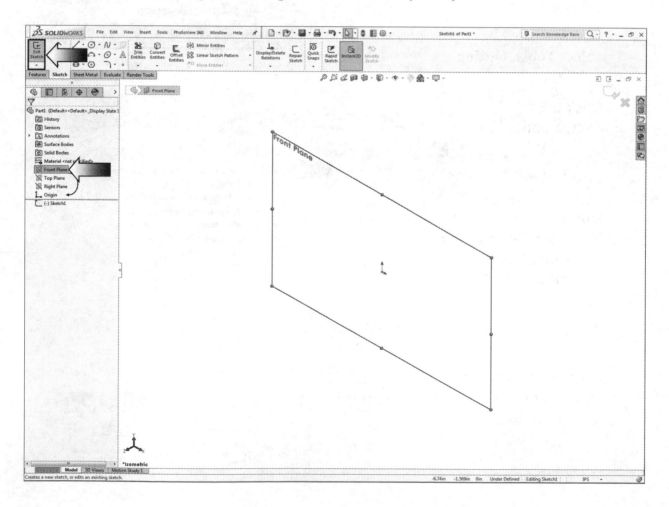

Select the **Line** command 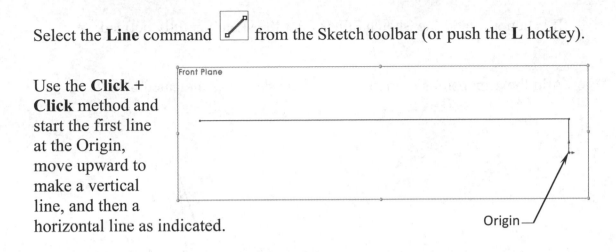 from the Sketch toolbar (or push the **L** hotkey).

Use the **Click + Click** method and start the first line at the Origin, move upward to make a vertical line, and then a horizontal line as indicated.

Continue with sketching the rest of the lines as shown below.

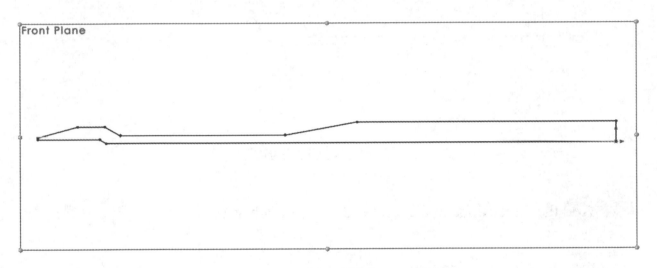

Select the **Smart Dimension** command ; add the dimensions shown. More...>

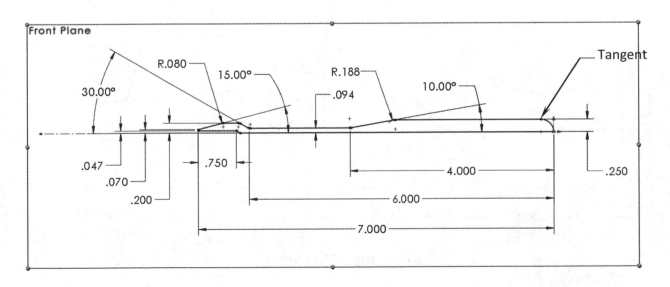

It is better to add the Sketch Fillets <u>after</u> the sketch is fully defined.

Add the dimensions to the left end of the sketch as indicated below.

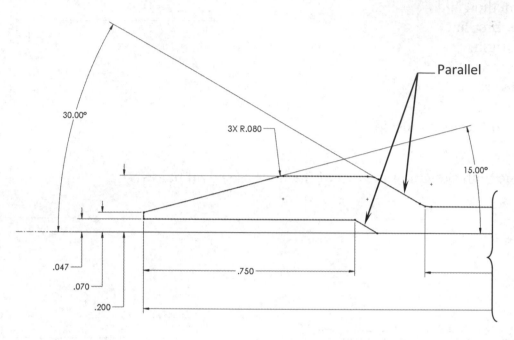

<u>Left end of the sketch</u>

Add the dimensions to the right end of the sketch as shown.

The dimension R.250 is a reference dimension and shown in gray color.

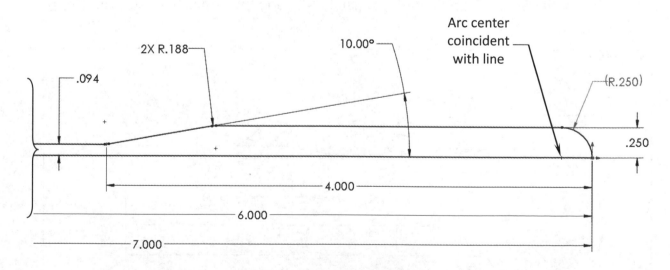

<u>Right end of the sketch</u>

4. Revolving the parent sketch:

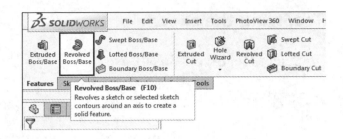

Switch to the **Features** tab.

Select the revolved **Boss/Base** command.

Use the default **Blind** type and **360°** angle.

Click **OK**.

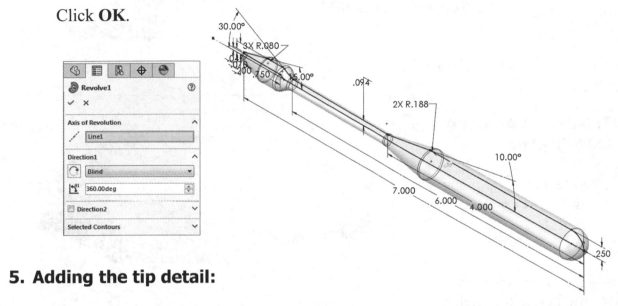

5. Adding the tip detail:

Select the <u>Top</u> plane and open a **new sketch**.

Sketch the profile shown on the right.

Sketch a horizontal centerline to use as the revolve line.

Add the relations and the dimensions as indicated.

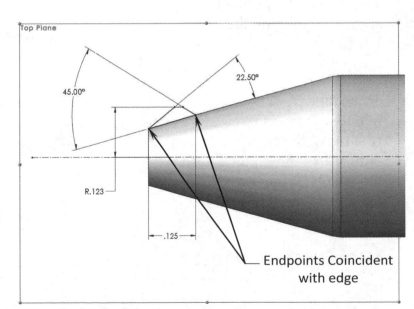

Be sure to fully define the sketch before revolving it.

6. Revolving the sketch:

Select **Revolved Boss/Base**.

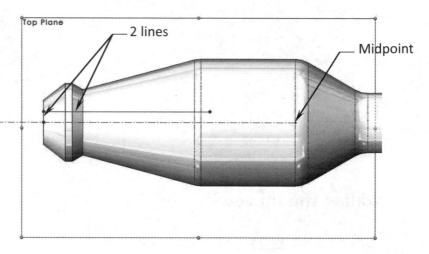

Use the default **Blind** type.

Revolved Angle **360°**.

Click **OK**.

7. Transitioning from Line-to-Arc:

Select the Top plane and open a **new sketch**.

Use the **click-click** method and sketch a short vertical line and a horizontal line starting at the midpoint on the left side.

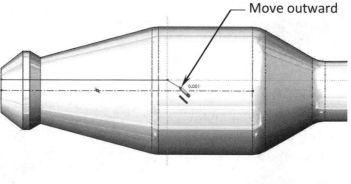

To change from a line to a Tangent Arc first move the cursor outward as shown then push the **A** key on the keyboard; the tangent arc appears.

Make an arc as shown.

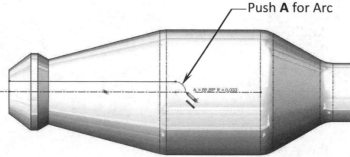

Push the **A** key again to switch back to the line.

8. Mirroring in sketch mode:

Push **Esc** to exit the line command.

<u>Box-Select</u> all sketch entities and click the **Mirror Entities** command.

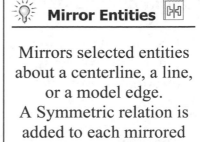

Mirror Entities

Mirrors selected entities about a centerline, a line, or a model edge.
A Symmetric relation is added to each mirrored entity.

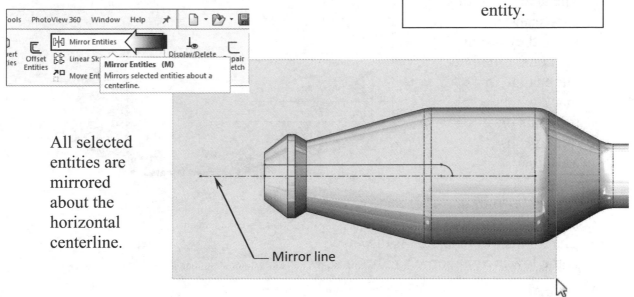

All selected entities are mirrored about the horizontal centerline.

Mirror line

A **Symmetric** relation is added to each mirrored entity.

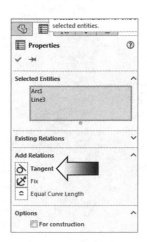

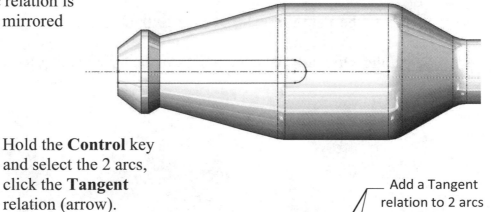

Hold the **Control** key and select the 2 arcs, click the **Tangent** relation (arrow).

Add a Tangent relation to 2 arcs

Click **OK**.

9. Adding dimensions:

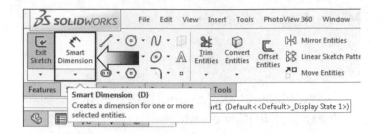

Use the Smart Dimension tool to specify the size and location for each sketch entity.

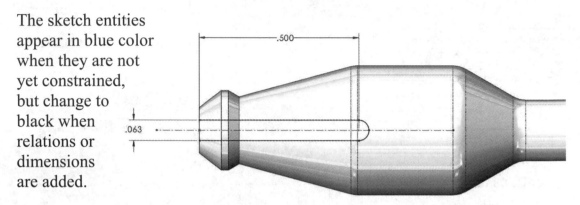

The sketch entities appear in blue color when they are not yet constrained, but change to black when relations or dimensions are added.

Add the two dimensions shown.

The Status of the sketch is displayed at the lower right corner of the screen.

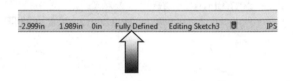

10. Extruding a cut:

Switch to the **Features** tab and click **Extruded Cut**.

Select the **Through All** condition from the list and click **Reverse** direction.

Click **OK**.

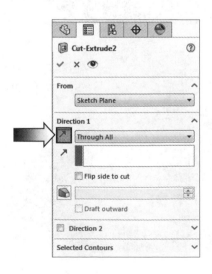

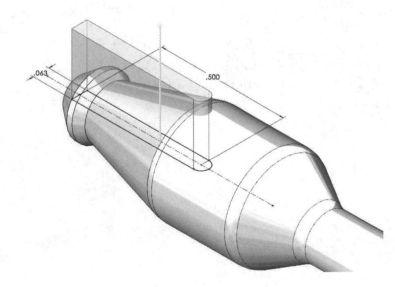

11. Creating a Circular Pattern:

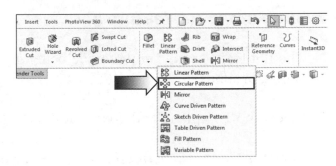

The Circular pattern command creates multiple instances of one or more features and spaces them uniformly around an axis.

Select the **Circular Pattern** command below the Linear Pattern drop-down list.

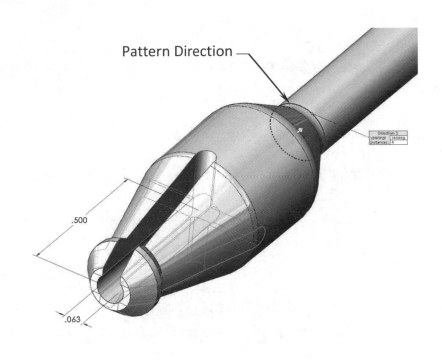

Pattern Direction

.500

.063

For Pattern Direction, select one of the **circular edges** in the model.

Enable the **Equal Spacing** checkbox (arrow).

Enter **4** for Number of Instances.

Select the **Cut-Extrude1** either from the Feature tree or from the graphics area. The preview graphics of the four instances appears.

Click **OK**.

12. Adding other cut features:

The flat features not only provide better grips but also help keep the device from rolling around.

Select the <u>Top</u> plane and open a **new sketch**.

Sketch **2 Corner Rectangles** at both ends of the model.

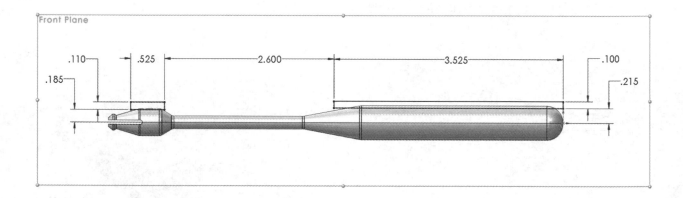

Add the dimensions shown to fully define the sketch.

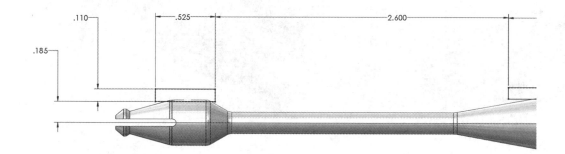

<u>Left end of the sketch</u>

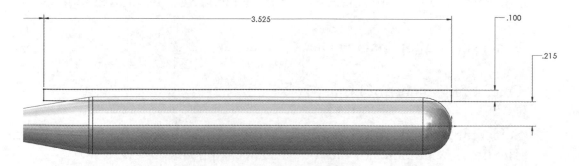

<u>Right end of the sketch</u>

Switch to the **Features** tab and click the **Extruded Cut** command.

Select the **Through All – Both** from the list to cut through both directions.

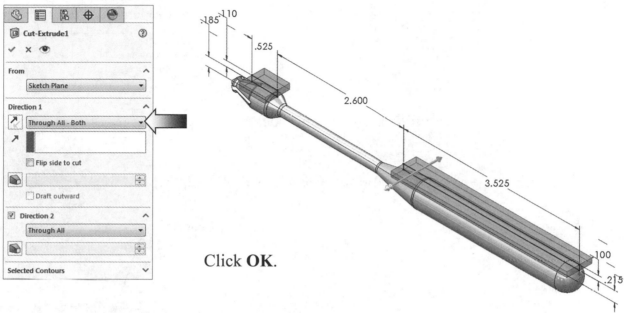

Click **OK**.

13. Creating another Circular Pattern:

Select the **Circular Pattern** command below the Linear Pattern option.

For Pattern Direction, select one of the **circular edges** of the model.

Enable the **Equal Spacing** checkbox.

Enter **4** for Number of Instances.

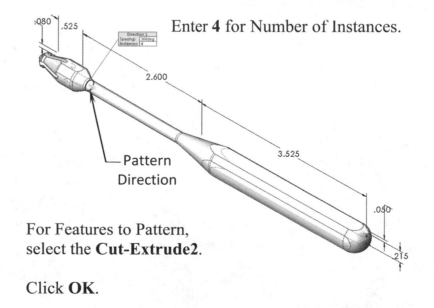

— Pattern
Direction

For Features to Pattern,
select the **Cut-Extrude2**.

Click **OK**.

14. Adding a .032" Constant Size Fillet:

Select the **Fillet** command 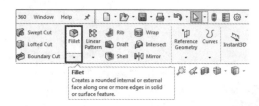 from the Features tool tab.

The **Constant Size Radius** is the default type.

Select one edge of each flat feature, select a total of **8 edges**.

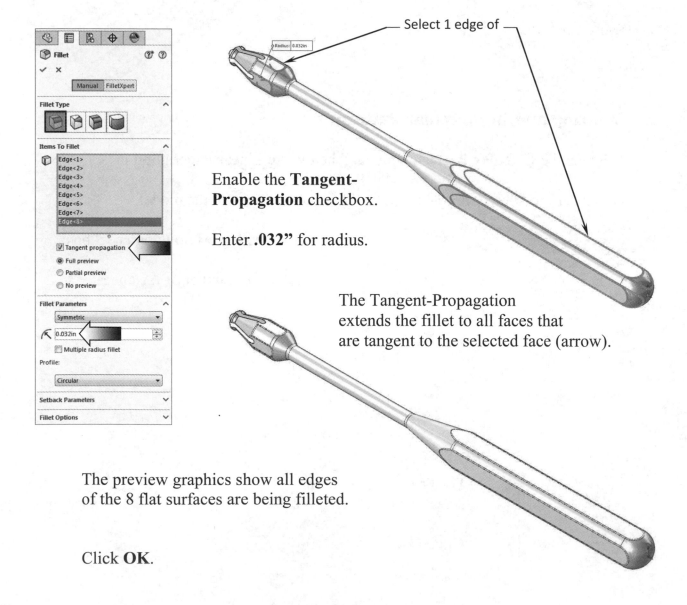

Enable the **Tangent-Propagation** checkbox.

Enter **.032"** for radius.

Select 1 edge of

The Tangent-Propagation extends the fillet to all faces that are tangent to the selected face (arrow).

The preview graphics show all edges of the 8 flat surfaces are being filleted.

Click **OK**.

15. Assigning material to the model:

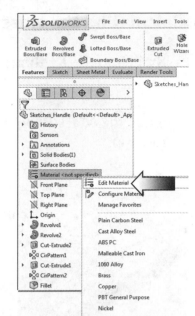

SOLIDWORKS has two sets of properties, visual and physical (mechanical). The response of a part when loads are applied to it depends on the material assigned.

Right click **Material** on the FeatureManager tree and select **Edit Material**.

In the SOLIDWORKS Materials folder, expand the **Steel** sub folder.

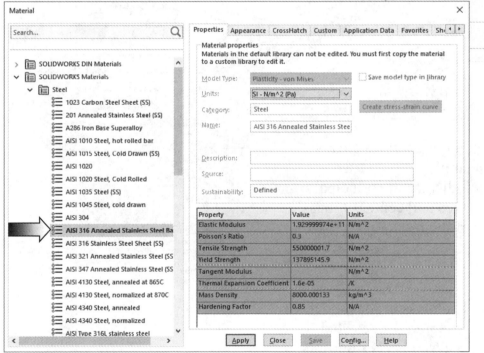

Select the material **AISI Type 316L Stainless Steel Bar** from the list (arrow).

Click **Apply** and **Close**.

Enable the **RealView Graphics** option (under View Settings) to enhance the model display if applicable.

16. Calculating the Mass of the model:

Switch to the **Evaluate** tab.

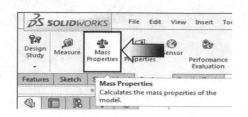

Click the **Mass Properties** command (arrow).

— Center of mass

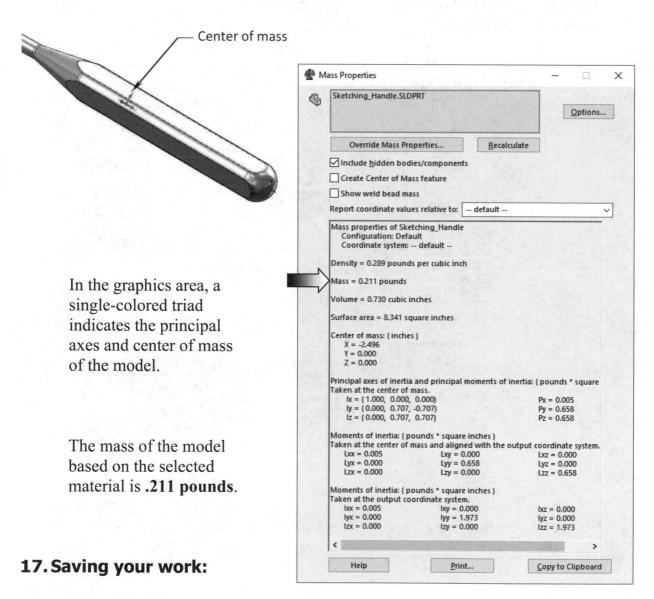

In the graphics area, a single-colored triad indicates the principal axes and center of mass of the model.

The mass of the model based on the selected material is **.211 pounds**.

17. Saving your work:

Select File / Save As.

Enter **Sketching_Handle.sldprt** for the name of the part.

Click **Save**.

Working with Sketch Pictures

Working with Sketch Pictures

Pictures and images that were saved as one of the file formats supported by the Windows operating system can be inserted into SOLIDWORKS, and used as an underlay for creating 2D sketches (raster data to vector data).

The supported formats are .jpg, .tif, .bmp, .gif, .png, .wmf, and .psd. The source image should be hi-resolution, with a minimum of 300dpi. The line art should be pen on paper (not pencil), and with precise contours and high contrast. The current supported resolution is limited to 4096 x 4096.

There are a few things to keep in mind when working with sketch pictures:

> The picture will be embedded in the document, but if the original image is changed, the sketch picture <u>will not</u> update.

> If you sketch over the picture, there is <u>no snap</u> to picture, inferencing, or auto tracing capability. If the image is moved, or deleted and replaced, the sketch <u>will not</u> update. And if the sketch is hidden, the picture will be hidden as well.

If the sketch or the picture becomes inactive, simply double-click the picture to reactivate it. The values in the PropertyManager allow the picture to be moved, rotated, or scaled either proportionally or un-proportionally.

There is an Auto-Trace option in SOLIDWORKS Add-Ins, but it only works well if the image's outline is sharp and the background is clear or white. This lesson discusses a different approach, where a jpeg file gets converted to a SOLIDWORKS 2D sketch, by tracing its outline with the sketch tools, and then revolves it into a 3D model.

Working with Sketch Pictures

| Dimensioning Standards: **ANSI** |
| Units: **INCHES** – 3 Decimals |

Tools Needed:

Insert Sketch   Sketch Picture   Centerline   Line

Tangent Arc   Sketch Fillet   Dimension   Revolve Boss/Base

1. Starting with a layout sketch:

Most of the time, a sketch picture will get inserted into SOLIDWORKS with a wrong size or scale. It would be quite helpful to have an overall full-size construction box laid out ahead of time, to use as a guide to scale the picture.

<u>*NOTE:*</u> *There will be no links or snaps between the layout sketch and the picture. If the picture is moved or changed, the sketch <u>will not</u> update.*

Click **File / New / Part**. Set the Units to **IPS**, 3 decimals. Select the <u>Front</u> plane and open a **new sketch**.

Sketch **eight centerlines** and add the dimensions as shown below to fully define the sketch. The centerlines will be used to help scale the image to size.

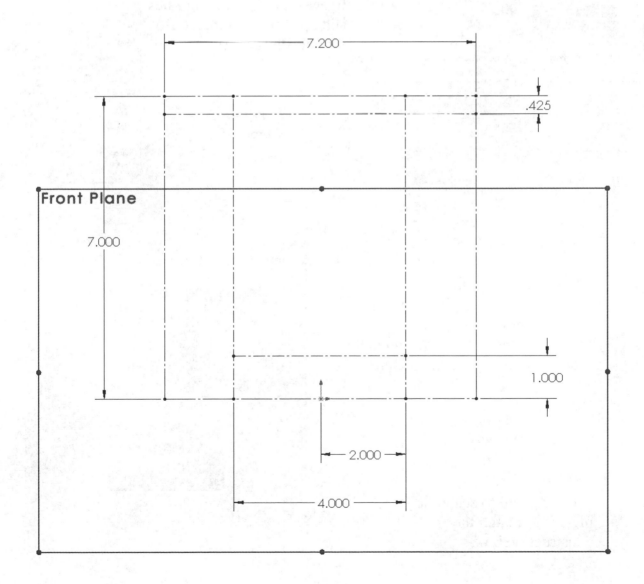

2. Inserting the picture:

Click **Tools / Sketch tools / Sketch Picture**.

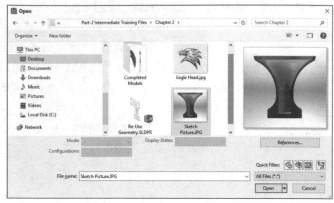

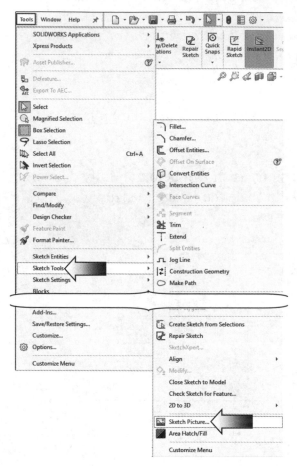

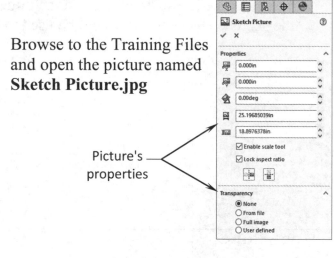

Browse to the Training Files and open the picture named **Sketch Picture.jpg**

Picture's properties

The lower left corner of the picture is placed on the origin by default, and the picture's properties appear displaying the options for moving, rotating, and scaling the picture.

The lower left corner is placed on origin

We will need to scale the picture to match the construction box.

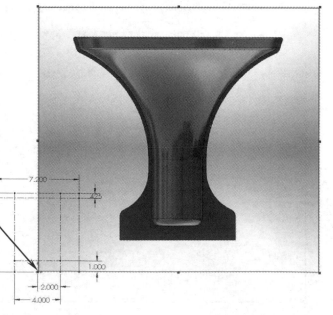

3. Scaling the picture:

Using the Properties tree enter the following:

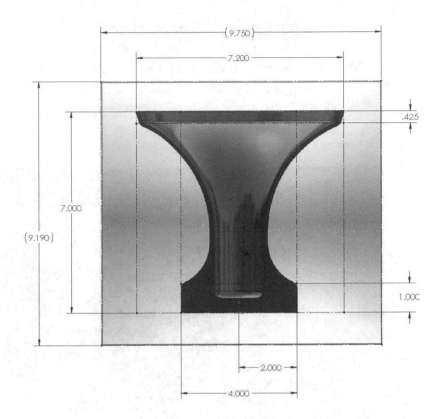

Location:
X = -4.835in.   **Y = -1.135in**.

Size:
Horizontal = 9.750in.   **Vertical = 9.190in**.

Keep the option **Lock Aspect Ratio** checked and clear the **Enable Scale Tool** (arrows).

Click **OK**.

4. Tracing the picture:

The geometry in this picture can be traced with the line and arc commands.

Sketch a line approx. as shown and connect it with a tangent arc.

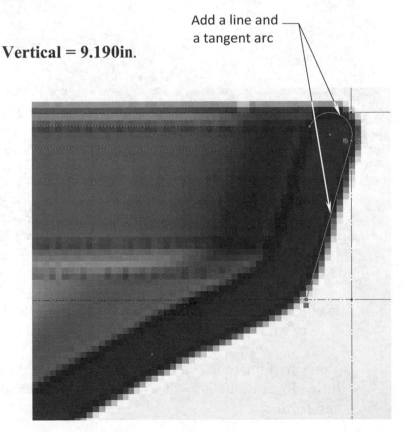

Add a line and a tangent arc

Continue tracing the outline of the picture using the sketch tools as noted.

Keep the corners sharp; they will be filleted at the end. Sketch only one half of the picture; the horizontal line at the bottom should stop right at the origin. The sketch will get revolved once completed.

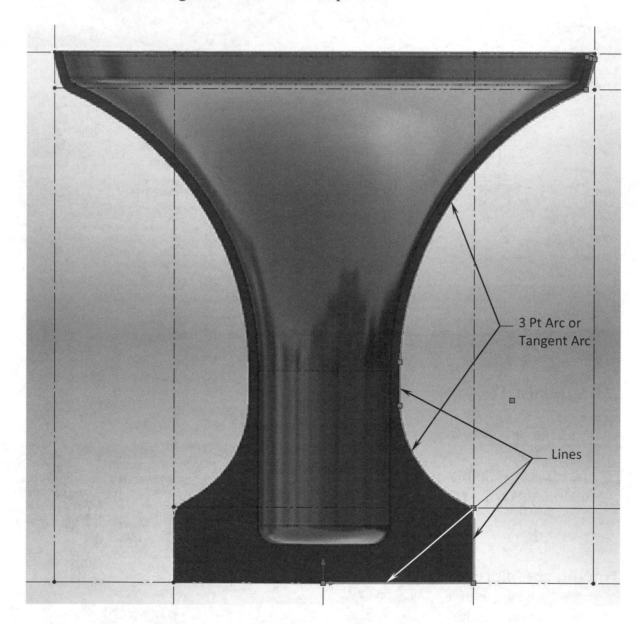

If you accidently got out of the sketch mode and the picture becomes in-active, simply double click the picture to reactivate it.

You can adjust the sketch afterwards. When the sketch is completed, zoom in a little closer to the area that needs adjustment and drag the sketch entities back and forth.

5. Creating the offset entities:

Click the **Offset Entities** command from the Sketch toolbar.

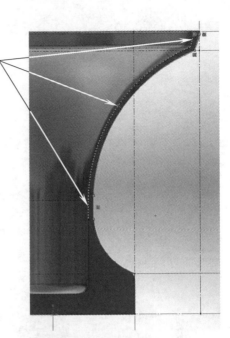

Select 3 entities

Select the 3 entities as noted.

Enter **.125in** for offset distance.

Verify that the offset is showing on the left side, and click the reverse checkbox if needed.

Click **OK** to close the offset command.

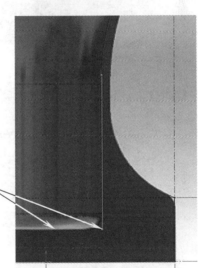

Drag the endpoint of the vertical line downward as indicated.

Drag the endpoint until it lines up with the horizontal edge

6. Closing off the sketch profile:

Add **2 more lines** to close off the sketch profile.

Snap the end of the last line to the origin. At this point, the sketch profile should be closed. To verify that all entities are in fact connecting with one another, right click one of the lines and pick **Select Chain**.

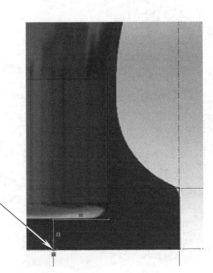

Stop at origin

If the entire sketch highlights, it indicates that all entities are connected properly.

Add a **Tangent** relation between the Arc and the line as indicated.

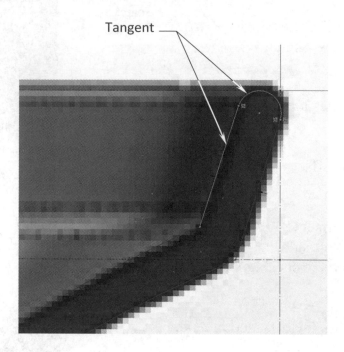

Tangent

7. Adding the sketch fillets:

It would be easier to create the fillets within the same sketch. That way we can see if they are going to look right while the sketch still overlays with the picture.

Click the **Sketch Fillet** command from the Sketch toolbar.

Enter **.250"** for radius.

Select the <u>3 vertices</u> as noted, and keep the **Constraint Corners** option checked.

The preview fillets should look similar to your image.

Click **OK**.

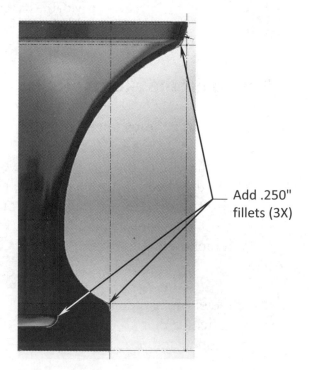

Add .250" fillets (3X)

Change the fillet size to **.125"**.

Select the <u>vertex</u> as noted to apply the new fillet.

Keep the option **Constraint Corners** checked.

Add .125" fillet (1X)

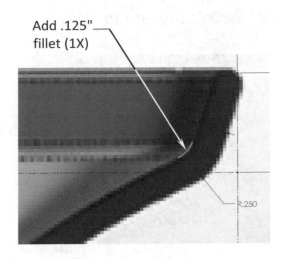

Click **OK** twice to exit the fillet command.

Add a <u>vertical centerline</u> as shown. It will be used to revolve the sketch profile.

Add a centerline

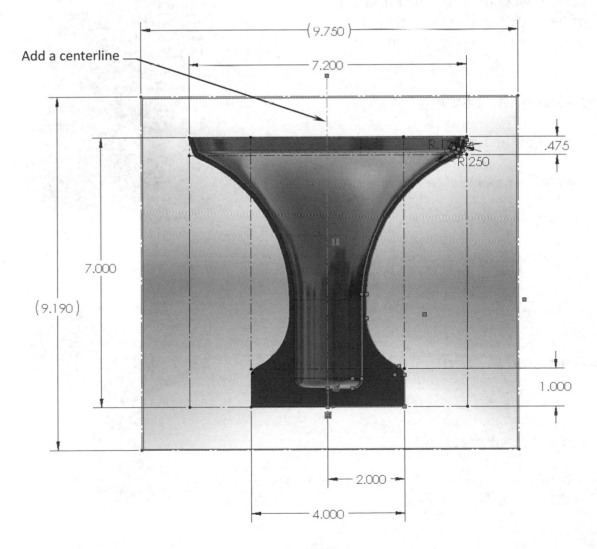

8. Revolving the profile:

Select the vertical centerline and click the **Revolve Boss/Base** from the Features toolbar.

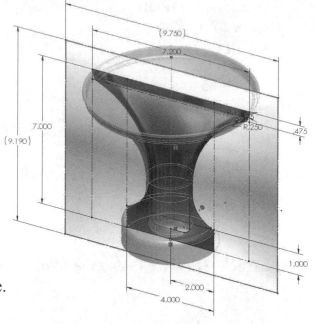

Use the default **Blind** type and the **360°** angle.

Click **OK**.

Expand the Revolved1 feature and <u>suppress</u> the Sketch Picture.

Use the Front plane and create a section view to verify the thickness of the revolved part. Exit the section view when you are done viewing.

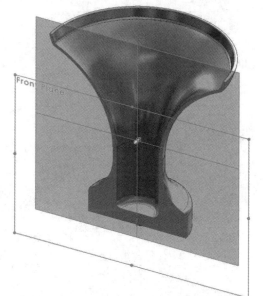

9. Measuring the Mass:

Switch to the **Evaluation** tool tab.

Change the material to **Aluminum Bronze**.

Click the **Mass Properties** command.

Using 3 decimals enter the mass here:

_____ (6.35lbs +/- .15 lbs.)

10. Saving your work:

Save your work as **Sketch Picture Completed**.

Close the part document.

NOTE: *Refer to the completed part saved in the Training Files folder for reference, or to compare your results against it.*

<u>Exercise 1</u> Working with Sketch Pictures & using the Spline tool

A digital image can be used as a reference to model a part. Each image can be placed on its own plane, so several images can be inserted to help define the shape of the model from different orientations.

Formats such as jpg, tif, bmp, gif, png, etc., are supported in SOLIDWORKS. They can be inserted and converted into a sketch, so that a feature can be made from it.

When scanning or saving the digital images, it is best to use fine resolution and high contrast pictures. Sketching over the sharp edges would be much easier than the blurry edges.

Splines are often used to do the tracing of the images due to their flexibility in manipulating the shapes, and splines offer a set of control tools to assist you with creating and maintaining the smoothness of the curves.

The digital or scanned image can be scaled to size and repositioned with reference to the origin so that dimensions can be added for accuracy.

1. Inserting the scanned image:

The scanned image should be inserted onto an active sketch.

In a <u>new part</u> file, select the **Front** plane and open a new sketch.

From the **Tools** menu, click **Sketch Tools / Sketch Picture** .

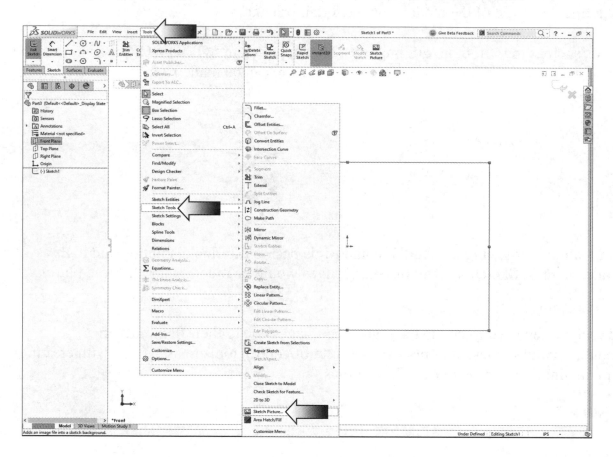

Browse to the Training Files folder and select the file named
Eagle Head.jpg and open it.

The lower left corner
of the image is placed
on the origin.

The image size and
locations appear on
the properties tree;
we will modify those
dimensions in the
next step.

2. Positioning and sizing the scanned image:

Sketch a **vertical centerline** to help center the image.

Double click the image to activate it.

Enter the following dimensions to re-position and re-size the image:

* X = **-1.825in**.
* Y = **-0.00in**.
* Angle = **0deg**.
* Width = **3.600in**.
* Height = **3.000in**.

Be sure to enable the **Lock Aspect Ratio** checkbox (arrow).

Click **OK**.

Splines:

A spline is a sketch entity that gets its shape from a set of spline points. This tool is great for modeling free-form shapes that require a few more "flexibilities" than other curve tools.

During the creation of a spline, each click creates a spline point and these points can be added or deleted when needed.

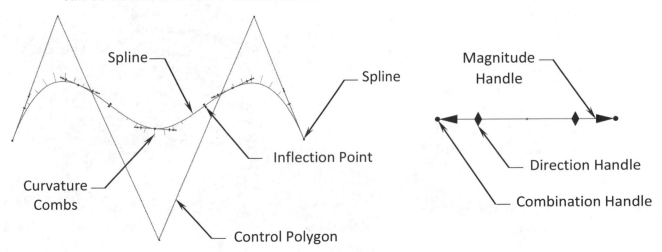

Try to use as few spline points as possible in the general, long curving areas.

Only use more spline points on tighter, smaller radiuses.

Use the spline handles to drag freely or hold the ALT key to drag symmetrically. The spline handles are used to change the direction and magnitude of the tangency at a spine point.

Use the Control Polygons in place of the spline handles. Drag its control points to manipulate the spline.

The Curvature Combs display the curvature of the spline in a form of a series of lines called a comb. The length of the lines represents the curvature. The longer the line, the larger the curvature, and the smaller the radius.

Inflection Points or Markers are used to show the inflection changes in a spline, whether it is convex or concave.

3. Tracing the image with the spline tool:

The sketch should still be active at this time; select the **Spline** command from the Sketch toolbar.

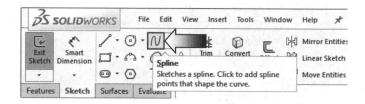

Keep in mind that the simpler the spline, the easier it is to manipulate it. So, we are going to create one spline with two or three spline points each time, and then adjust it to match the outline of the image as close as possible. (Zoom in a little closer.)

Start at "**point 1**," and "**point 2**," then "**point 3**" as indicated.

Push the **Escape** key when done.

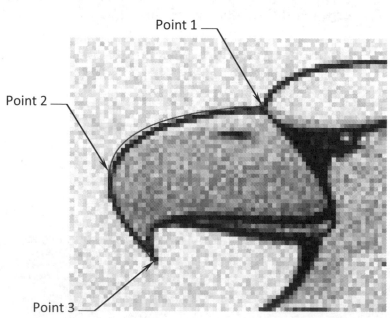

Zoom in even closer so that you can adjust the spline a little easier.

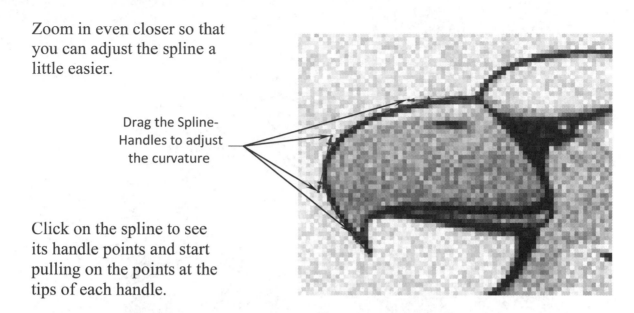

Drag the Spline-Handles to adjust the curvature

Click on the spline to see its handle points and start pulling on the points at the tips of each handle.

It may take some getting used to, so work on a small area each time. Create only one spline each time, and each spline should have two or three points only.

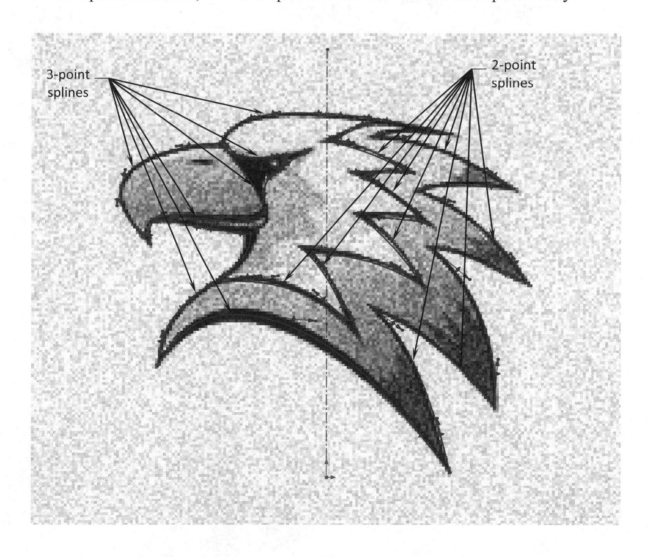

3-point splines

2-point splines

There should be a small gap
around the sketch (as noted)
so that a single sketch
can be extruded
into a feature.

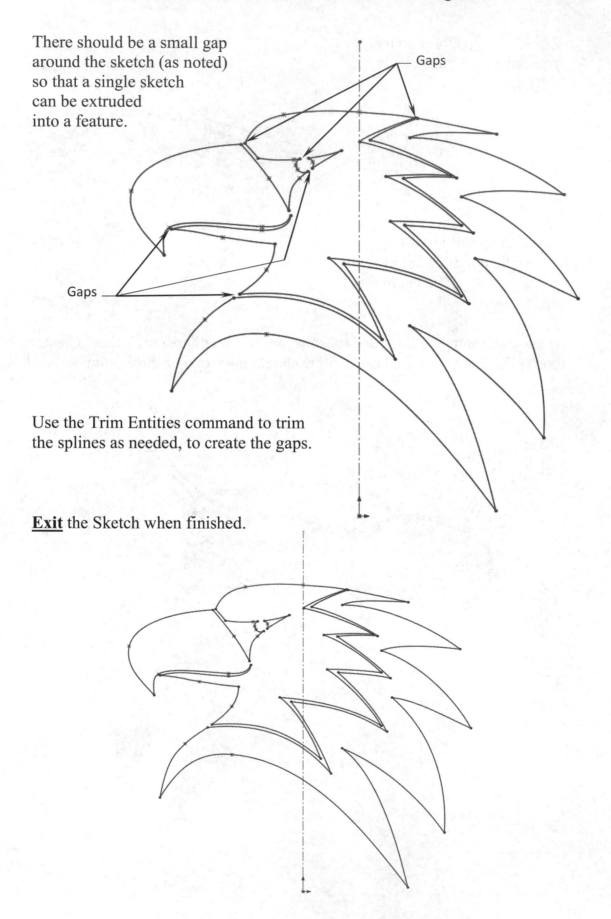

Gaps

Gaps

Use the Trim Entities command to trim
the splines as needed, to create the gaps.

Exit the Sketch when finished.

4. Extruding the traced sketch:

Click **Extruded Boss-Base**.

Use the default **Blind** type
and enter **.125"** for thickness.

Click **OK**.

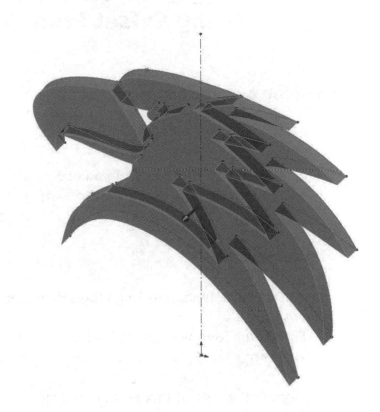

5. Optional:

Use **Photoview 360** and render the model with the following settings:

Appearances: **Glass / Clear Thick Gloss / Clear Thick Glass**.

Scene: **Studio Scenes / Reflective Floor Black**.

Lighting: **Green**
         Brown
         Blue

Output Image Quality:
    1280 X 1024

6. Saving your work:

Save your work as
Eagle Head_Sketch Picture.

Exercise 2 – Working with Sketch Picture Using Offset From Surface option

1. Opening a part document:

Select **File, Open**.

Browse the <u>Training Folder</u> and open a part document named: **Sketch Picture.sldprt**.

2. Making a block:

<u>Edit</u> the sketch named: **Logo Sketch** and press <u>Control+6</u> (bottom view).

The sketch logo was created using several splines and it has not been fully defined.

Moving the logo at this point will distort the entities. One quick way to overcome that is to make a block out of it. That way the entire sketch can be moved or scaled as one entity.

<u>Box-select</u> the entire sketch and select: **Make-Block**, or select: **Tools, Block, Make**.

Click **OK**.

All selected entities are joined into a single entity.

3. Extruding the sketch:

Switch to the **Features** tab and click: **Extruded Boss-Base**.

Select 9 contours

Under Selected Contours, select the 9 contours as noted.

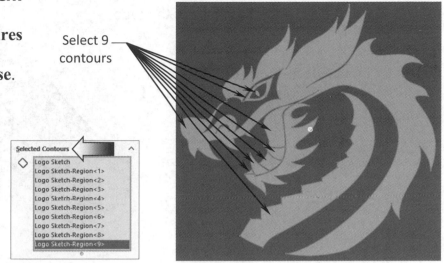

For Extrude From, select: **Surface/Face/Plane** from the drop-down list.

Select face for Extrude From

For Direction 1, select the **Offset From Surface** option.

Select the face to offset from, as indicated.

Select face to Offset From

For Offset Distance enter **.020in**.

Enable the **Reverse Offset** and **Translate Surface** checkboxes.

Click **OK**.

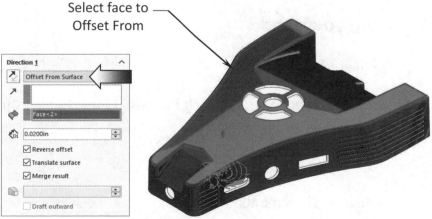

Optionally, change the color of the logo to white.

Click the **Boss-Extrude1** feature and select **Appearance**, **Boss-Extrude1**.

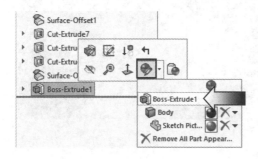

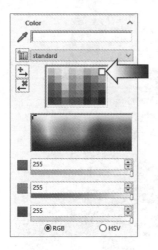

Select the **White** color (arrow).

Click **OK**.

4. Saving your work:

Select **File, Save As**.

For file name, enter: **Sketch Picture_Completed.sldprt**.

Click **Save**.

Close all documents.

Re-using the Geometry

The Contour Select Tool allows you to select sketch contours and model edges and apply features to them. The same sketch can be reused over and over again.

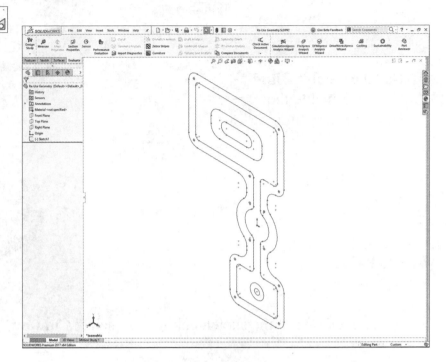

Contour selection is also restricted as follows:

* When reusing a sketch, you can select only on the original face. If, for example, part of the face has been extruded, the tool does not recognize the new face.

* You can select contours only on the face with the sketch. If, for example, the face with the sketch is cut by a solid object (as shown below), the tool can select the part of the face still visible but does not recognize the solid object.

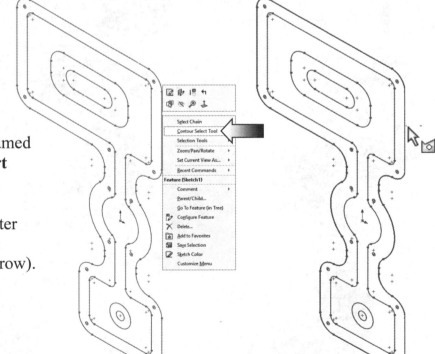

1. Open a part document named **Re-Use Geometry.sldprt**

2. Right click one of the outer sketch entities and select **Contour Select Tool** (arrow).

The entire outer contour is selected; it will be extruded first to create the parent feature.

3. Switch to the Features tool bar and click **Extruded Boss Base**.

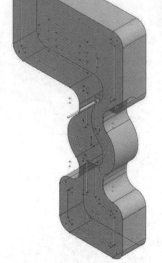

4. Use the default **Blind** type and enable the **Reverse** direction.

5. Enter a thickness of **1.500in**.

6. Click **OK**.

7. Right click one of the sketch entities and select **Contour Select Tool** again (arrow).

8. Hold the **Control** key and select the **3 contours** as indicated below.

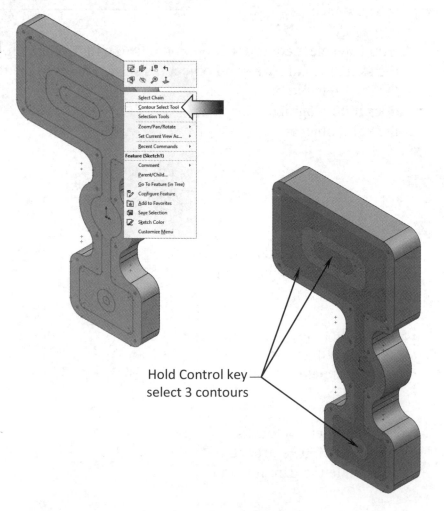

Hold Control key select 3 contours

9. Switch to the Features toolbar and click **Extruded Cut**.

10. Use the default **Blind** type.

11. Enter a depth of **1.000in**.

12. Click **OK**.

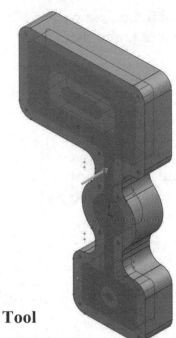

13. Right click one of the circles and select **Contour Select Tool** once again (arrow).

14. Hold the **Control** key and select all **13 circles** as indicated below.

15. Click **Extruded Cut**.
For Direction 1, select **Through All**.

16. Click **OK**.

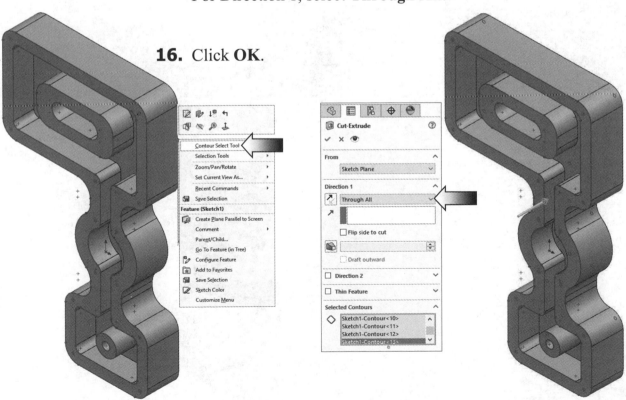

Expand the features to see the sketches under them.

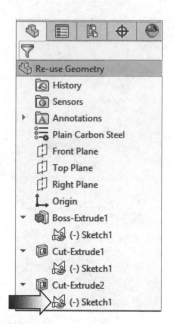

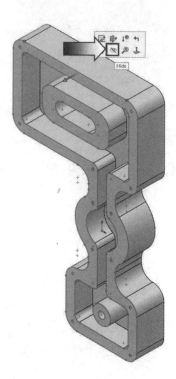

The symbol means the sketch has been used several times to create multiple features.
It is called **Shared Sketch**.

Click one of the sketch entities and select **HIDE**.

17. Save and close the part document.

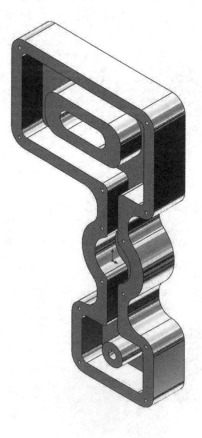

CHAPTER 3

Creating Multibody Parts

Creating Multibody Parts
Wooden Crate

Part documents can contain multiple solid bodies.
Multibody parts should not replace the use of assemblies.

A general rule to keep in mind is that one part, multibody or not, should represent
one part number in a Bill of Materials. A multibody part consists of multiple solid
bodies that are not dynamic. If dynamic motion among bodies is needed it should
be done in an assembly instead. Tools such as Move Component, Dynamic
Clearance, and Collision Detection are available only with assembly documents.

Multibody solids can be manipulated the same way you manipulate single solid
bodies. For example, you can add and modify features, and change the names and
colors of each solid body.

A folder named Solid Bodies 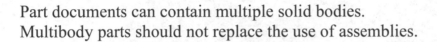 appears in the FeatureManager design tree when
there are solid bodies in a single part document. The number of solid bodies in the
part document is displayed in parentheses next to the Solid Bodies folder. The
solid bodies can be organized and managed in the following ways:

* Group bodies into folders in the Solid Bodies folder
* Select commands to apply to all bodies within a folder
* List features that belong to each body

Multiple solid bodies can be created from a single feature with the following
commands:

* Extrude boss and cut	* Revolve boss and cut (including thin features)
* Surface cut	* Sweep boss and cut (including thin features)
* Boss and cut thicken	* Cavity.

Creating Multibody Parts
Wooden Crate

View Orientation Hot Keys:

Ctrl + 1 = Front View
Ctrl + 2 = Back View
Ctrl + 3 = Left View
Ctrl + 4 = Right View
Ctrl + 5 = Top View
Ctrl + 6 = Bottom View
Ctrl + 7 = Isometric View
Ctrl + 8 = Normal To Selection

Dimensioning Standards: **ANSI**

Units: **INCHES** – 3 Decimals

Tools Needed:

Insert Sketch	Center Rectangle	Extruded Boss/Base
Dimension	Add Relations	Linear Pattern
Move/Copy	Exploded View	Mirror

1. Starting a new part document:

The Multibody-Parts topic is best learned by creating a new model from scratch. You can see step-by-step how each body gets created and how the bodies get organized and managed as the model is built.

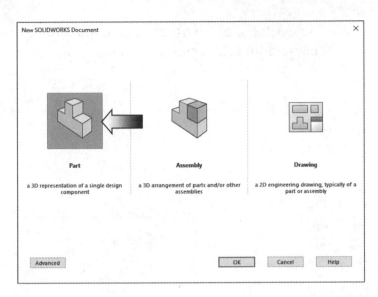

Select **File / New / Part**.

Select the **Part** template and click **OK**.

Select the <u>Front</u> plane from the FeatureManager tree (arrow) and open a **new sketch** by clicking on the **Sketch** icon (arrow).

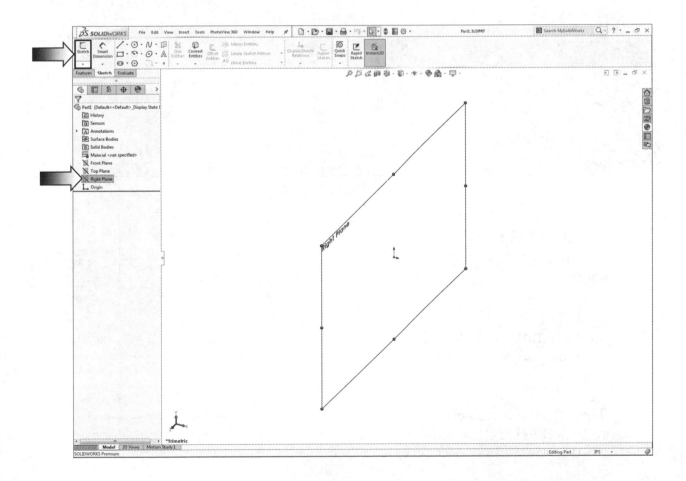

2. Sketching the first body profile:

Select the **Center Rectangle** command (arrow) and sketch a rectangle that is centered on the origin.

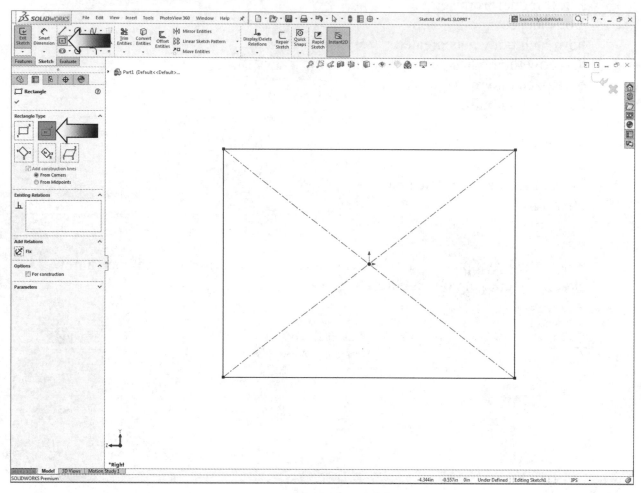

Add the vertical and the horizontal dimensions shown.

Vertical dim. = **12.000"**
Horizontal dim. = **14.250"**

The sketch should become fully defined at this point.

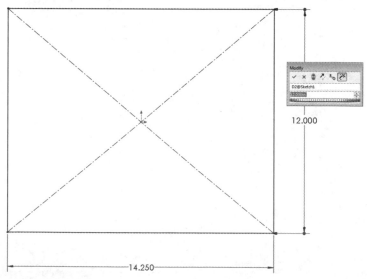

Click the **Straight-Slot** command; <u>clear</u> the **Add Dimensions** checkbox (arrows).

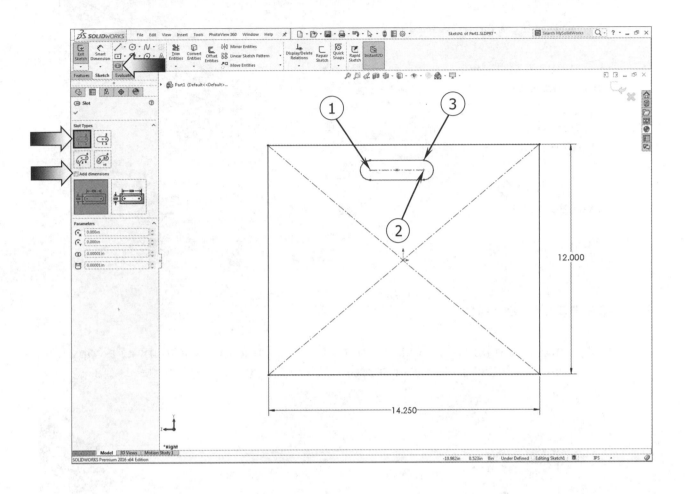

Sketch the slot by clicking the 3 points in the order noted above.

Add the 3 dimensions as indicated to position the slot.

Add the vertical relation to fully define the slot.

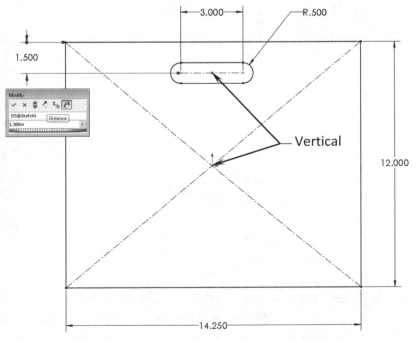

3. Extruding the first body:

Switch to the **Features** tool tab.

Click the **Extruded Boss-Base** command.

Use the default **Blind** extrude type.

Enter **.750in** for extrude depth.

Click **OK**.

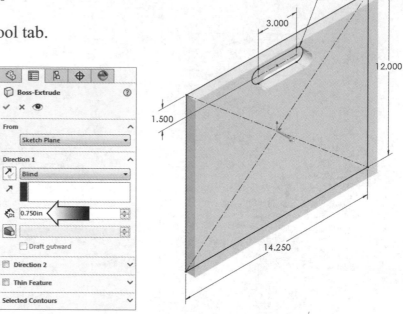

4. Copying the body:

Since the second body is identical to the first one, it is quicker to make a copy than to recreate it. Select **Insert / Features / Move/Copy** (arrow).

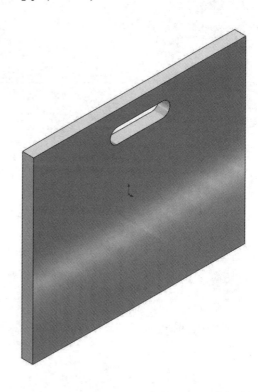

To add command: Select **Tools / Customize / Commands / Features** and drag/drop the **Move/Copy** command to your Features tool tab.

Click the **Translate/Rotate** button at the bottom of the tree to switch from the Constraints mode to Move mode.

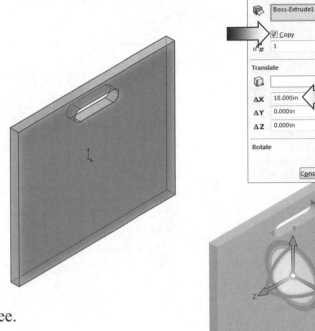

For Bodies to Move/Copy select the **Boss-Extrude1** either from the graphics or from the FeatureManager tree.

Enable to **Copy** checkbox (arrow).

Enter **18.00in**. in the **Delta X** dimension field.

Click **OK**.

5. Creating the upper plank:

Only one of the planks will be made. It will then be replicated and mirrored a few times to create the front and rear planks.

Select the left side face and open a **new sketch**.

Select sketch face

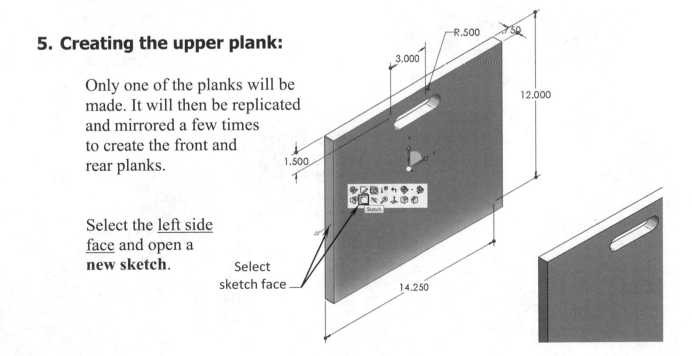

Change to the front orientation (Control+1).

Sketch a **Corner Rectangle** from the upper left corner of the 1st body to the outer edge of the second body.

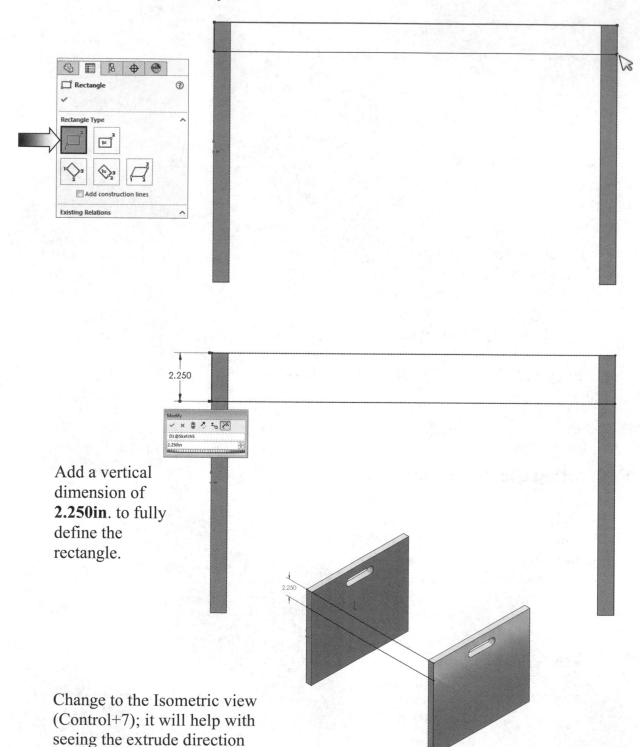

Add a vertical dimension of **2.250in**. to fully define the rectangle.

Change to the Isometric view (Control+7); it will help with seeing the extrude direction more clearly in the next step.

Switch to the **Features** tool tab and click **Extruded Boss/Base**.

Use the default **Blind** type, extrude away from the bodies.

Enter **.500in**. for extrude depth.

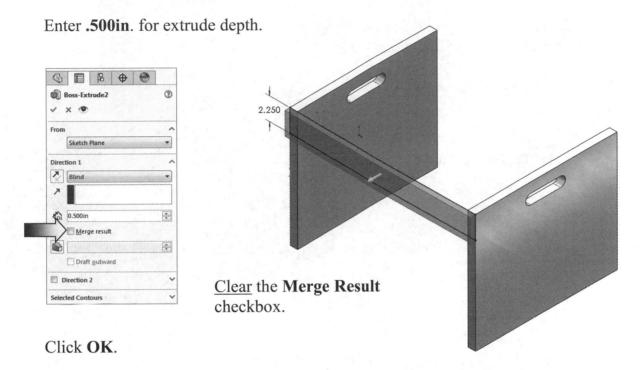

Clear the **Merge Result** checkbox.

Click **OK**.

6. Creating the first linear pattern:

Click the **Linear Pattern** command from the **Features** tool tab.

Select the <u>edge</u> as noted for Pattern Direction.

Enter **3.25in** for spacing.

Enter **4** for Number of Instances.

Expand the **Bodies** section (arrow) and select the plank (**Boss-Extrude2**) for Bodies to Pattern.

Click **OK**.

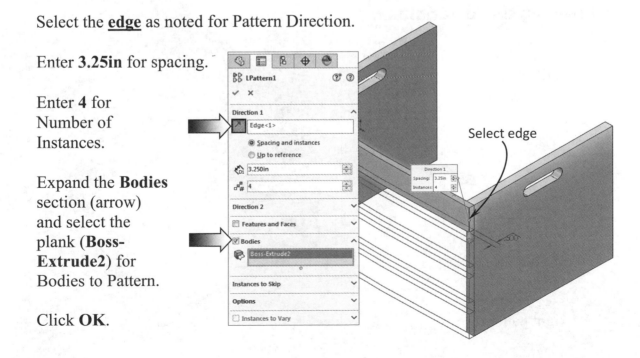

Select edge

7. Mirroring the planks:

Click the **Mirror** command from the Features tool tab.

Select the **Front** plane from the Feature tree for Mirror Face/Plane.

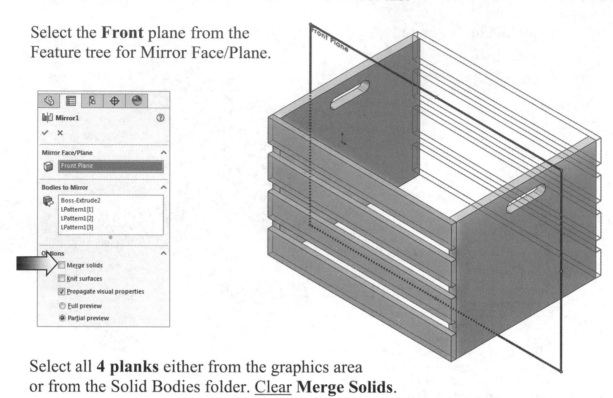

Select all **4 planks** either from the graphics area or from the Solid Bodies folder. <u>Clear</u> **Merge Solids**.

Click **OK**.

8. Creating the lower plank:

There should be a total of **10 Solid bodies** at this point.

Open a **new sketch** on the <u>bottom face</u> of the right panel.

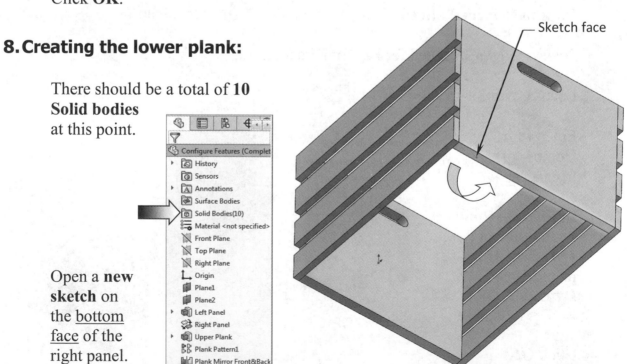

Change to the bottom view orientation (Control+6).

Sketch a **Corner Rectangle** approximately as shown.

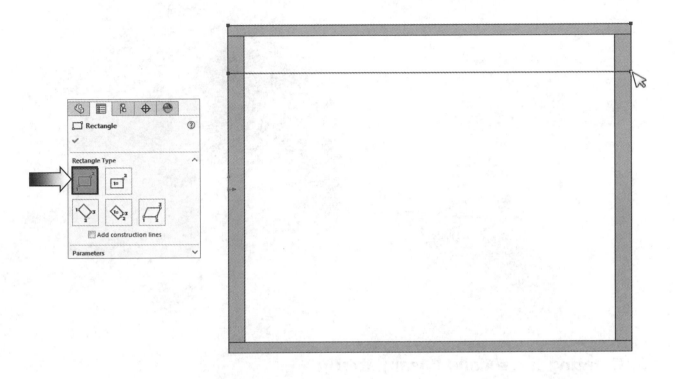

Add a vertical dimension of **2.250in.** to match the height of the other planks.

Change to the Isometric view (Control+7) before extruding the sketch. Switch to the **Features** tool tab and click **Extruded Boss/Base**.

Use the default **Blind** type and extrude away from the bodies.

Enter **.500in**. for extrude depth.

Click **OK**.

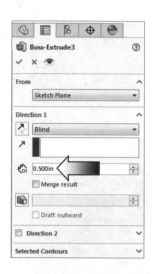

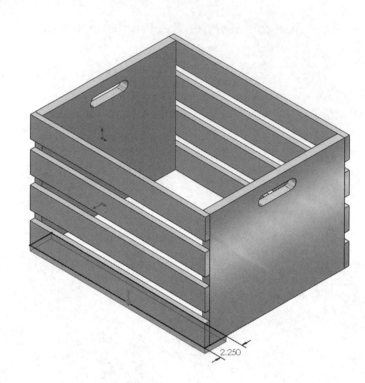

9. Creating the second linear pattern:

Click the **Linear Pattern** command from the **Features** tool tab.

Select the **edge** as noted for pattern direction.

Enter **3.250in** for Spacing.

Enter **5** for Number of Instances.

Expand the **Bodies** section (arrow) and select the bottom plank (**Boss-Extrude3**) for Bodies to Pattern.

Click **OK**.

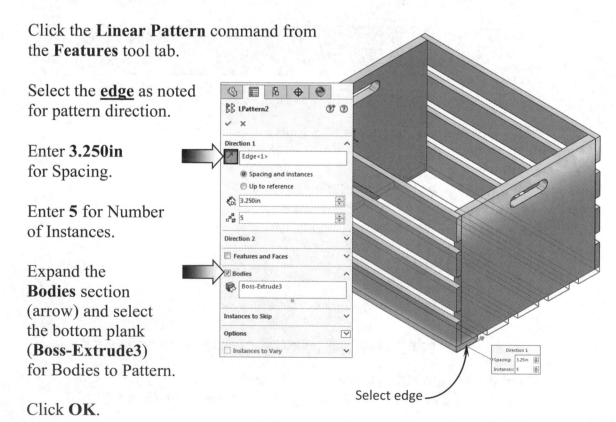

10. Creating an exploded view:

An exploded view is created by selecting and dragging a triad arm in the graphics area to show the solid bodies spread out, creating one or more explode steps.

Select **Insert / Exploded View**.

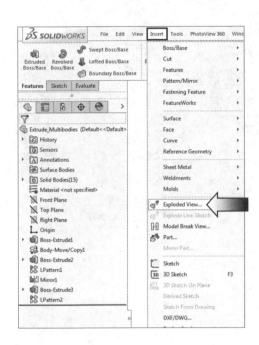

Select the **4 solid bodies** as indicated.

A triad appears in the graphics area.

Click the **Z direction** arrow of the triad.

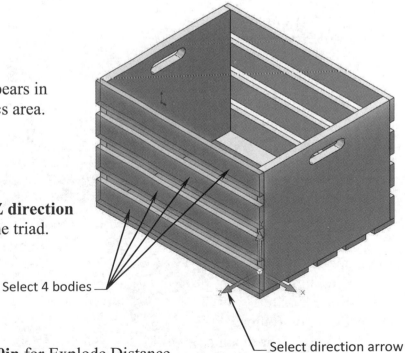

Select 4 bodies

Select direction arrow

Enter **15.00in** for Explode Distance.

<u>Clear</u> the **Auto-Space** checkbox.

Click **Apply** to move the selected bodies.

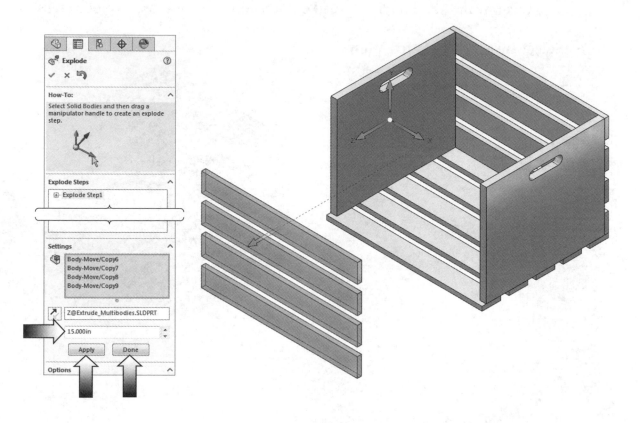

Click **Done**.

The 4 bodies are moved
15 inches along the
Z direction.

Repeat the
same step and
move the other
4 bodies along
the opposite
direction (enter
-15.00in. for distance).

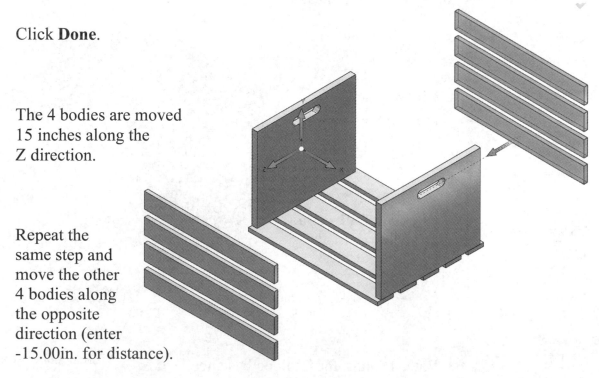

Create 3 additional steps to explode the left panel, the right panel, and the bottom planks similar to the image shown here.

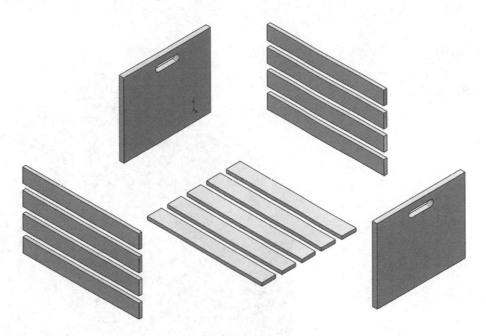

11. Collapsing the view:

The exploded view can be collapsed, edited, or deleted.

Switch to the **ConfigurationManager** tree (arrow).

Expand the **Default** configuration.

Right-click on **ExplView1** and select **Collapse**.

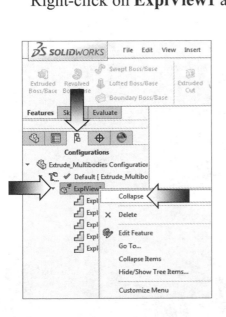

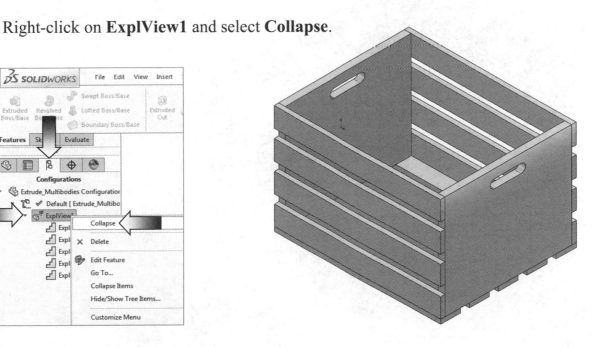

12. Saving your work:

Select **File / Save As**.

Enter **Multibody Parts** for the file name.

Click **Save**.

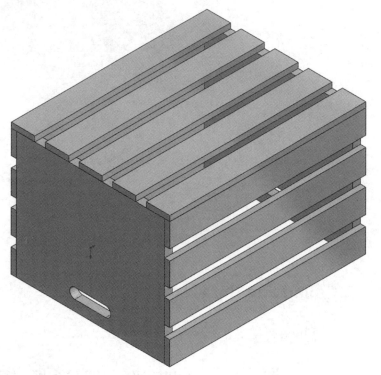

Exercise: Combining Multibodies

1. Opening a part document:

Click **File / Open**.

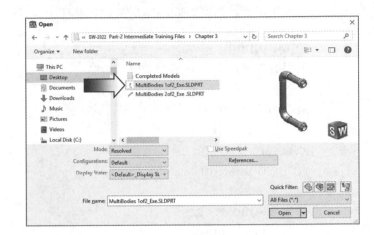

Browse to the Training Files folder and open the part document named **MultiBodies 1of 2_Exe**.

2. Inserting a solid body:

Using the pull-down menus, select **Insert / Part**.

Locate the Training Files folder, and select the document named: **MultiBodies 2of 2_Exe** and open it.

A preview of the selected part is attached to the mouse cursor; do not click OK. Enable the two check boxes: **Locate Part with Move / Copy Feature** and **Break Link to Original Part** (arrows).

Click **OK** to place the part on the **Origin**.

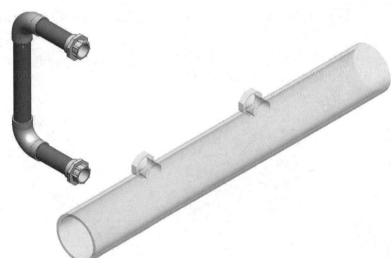

The **Locate part** dialog appears. We will use the Rotate options to reorient the large tube to the upright position.

Click the **Translate/Rotate** button `Translate/Rotate` .

3. Rotating a solid body:

Click the **Rotate** tab (arrow) to activate its options.

Enter **90.00deg** in the **X** direction.

Select the large tube and press **Enter** to rotate it to the vertical position.

This is one of the few areas where the Reverse or Flip direction is not available; use either a 90° or -90° to rotate the body either to the left or to the right.

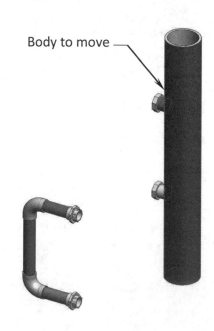

Click **OK** to exit the Locate Part mode.

4. Creating new mates:

Using the pull-down menus, select **Insert / Features / Move / Copy** (arrow).

Under the Bodies to Move section, select the large tube as noted.

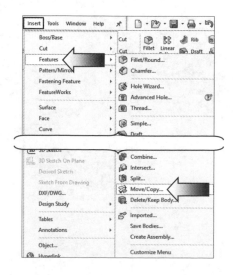

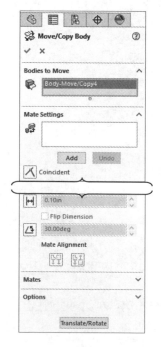

Body to move

Click the **Constraints** button `Constraints` (at the lower left side of the tree) and click inside the **Mate Settings** section box to activate this option.

First, select the 2 cylindrical faces of the 2 bodies as indicated.

A **Concentric** mate is created and the large tube moves to the new position.

Click **Add** to accept the mate.

This mate is stored in the Mate section near the bottom of the dialog box. It can be edited or deleted if needed.

Next, select the 2 planar faces of the 2 bodies as noted below.

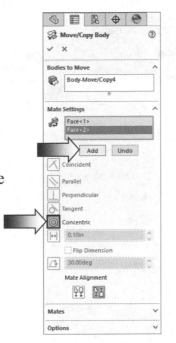

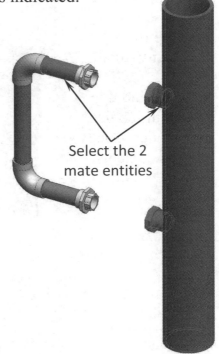

Select the 2 mate entities

A **Coincident** mate is created. The large tube moves forward and touches the fixed body. Click **Add** to accept the mate.

Click **OK**.

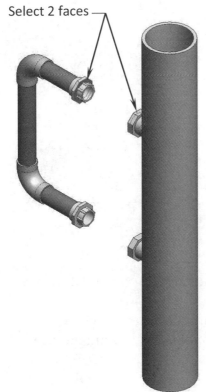

Select 2 faces

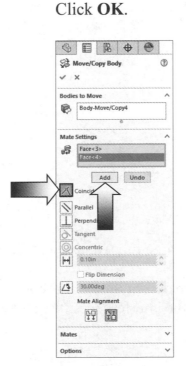

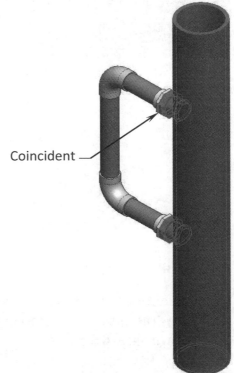

Coincident

5. Checking the overlapped geometry:

After the two solid bodies are positioned, they cause some interferences. One of the easier ways to view the interference is to change the Body2 to transparent.

Expand the Multibodies folder to see its contents.

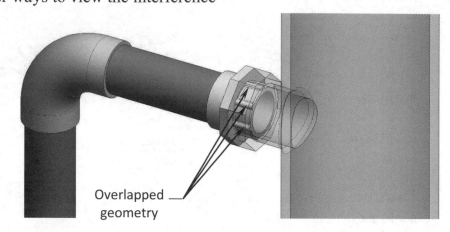

Overlapped geometry

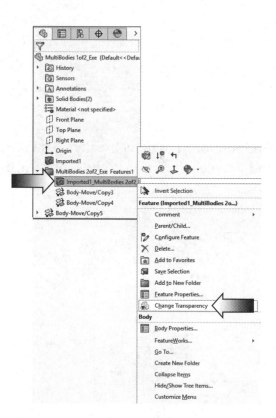

Right-click the **Body2** (the large tube) and select **Change Transparency** (arrow).

Examine the interferences between the 2 solid bodies in both areas. There are several options to remove the interferences; the method that we are going to use is **Combine-Add**.

6. Combining the solid bodies:

Using the drop-down menus once again; select:
Insert / Features / Combine.

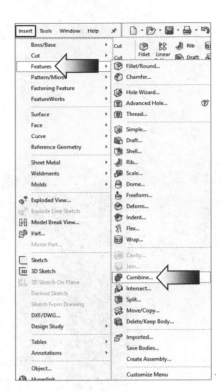

As mentioned earlier in this chapter, the Combine command offers three different options:

* **Add** = Combines multiple bodies into a single body.

* **Subtract** = Subtracts overlapping material from a selected main body.

* **Common** = Removes all material except that which overlaps.

Under the Operation Type, select the **Add** option (arrow).

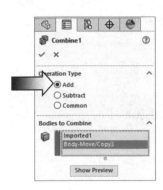

Select the **2 solid bodies** either from the graphics areas or from the FeatureManager tree.

Click the **Show Preview** button and zoom in on one of the areas that has the interference.

The interferences are consumed by the Combine-Add operation and the 2 flanges are joined as one.

Examine the result of the Combine-Add.

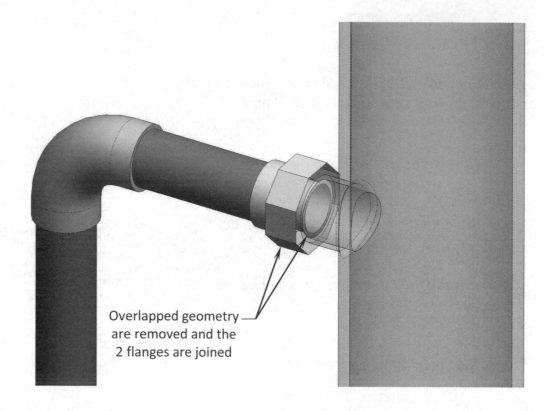

Overlapped geometry
are removed and the
2 flanges are joined

7. Saving your work:

Click **File / Save As**.

Enter **Combining Multibodies_Exe** for the name of the file.

Press **Save**. Close all documents.

CHAPTER 4

Working with Multibody Parts

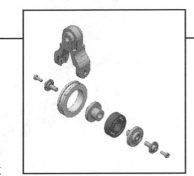

Working with Multibody Parts
Creating Mates & Exploded Views

Use the **Insert Part** command to insert a base part into another part document.

When you insert a part into another part document, it becomes a Multibody Part, and the part you inserted becomes a Solid Body.

If the inserted part contains mate references, you can use them to position the inserted part.

There are 3 options available to help locate a solid body: Constraints, Translate, and Rotate. You can move or rotate the body after it is placed in the graphics area. A feature Body-Move/Copy is then added to the FeatureManager design tree.

The inserted body can also be constrained with the mate options. Some of the standard mates are available in the part mode to help position the solid bodies.

An exploded view is created after all solid bodies are positioned. Exploded views are stored in the ConfigurationManager. They can be edited, deleted, or more than one exploded view can be created, if needed.

This chapter will guide you through the use of multibodies parts, where an existing part document with a number of solid bodies is used. Other parts will then get inserted into the same document and constrained with some mates. After all solid bodies are positioned, an exploded view will be created in the same part mode, and the Solid Bodies folder is used to help keep track of the number of solid bodies existing in the part document.

Creating Mates & Exploded Views
Working with Multibody Parts

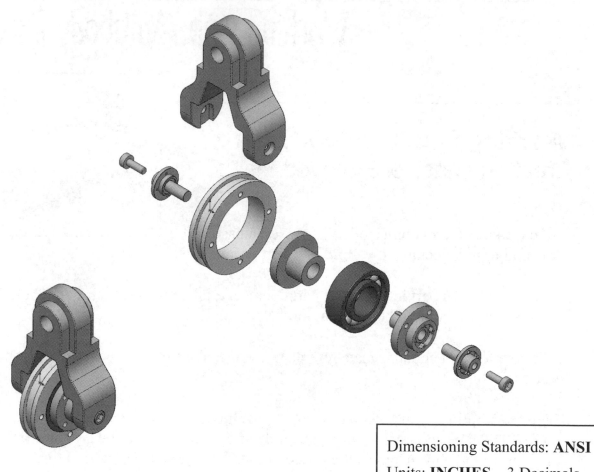

Dimensioning Standards: **ANSI**

Units: **INCHES** – 3 Decimals

Tools Needed:

Insert Part	Locate Part	Plane
Mirror	Exploded View	Mates
Concentric	Coincident	Mate alignment:

1. Opening a part document:

The **3D Interconnect** option should be <u>disabled</u> for this exercise. Go to:
Tools, Options, System Options, Import, <u>clear</u> **3D Interconnect** checkbox.

Go to **File / Open**.

Browse to the Training Files folder. Locate the part document named
MultiBody Parts and open it.

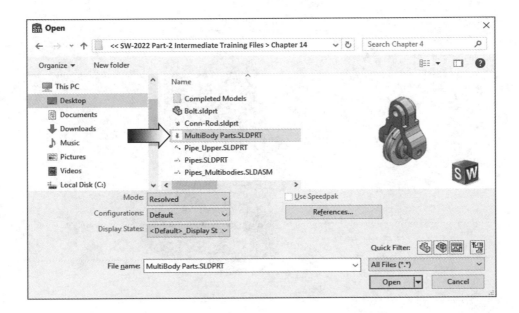

When opening a non-SOLIDWORKS native document, the program will prompt
you with a couple of options such as Import Diagnostics or Features Recognition.
We will not discuss these options in this chapter but focus on the use of
MultiBodies instead.

<u>If prompted</u>, click **No** to cancel the Import Diagnostics option.

Click **No** once again to cancel the Feature Recognition options.

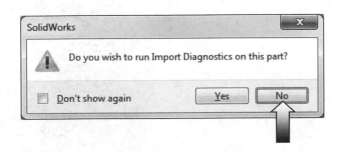

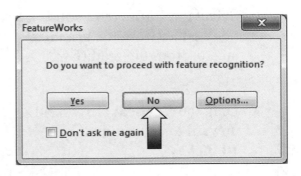

2. Creating an exploded view:

An exploded view needs to be created so that the collapse / explode stages can be toggled back and forth in the future.

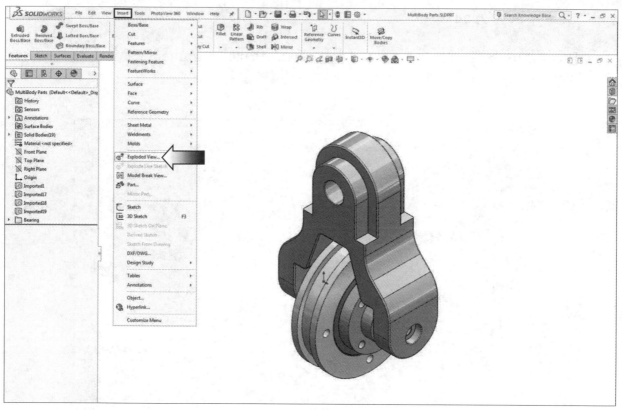

Select **Insert / Exploded View**.

The Exploded dialog appears similar to the one found in the assembly mode.

Select the **Imported2** solid body either from the graphics area or from the feature tree.

Drag the **Vertical** arrowhead upward to approximately **10 inches**.

Drag the Vertical arrowhead upward, about 10.00"

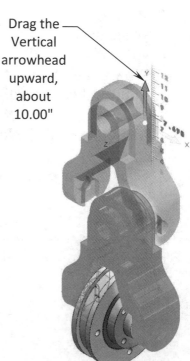

Right-click anywhere
in the graphics area
and select **Box Selection**.

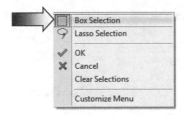

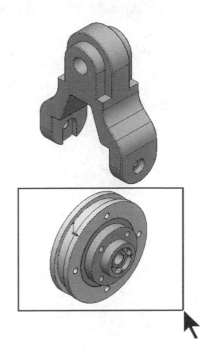

Drag a selection-box around the solid bodies as
shown. All solid bodies within the selections box
are selected.

The outer pulley (Imported1) should not be part
of this group, so we are going to unselect it before
exploding this group of solid bodies.

Either hold the Control key and click the **Imported1** body to exclude it from the
group, or delete it from the list under the Settings section.

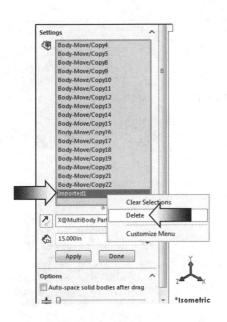

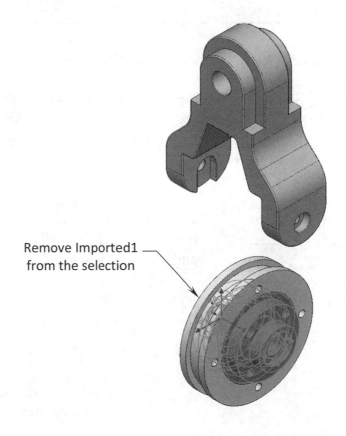

Remove Imported1
from the selection

An alternative method is to
select the bodies-to-explode
from the Solid Bodies folder,
on the FeatureManager tree.

Explode the solid bodies similar to the image shown below.

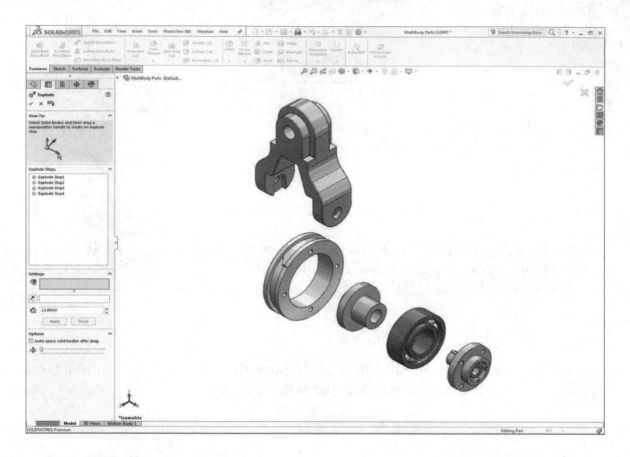

3. Collapsing the solid bodies:

Switch to the **ConfigurationManager** tree (arrow).

Expand the Default configuration to see the ExpView1. Right-click on ExpView1 and select **Collapse** (arrow).

The solid bodies are collapsed to their initial stage. We will toggle back and forth between the Collapse and Explode modes while adding more parts in the next few steps.

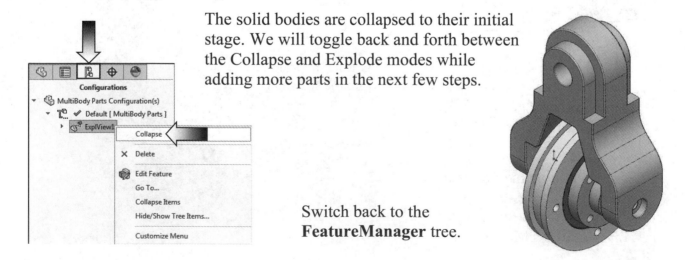

Switch back to the **FeatureManager** tree.

4. Inserting another part:

Other parts can be inserted into an existing part document. Once inserted they will become Solid Bodies, and the original part becomes a Multibody Part.

Click **Insert / Part**.

Select the part document named: **Conn-Rod.sldprt** from the same training files folder and open it.

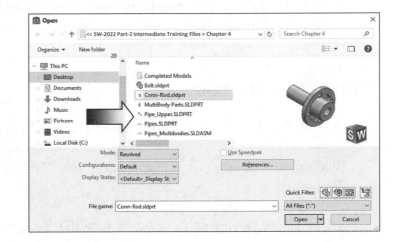

Place the new part on the right side of the others, approximately as shown below.

The **Insert Part** dialog box appears.

Select the following options:

    * **Solid Bodies**

    * **Planes**

    * **Locate Part with Move/ Copy feature**

Clear all other check-boxes.

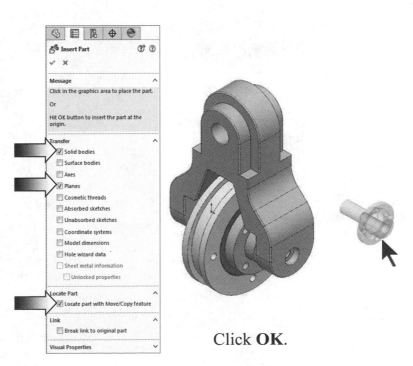

Click **OK**.

5. Constraining the solid bodies:

By clicking OK from the previous step the **Locate Part** section is activated.

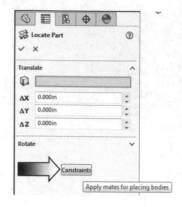

Click the **Constraints** button to switch to the Mate Settings mode (arrow).

(Click the Translate / Rotate button to change to the Constraints mode if this option is not visible on the tree.)

Adding a Concentric mate:

Select the **shaft body** of the bolt and the **cylindrical face** of the hole.

A **Concentric** mate is created, and the bolt moves to the same center with the hole.

Click the **Add** button (arrow) to save the mate in its mate group below.

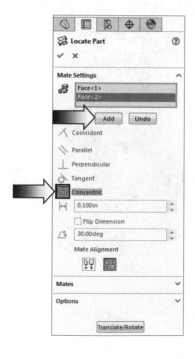

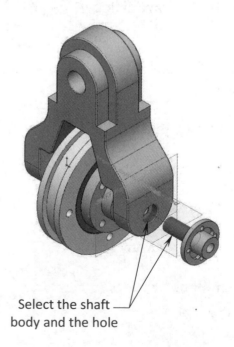

Select the shaft body and the hole

Adding a Coincident mate:

Select the 1st **planar face** (face1) as indicated.

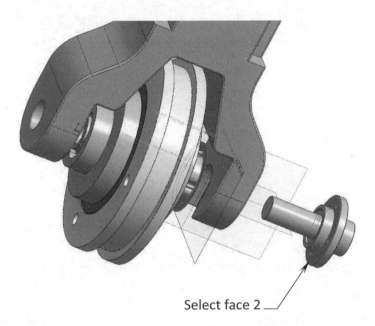

Select face 1

Select the 2nd **planar face** (face2) as noted).

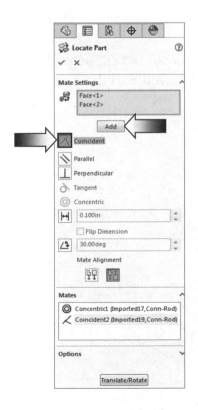

A **Coincident** mate is created and the 2 selected faces are now coinciding or occupying the same place.

Select face 2

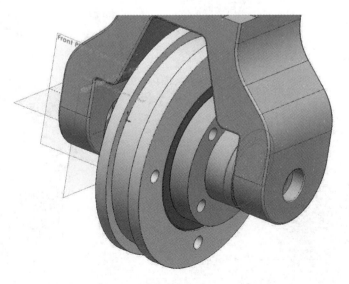

To help keep the file size smaller, we are going to leave the 3rd degree of freedom open. The last mate is not really needed for the purpose of the lesson.

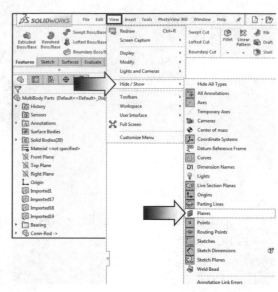

Click: **View / Planes** to hide all the planes in the Imported1 solid body (arrow).

6. Creating a new mirror plane:

Switch to the **Features** tool tab and
select the **Plane** command from the Reference Geometry drop-down arrow.

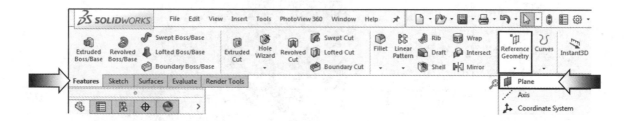

To create a Mid Plane, two **planar surfaces** are needed.

Select the outer most **left** and **right** faces of the main housing as noted. A Mid Plane is created in the middle of the 2 selected faces.

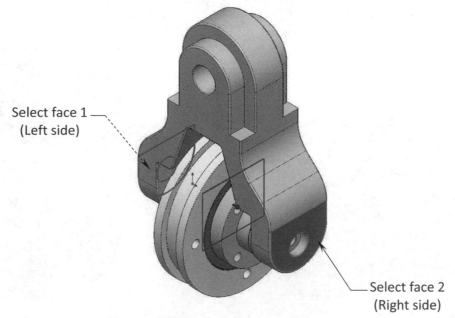

Select face 1
(Left side)

Select face 2
(Right side)

7. Creating a mirror body:

Click the **Mirror** command from the **Features** tab.

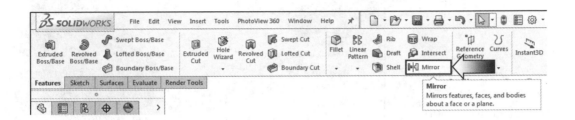

For **Mirror Face / Plane**, select the **new plane** (arrow).

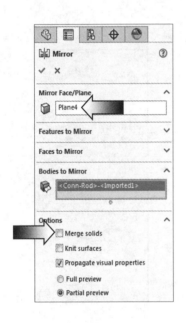

Expand the **Bodies to Mirror** section.

Expand the **Solid Bodies folder** from the FeatureManager tree and select the **Conn-Rod** from the list.

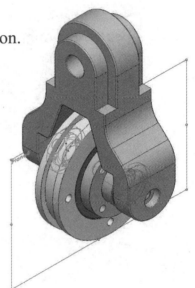

Clear the **Merge Solids** checkbox and click **OK**.

Switch back to the **Configuration** tree, **Edit** the Exploded view and **move** the solid bodies to the positions approx. as shown.

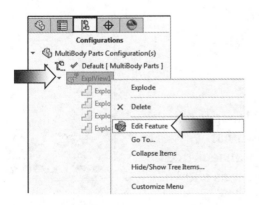

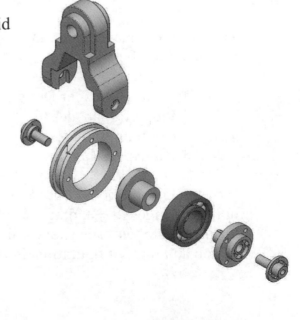

Right-click on
ExplView1 and
select **Collapse**.

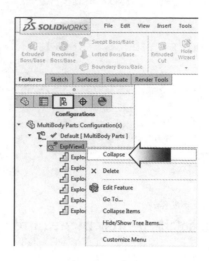

We will now insert
the **Bolt** component
into this document.

Switch back to the
FeatureManager
tree (arrow).

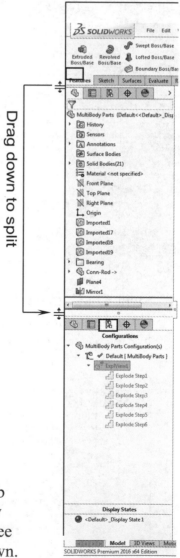

Optional:

The FeatureManager tree can be split to display any
combination of:

* FeatureManager Design tree
* PropertyManager tree
* ConfigurationManager
* DimXpertManager

Move the mouse pointer under the FeatureManager tab
until it changes to ⯐, drag the split bar down halfway
to split it into two panes. Show the FeatureManager tree
on top and the ConfigurationManager below it as shown.

8. Inserting another part:

The Bolt is going to be inserted into the same part document; it will become another solid body, at this point, the part document is called: Multibody Part.

Click **Insert / Part**.

Select the part **Bolt.sldprt** from the previous training files folder.

Place the new part on the right side of the main housing.

Click **YES** to accept the units of the new part to be the same as the units of the MultiBody part.

Select the **3 checkboxes** as noted and click **OK**.

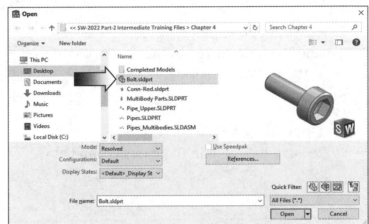

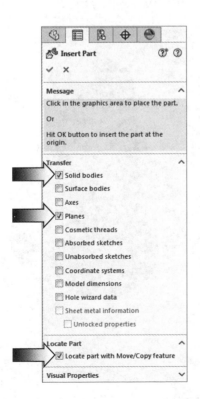

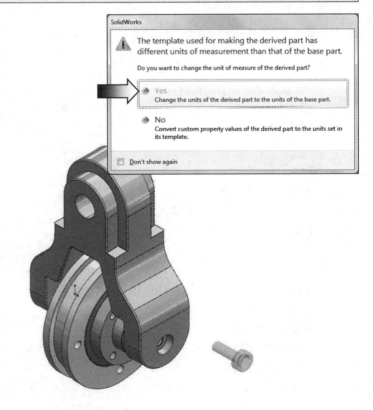

The **Locate Part** options appear.

Adding a Concentric mate:

Select the **shaft body** of the bolt and the **cylindrical face** of the hole.

A **Concentric** mate is created and the bolt moves to the same center with the hole.

Click the **Add** button (arrow) to save the mate in its mates group below.

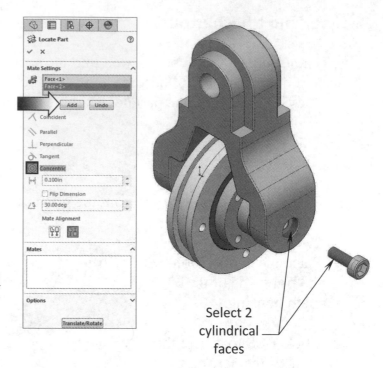

Select 2 cylindrical faces

Adding a Coincident mate:

Select the 2 **planar-faces** as indicated and click the **Add** button.

A **Coincident** mate is created and the 2 selected faces are now coincident.

The bolt is mated to the hole and still has 1 degree of freedom left. (It is good practice to add only 2 mates to each fastener to help keep the file size down).

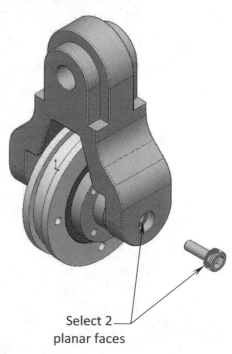

Select 2 planar faces

9. Adding another instance of the bolt:

A second instance of the bolt is needed on the opposite side of the main housing.

Click **Insert / Part**.

Select the **Bolt.sldprt** once again from the previous file location.

Rotate the view and place the new part on the left side of the main housing.

Adding a Concentric mate:

Select the body of the **bolt** and the side **hole** of the main housing to add a **Concentric** mate.

Click the **Anti-Aligned** button to flip the bolt 180 degrees (arrow), and then click the **Add** button to save the new mate in its Mates group.

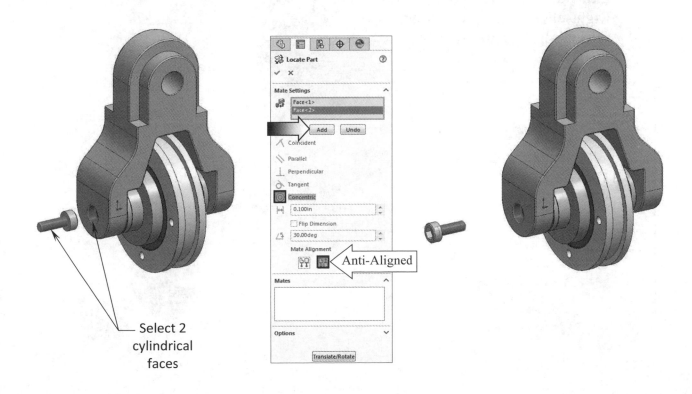

Select 2 cylindrical faces

Anti-Aligned

Adding a Coincident mate:

Select the 2 **planar faces** as indicated.

The 2 selected faces are constrained to share the same location.

Click the **Add** button to save the new mate.

Click **OK** to exit the Locate Part mode.

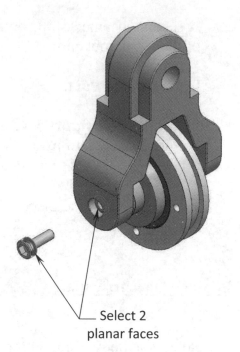

Select 2
planar faces

10. Editing the exploded view:

Switch to the **ConfigurationManager** tree.

Expand the **Default** configuration. Right-click the **ExpView1** and select **Edit Feature** (arrow).

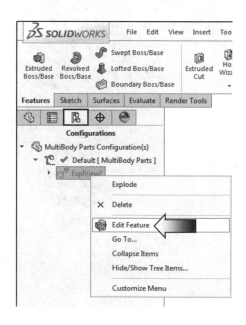

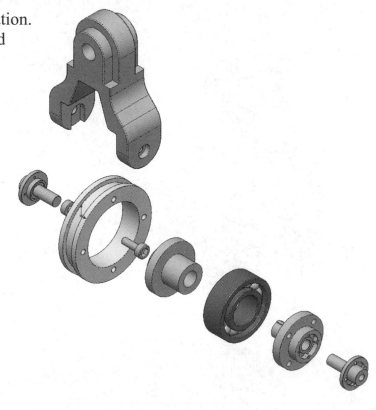

Select the **Bolt** on the right.

Drag the head of the **horizontal arrow** to move the bolt along X direction.

NOTE: Enter a dimension*
if a precise distance is preferred.

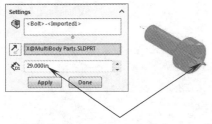

* If distance is used: Select a direction
arrow (red), enter a dimension
then click Apply and Done.

Select the bolt on the left.

Drag the head of the
horizontal arrow to the
left to move the bolt along
the -X direction.

Click **OK** to exit the exploded View mode.

11. Saving your work:

Click **File / Save As**.

Enter: **MultiBody Parts** for the name of the file.

Click **Save**.

Close all documents.

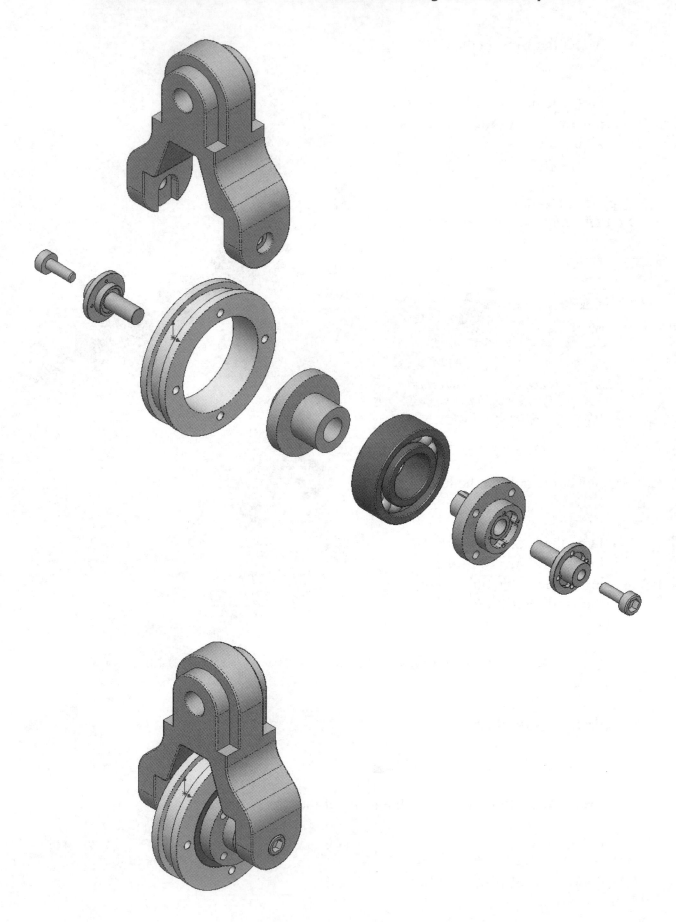

Working with MultiBodies
Pipes with Flanges

Working with Multibodies

Part documents can now have multiple solid bodies. For example, when you model a laptop power charger, you know the dimensions for the power supply but do not know the type of cable or the dimensions for the outlet adapters. With multibody parts, you can create the power supply unit first, then create the outlet adapters later when their data becomes available.

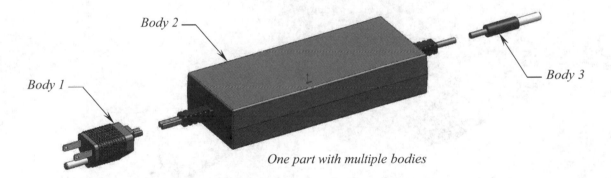

Body 2

Body 3

Body 1

One part with multiple bodies

A folder named **Solid Bodies** appears in the FeatureManager design tree when there are solid bodies in a single part document. The number of solid bodies in the part document is displayed in parentheses next to the **Solid Bodies** folder.

In this lesson, the two pipes were modeled as two separate bodies and saved as an assembly document, but they will need to be reoriented and mated in the part level.

After being fully positioned, one of the bodies will have features that would cause interferences between the two bodies, and the intention is not only to remove those interferences but to combine the two bodies into a single solid.

The assembly document will then be updated accordingly, and its mass properties can also be calculated.

Working with MultiBodies

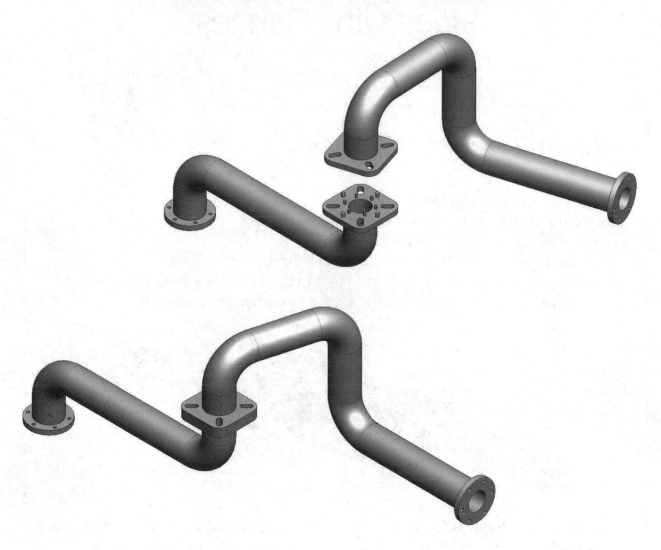

Dimensioning Standards: **ANSI**

Units: **INCHES** – 3 Decimals

Tools Needed:

 Edit Part/Assembly   Move/Copy Bodies

 Combine   Mass Properties

1. Opening an assembly file:

Click **File / Open**.

Browse to the Training Files folder, and open the assembly named **Pipes_Multibodies**.

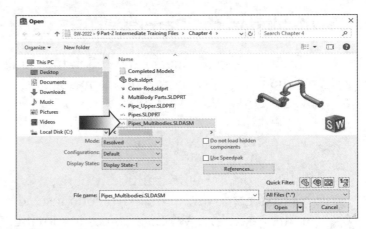

This assembly has only one component but with two solid bodies in it. Expand the component to see the Solid Bodies folder; it has the number (2) next to its folder indicating there are 2 bodies in this component.

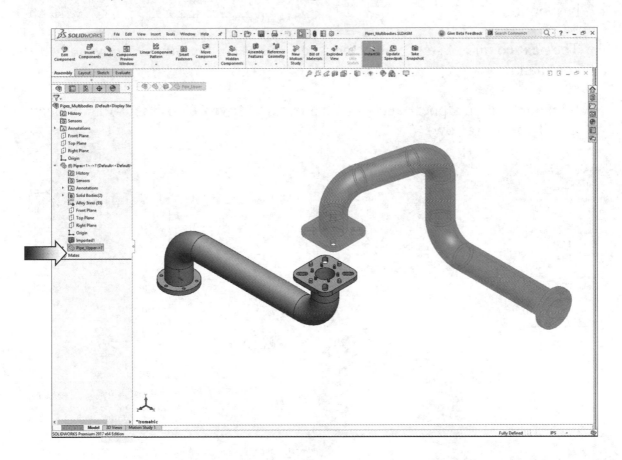

The <u>Pipes</u> is the name of the main component, and the <u>Pipe-Without-Pins</u> is the name of the inserted part, which is being treated as a body of the main part.

The intent is to reorient the Pipes-Without-Pins and mate it to the part Pipes, then combine the 2 solid bodies into one. After the two parts are merged as one, the small pins in the Pipes must be consumed by the combine operation, leaving no interferences between the bodies.

2. Switching to the Part level:

On the FeatureManager tree, click the name of the part **Pipes**, a pop-up menu appears, select **Open Part** (arrow).

The part Pipes is opened at the part level.

When working with multibody parts, it is a bit easier to work in the part mode instead of in the assembly. When it comes to moving, copying, rotating, or mating the bodies, the command Move / Copy Body would be your best option.

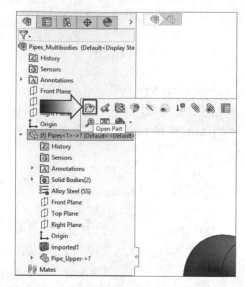

The next couple of steps will demonstrate the use of the Move / Copy Body command.

Expand the Pipe Upper, right-click the feature **Move / Copy Body1** and select: **Edit Feature** (arrow).

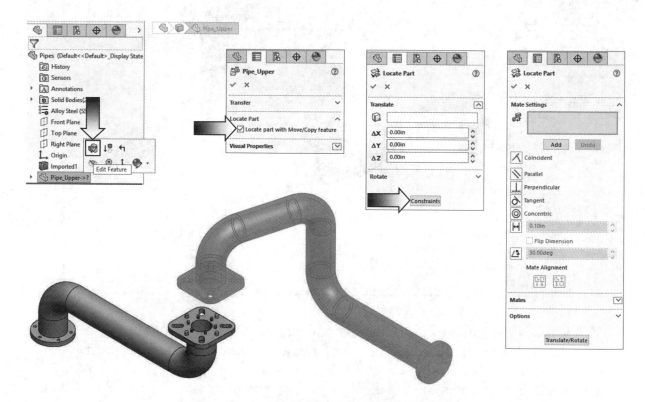

Enable the **Locate Part with Move/Copy feature** checkbox and click the **Green** checkmark. Click the **Constraints** button (arrow) to switch to the Mate Settings section.

3. Changing the mate:

Select the 2 faces as indicated.

The Coincident
mate is
selected
by default.

Click **Add**.

Click **OK** to
close the Locate Part command.

4. Adding new mates:

Select **Insert / Features / Move-Copy** (arrow).

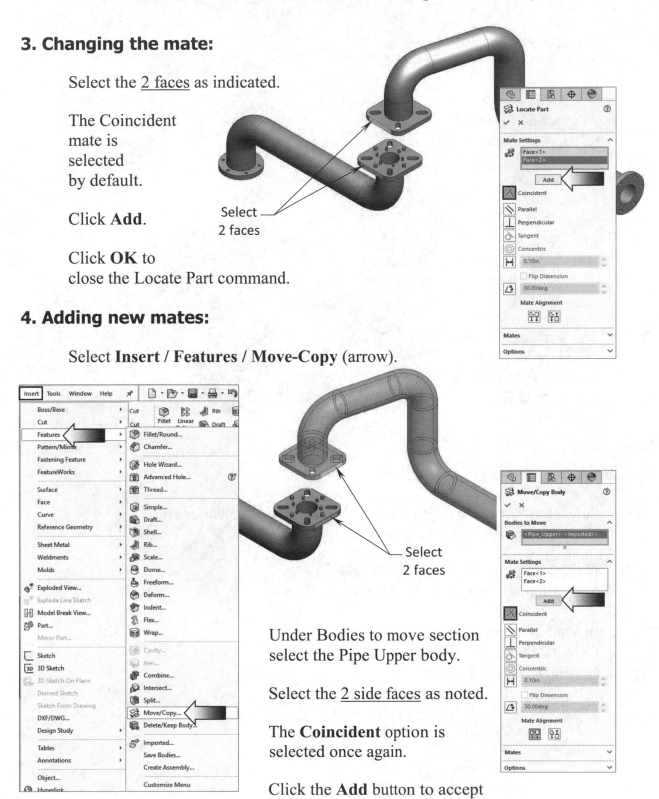

Select
2 faces

Select
2 faces

Under Bodies to move section
select the Pipe Upper body.

Select the 2 side faces as noted.

The **Coincident** option is
selected once again.

Click the **Add** button to accept
the new mate.

Keep in mind that it takes 3 mates to fully position a part or a body, so do not
close the Mate options just yet; one more mate needs to be added to the 2 bodies in
the next step.

Select the upper and lower faces of the 2 flanges.

Another **Coincident** relation is added to the flanges.

Click **Add**.

The Pipe Upper body becomes fully defined, or fully positioned.

The 3 mates are now saved in the Mates section. These mates can be modified or removed if needed.

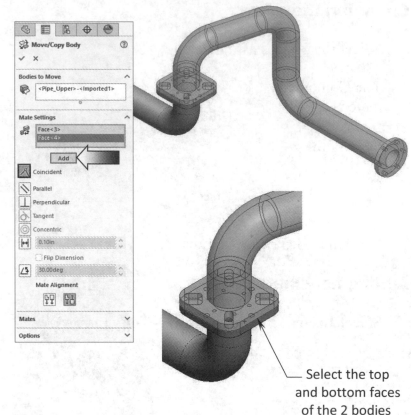

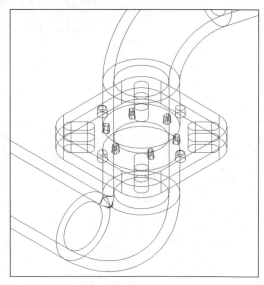

Select the top and bottom faces of the 2 bodies

5. Changing the display option:

Select the Wireframe option under the Heads-Up toolbar or click:
View / Display / Wireframe.

Notice the Pins in the Imported1 body are interfering with the Pipe Upper body.

Instead of deleting or cutting those pins from the part, we are going to try a different approach: **Combine Add**.

Combine is one of the unique features in SOLIDWORKS that offers three different options:

* **Add** = Combines multiple bodies into a single body.
* **Subtract** = Subtract overlapping material from a selected main body.
* **Common** = Removes all material except that which overlaps.

6. Combine Add the bodies:

Select **Insert / Features / Combine** (arrow).

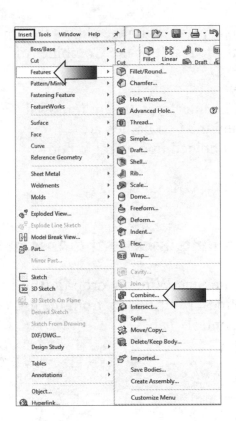

For Operation Type selection, click the **Add** option (arrow).

For Bodies to Combine selection, select <u>both bodies</u> from the graphics area.

Notice the pins in the lower body are still visible? They will be removed as soon as the operation is completed.

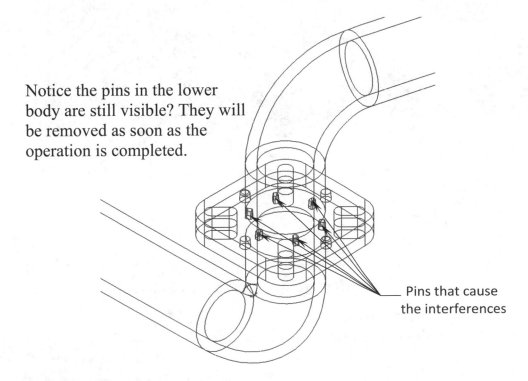

Pins that cause the interferences

Click the **Show Preview** button.

The preview image shows the pins are removed as the result of the Combine Add.

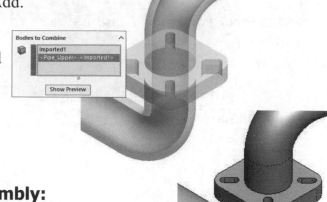

The interferences are removed and the two bodies are combined into a single body.

Click **OK**.

7. Switching back to the Assembly:

Click the Window drop-down menu and select the Pipes_Multibodies from the list, or press the hotkey Control+Tab.

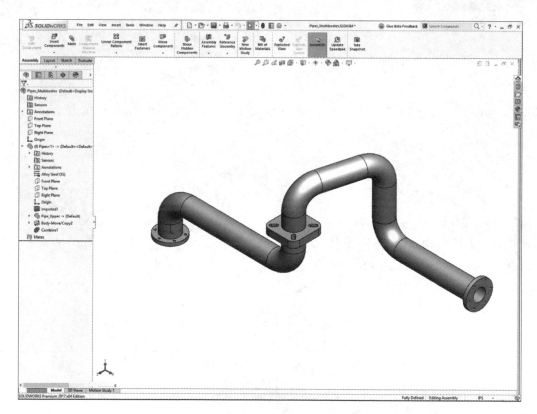

A dialog box appears asking to rebuild the assembly. Click **Yes**.

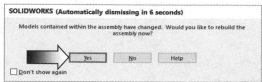

Resave your assembly and overwrite the existing document.

CHAPTER 5

Revolved & Thread Features

Revolved & Thread Features
Cylinder

The revolve command adds or removes material by revolving one or more profiles around a centerline. The revolve feature can be a boss/base, revolved cut, or revolved surface. The revolve feature can be a solid, a thin feature, or a surface.

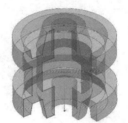   

Solid Feature   Thin Feature

The sketch for a thin or surface revolved feature can contain multiple open or closed intersecting profiles. The profile sketch must be a 2D sketch; 3D sketches are not supported for profiles. The Axis of Revolution can be a 3D sketch.

Profiles cannot cross the centerline. If the sketch contains more than one centerline, select the centerline you want to use as the axis of revolution. For revolved surfaces and revolved thin features only, the sketch cannot lie on the centerline.

When you dimension a revolve feature inside the centerline, you produce a radius dimension for the revolve feature.
When you dimension across the centerline, you produce a diameter dimension for the revolve feature.

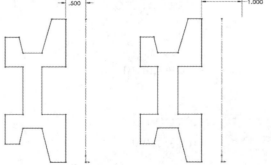

Revolved & Thread Features
Cylinder

Dimensioning Standards: **ANSI**

Units: **INCHES** – 3 Decimals

Tools Needed:

 Insert Sketch

 Circle

 Circular Sketch Pattern

 Dimension

 Geometric Relation

 Chamfer

 Revolved Boss/Base

 Thread

 Section View

1. Opening a part document:

Select **File / Open**.

Locate the part document named: **Revolved Features** and open it.

The sketch has been fully dimensioned to help focus on the revolved features.

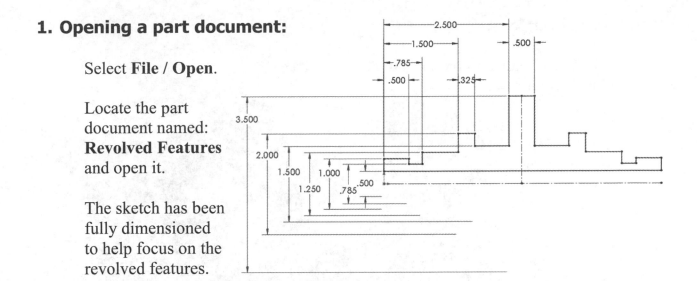

2. Creating a revolved feature:

Switch to the **Features** tool tab.

Click the **Revolved Boss/Base** command.

Select the **horizontal centerline** as noted for the Axis of Revolution.

Use the default **Blind** condition.

Keep the revolve angle at **360°**.

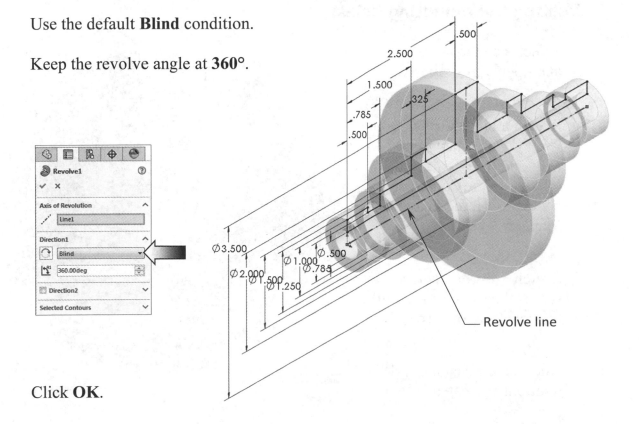

Revolve line

Click **OK**.

3. Adding chamfers:

Select the **Chamfer** command under the Fillet drop down arrow.

Enter **.032in** for the depth of chamfer.

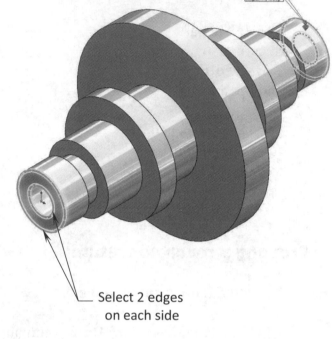

Keep the default angle **45.00 deg**.

Select the **4 edges** as noted.

Click **OK**.

Select 2 edges on each side

4. Creating the mounting holes:

Select the <u>face</u> as indicated and open a **new sketch**.

Sketch a **large circle** (Bolt-Circle*) and change it to **construction** geometry (click the For Construction checkbox on the Feature-Manager tree).

Sketch a **small circle** that shares the same center with the construction circle.

Add the dimensions and relations indicated.

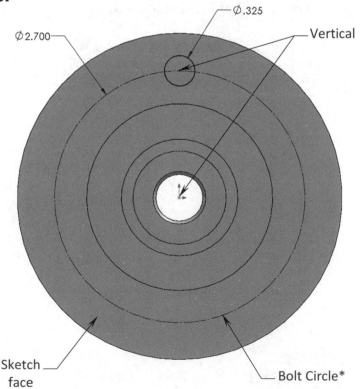

Ø.325

Vertical

Ø2.700

Sketch face

Bolt Circle*

Select the **small circle**.

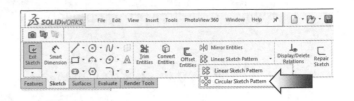

Expand the Linear Sketch Pattern and select the **Circular- Sketch- Pattern** command (arrow).

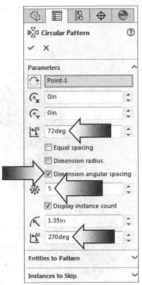

Enter **72deg** for spacing.

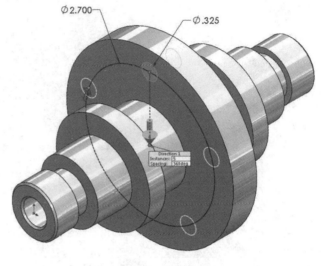

Enable the **Dim. Angular Spacing** checkbox.

Enter **5** for Number of Instances.

Click **OK**.

The circle is replicated 5 times but their centers are not yet constrained to the construction circle.

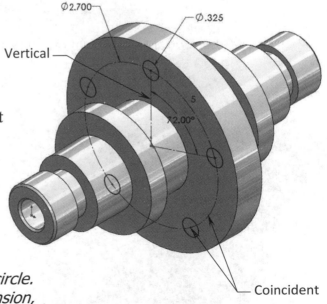

Add the **Vertical** and the **Coincident** relations as shown to fully define the sketch.

Bolt Circle: A theoretical circle on which the centerpoints of bolt holes lie when the bolt holes are positioned as equally spaced in a circle. It is often treated as a basic dimension, when true position for the centers is specified using GD&T.

Switch to the **Features** tool tab.

Click the **Extruded Cut** command.

Select the **Up-To-Next** condition for Direction 1.

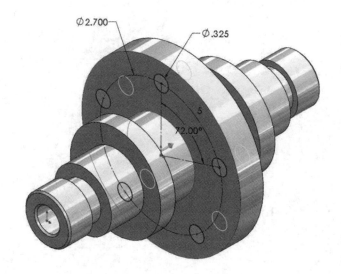

Click **OK**.

Rotate the model to different orientations to inspect the cut feature.

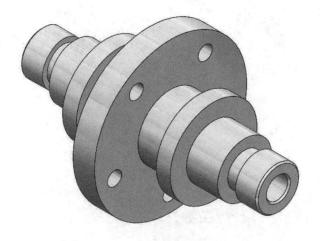

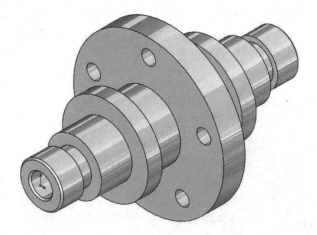

5. Adding threads:

Threads can be created with the "traditional method" by sweeping a thread profile along a helical path or by using the new Thread feature.

The Thread tool does not autosize the thread to the model.

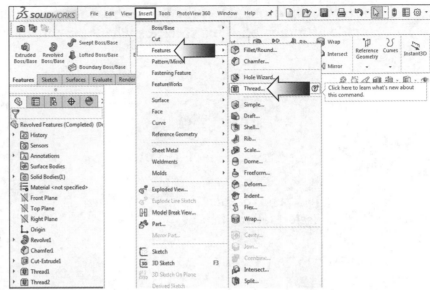

This tool accelerates creating an extrude or cut thread on a hole or shaft, based on the selected profile.

The next couple of steps will discuss the use of the Thread feature.

Select **Insert / Features / Thread** (arrow).

For Thread Location, click the **Edge** as noted.

For Start Location, click the **Face** shown.

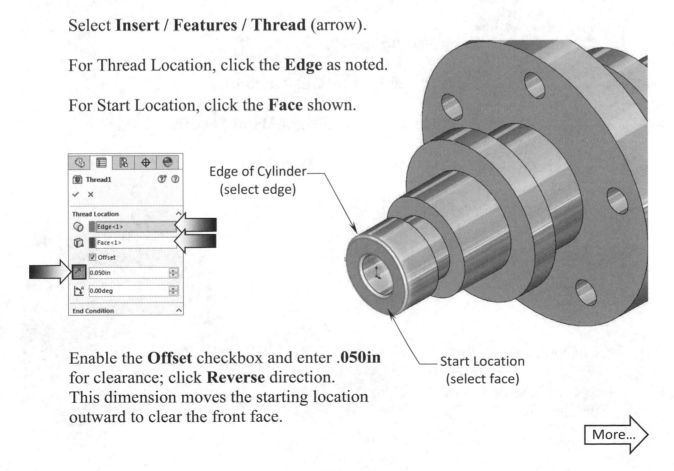

Edge of Cylinder
(select edge)

Start Location
(select face)

Enable the **Offset** checkbox and enter **.050in** for clearance; click **Reverse** direction. This dimension moves the starting location outward to clear the front face.

More...

For End Condition, select the **Up To Surface** option and click the <u>**face**</u> on the <u>far side</u> as indicated. This specifies where the threads will end.

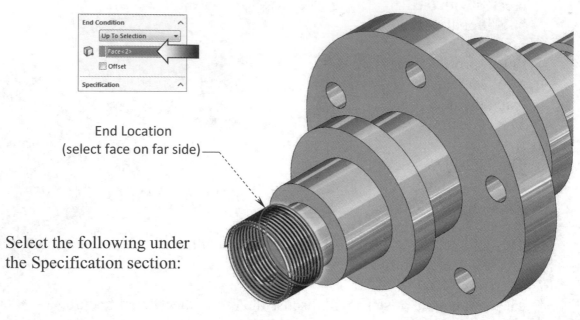

End Location
(select face on far side)——

Select the following under
the Specification section:

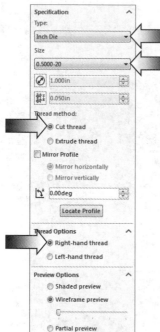

* Type: **Inch Die**.

* Size: **0.5000-20**.

* Thread Method: **Cut Thread**.

* Thread Options: **Right-Hand Thread**.

* Preview Options: **Wireframe**.

* Keep other settings at their
default values.

Click **OK**.

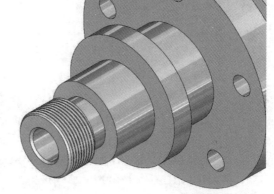

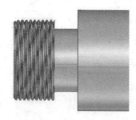

Change to the **Right**
orientation (Control+4)
to verify the details of
the threads.

6. Adding threads to the opposite end:

Switch back to the isometric view (Control+7) and press the hotkey **Shift+Up Arrow** <u>twice</u>. This shortcut will rotate the view orientation precisely 180°.

Select: **Insert / Features / Threads** once again.

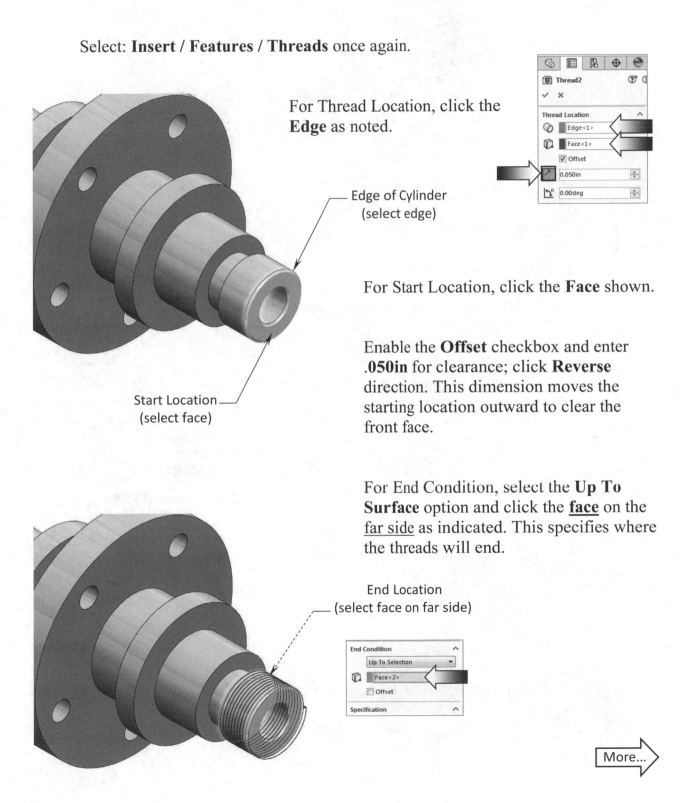

For Thread Location, click the **Edge** as noted.

Edge of Cylinder
(select edge)

For Start Location, click the **Face** shown.

Enable the **Offset** checkbox and enter .050in for clearance; click **Reverse** direction. This dimension moves the starting location outward to clear the front face.

Start Location
(select face)

For End Condition, select the **Up To Surface** option and click the <u>**face**</u> on the <u>far side</u> as indicated. This specifies where the threads will end.

End Location
(select face on far side)

More...

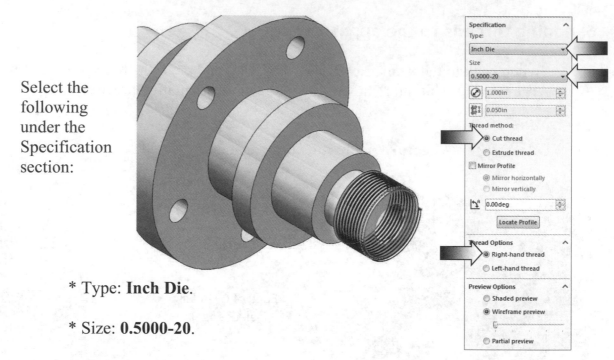

Select the following under the Specification section:

* Type: **Inch Die**.

* Size: **0.5000-20**.

* Thread Method: **Cut Thread**.

* Thread Options: **Right-Hand Thread**.

* Preview Options: **Wireframe**.

* Keep other settings at their default values.

Click **OK**.

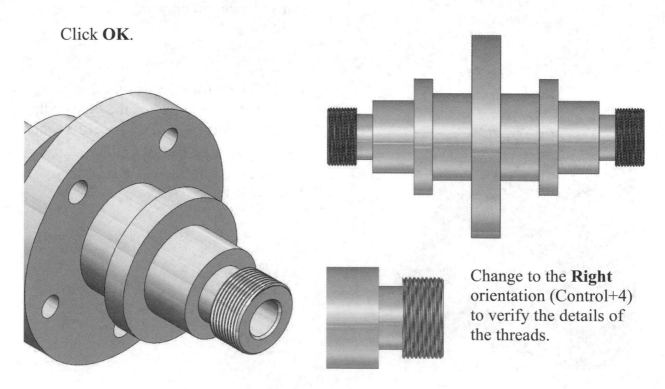

Change to the **Right** orientation (Control+4) to verify the details of the threads.

7. Creating a Zonal Section View:

You can create section views in a model by cutting away multiple areas of the model. These multiple areas are defined by "intersection zones."

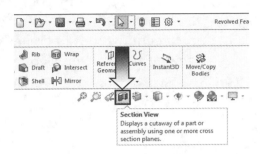

Click the Section View command from the **View (Heads-Up)** toolbar (arrow).

For Section Method, select the **Zonal** option.

For Section 1, select the **Top** plane.

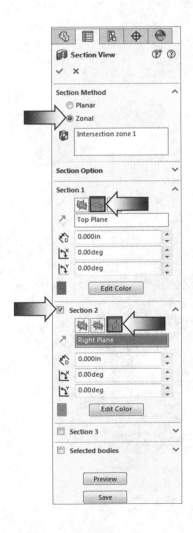

Enable the Section 2 checkbox and select the **Right** plane.

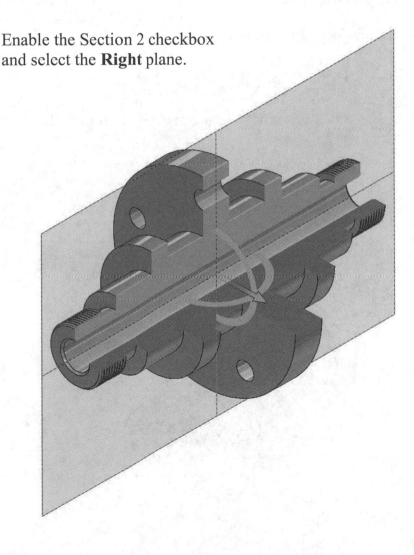

Click **OK**.

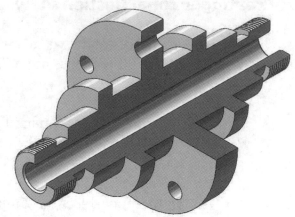

When you select section planes or faces, the bounding box of your selections is the intersection zones. You can use the intersection zones to create a section view of multiple areas of a model.

8. Saving your work:

Click **File / Save As**.

Use the <u>same file name</u> and press **Save**.

Overwrite the existing one when prompted.

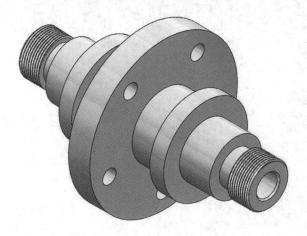

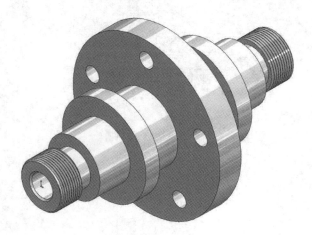

Exercise: Revolved & Threads

1. Opening a part document:

Browse to the Training Folder and open the part document named **Revolved & Threads.sldprt**

Edit the **Sketch1**.

The sketch has already been fully defined. Notice the "Virtual Diameter" dimensions? They were measured from the centerline to the entities that will become cylindrical features after the sketch is revolved.

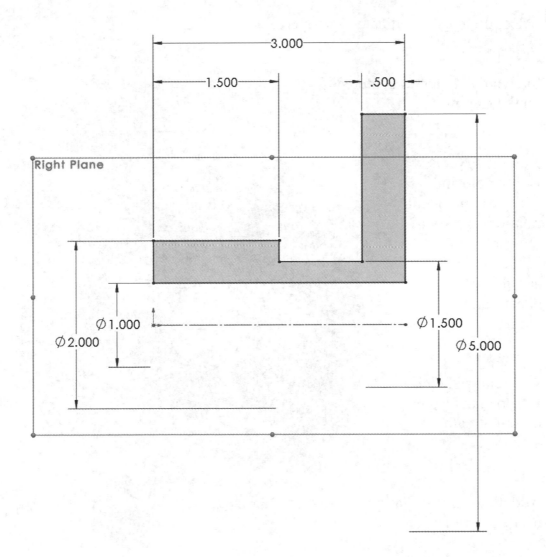

2. Revolving the sketch:

Switch to the **Features** tool tab and select:
Revolve Boss/Base.

Use the default **Blind** condition and revolve the sketch **360deg**.

Click **OK**.

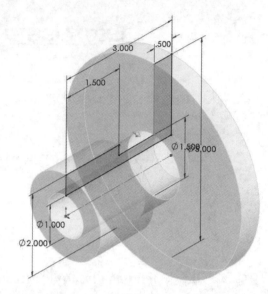

3. Creating the holes:

Select the <u>face</u> as indicated and open a **new sketch**.

Sketch the **4 circles** centered on the origin.

Add the Ø3.50 Bolt Circle and add the Coincident relations between the center points of the circles and the construction circle.

A Circular Sketch Pattern can also be used to create the 4 circles.

Add dimensions/relations needed to fully define the sketch.

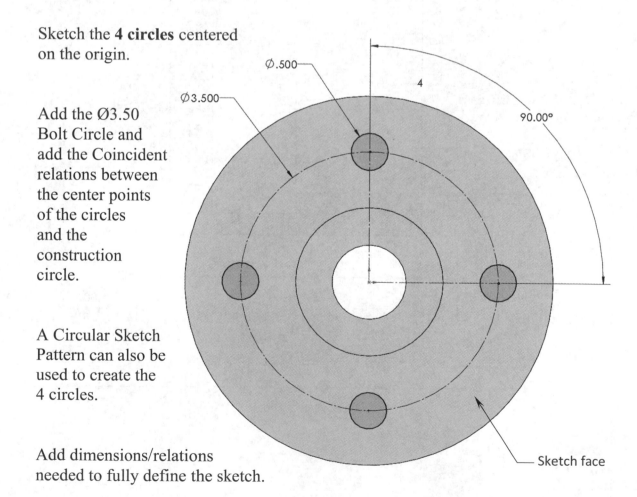

Sketch face

4. Creating an extruded cut:

Click the **Feature** tool tab.

Select **Extruded Cut**.

For Direction1, select **Through All**.

Click **OK**.

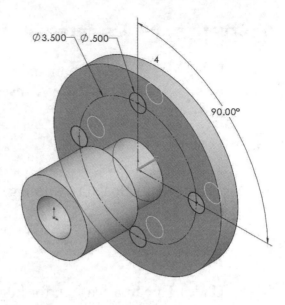

5. Adding chamfers:

Select the **Chamfer** command from the Fillet drop down list.

Click the **Angle and Distance** option (arrow).

For Distance, enter **.032in**.

For Angle, keep it at the default **45deg**.

Select the **edges** of the holes and the 2 cylindrical features on both ends.

There should be a total of **14 edges**.

Click **OK**.

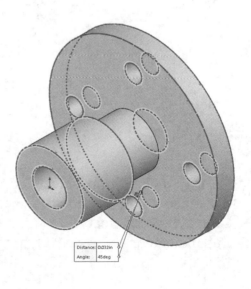

6. Creating the first internal threads:

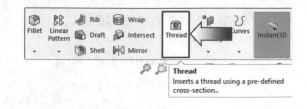

Select the **Thread** command on the Features tool tab – or click **Insert, Features, Thread**.

Click **OK** to close the warning dialog box.

Under Thread Location, for Edge of Cylinder select the <u>circular edge</u> of the center hole as noted.

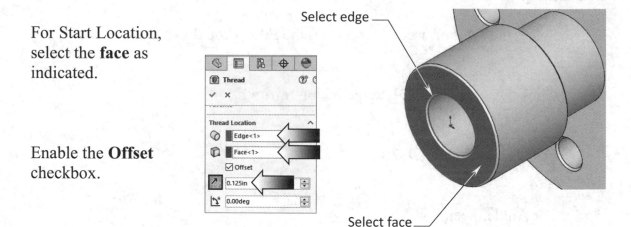

For Start Location, select the **face** as indicated.

Select edge

Enable the **Offset** checkbox.

Select face

For Offset Distance, enter **.125in** and click **Reverse**.

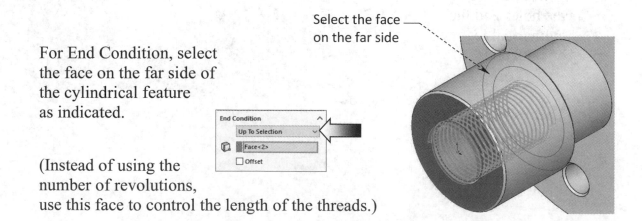

Select the face on the far side

For End Condition, select the face on the far side of the cylindrical feature as indicated.

(Instead of using the number of revolutions, use this face to control the length of the threads.)

For Specification, select:

Type: **Inch Tap**

Size: **1.000-12**

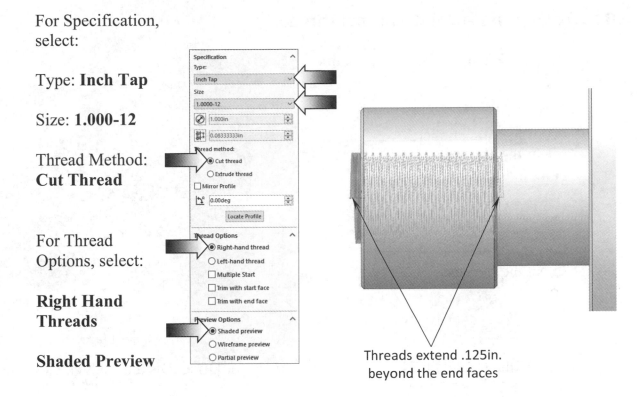

Thread Method: **Cut Thread**

For Thread Options, select:

Right Hand Threads

Shaded Preview

Threads extend .125in. beyond the end faces

Click **OK**.

7. Creating a section view:

Create the section view to see the thread details more clearly.

Click the Section View command from the **Heads-Up View** tools.

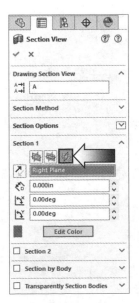

For Section 1, select the **Right** plane as the cutting plane.

Zoom in to inspect the threads and push **Esc** when finished.

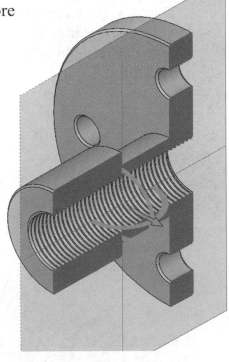

8. Creating the second internal threads

Select the **Thread** command again.

For Edge of Cylinder, select the **circular edge** indicated.

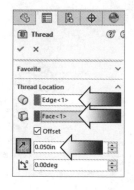

For Start Location, select the **face** as noted.

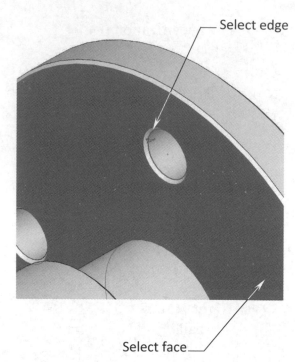

Select edge

Select face

Enable the **Offset** check box and enter **.050in** for Offset Distance. Click **Reverse**.

For End Condition, select **Up to Selection** and click the bottom face of the model.
The preview graphics show the threads extend beyond the 2 end faces by **.050in**.

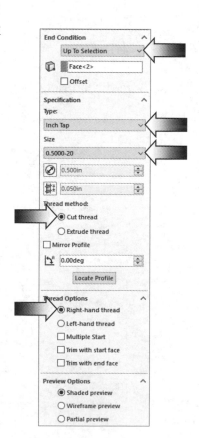

Set the following:

Inch Tap
0.5000-20
Cut Thread
Right hand Thread
Shaded Preview

Click **OK**.

Threads extend .125in beyond the end faces

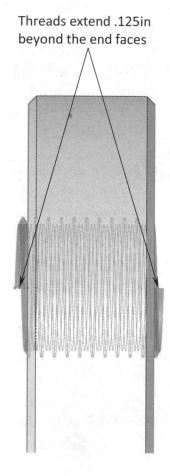

9. Creating a Section View:

Click the **Section View** command from the **Heads-Up View** tools.

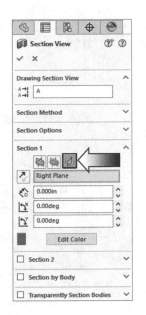

Select the **Right** plane as the Cutting plane.

Examine the threads and push **Esc** when finished.

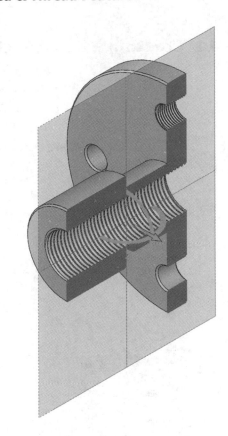

Create a **Circular Pattern** of the small hole with **4 instances**.

10. Assigning material:

Right-click the Material selection and select **Plain Carbon Steel** (arrow).

Enable **RealView Graphics** if applicable.

RealView gives the model and its material a more realistic look and dynamic representation without the need to render.

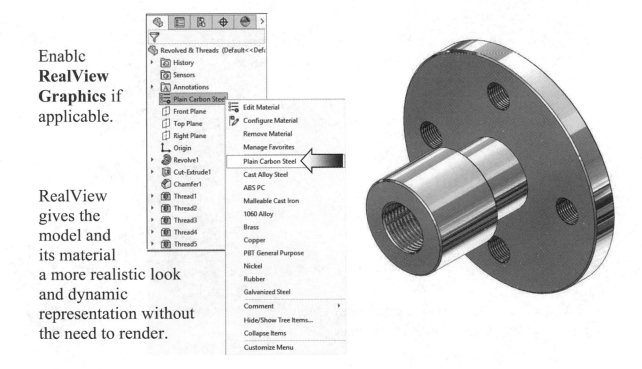

11. Creating a Zonal Section View:

You can create section views in a model by cutting away multiple areas of the model. These multiple areas are defined by "intersection zones."

Click the **Section View** command.

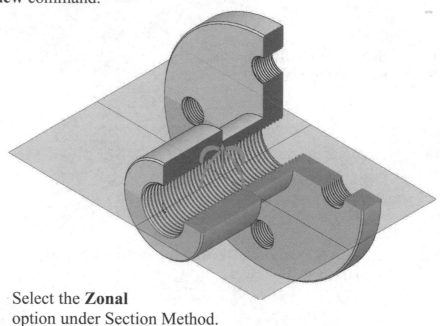

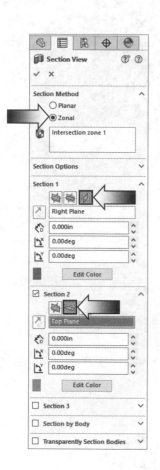

Select the **Zonal** option under Section Method.

For Section 1, select the **Right** plane for cutting plane.

Enable the Section 2 checkbox and select the **Top** plane.

Click **OK**.

12. Saving your work:

Click **File, Save As**.

Use the same file name and press **Save**.

Overwrite the previous document when prompted. Close all documents.

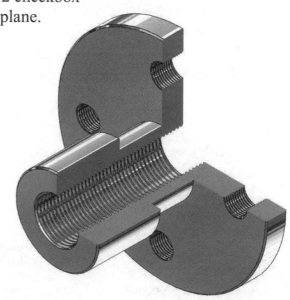

CHAPTER 6

Flex Bending

Flex Bending

Flex features are useful for modifying complex models
with predictable, intuitive tools for many applications including concepts
mechanical design, industrial design, stamping dies, molds, and so on.

There are 4 flex types available:

* **Flex Bending** Bends one or more solid or surface
  bodies about the bend axis (Red X-Axis). The neutral plane
  of the bend passes through the triad Origin and corresponds
  to the triad's X-Z plane. The arc length between the trim planes along
  the neutral plane stays constant throughout the bending operation.

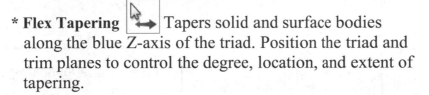

* **Flex Twisting** Twists solid and surface bodies
  around the blue Z-axis of the triad. Position the triad and
  trim planes to control the degree, location, and extent of
  twisting.

* **Flex Tapering** Tapers solid and surface bodies
  along the blue Z-axis of the triad. Position the triad and
  trim planes to control the degree, location, and extent of
  tapering.

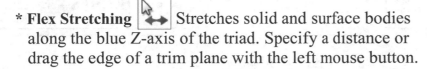

* **Flex Stretching** Stretches solid and surface bodies
  along the blue Z-axis of the triad. Specify a distance or
  drag the edge of a trim plane with the left mouse button.

The Flex feature calculates the extent of the part using a bounding box.
The trim planes are then initially located at the extents of the bodies, perpendicular
to the Z-axis of the triad. The Flex feature affects the region between the trim
planes only.

Flex Bending
Spanner

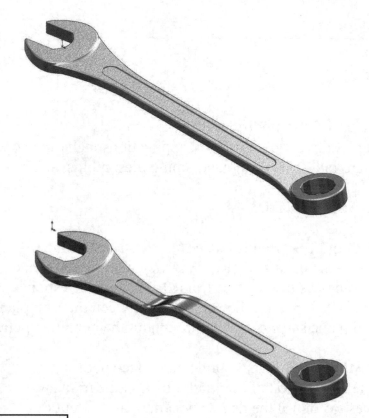

View Orientation Hot Keys:

Ctrl + 1 = Front View
Ctrl + 2 = Back View
Ctrl + 3 = Left View
Ctrl + 4 = Right View
Ctrl + 5 = Top View
Ctrl + 6 = Bottom View
Ctrl + 7 = Isometric View
Ctrl + 8 = Normal To Selection

Dimensioning Standards: **ANSI**

Units: **INCHES** – 3 Decimals

Tools Needed:

 Flex   Bending   Deform

Flex Bending – Part 1

1. Opening a part document:

Select **File / Open**.

Browse to the **Training Files** folder
and open a part document named:
Flex Bending 1.sldprt

2. Creating the first Flex-Bending feature:

The Flex features deform complex models in an intuitive manner. The tool is
useful for modifying complex models with predictable, intuitive tools for many
applications including concepts, mechanical design, industrial design, stamping
dies, molds, and so on. Flex features can change single body or multibody parts.

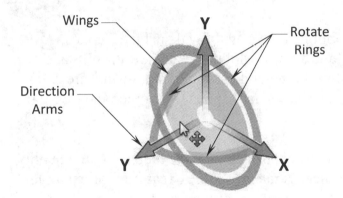

The Triad

The Flex command also uses the Triad to
facilitate manipulating various objects such
as 3D sketch entities, bodies, parts, certain
features, and components in assemblies.

The rings and wings are displayed when
rotation and dragging along the wings' planes
are possible.

Select **Insert / Features / Flex** (arrows).

Select the **model** in the graphics area for Flex Input.

Click the **Bending** option (arrow).

Enter **45deg** for Angle; the Radius is adjusted automatically.

Leave all other settings at their **default** values.

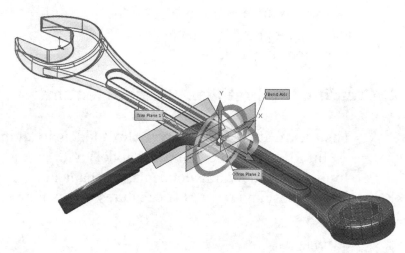

The **Hard Edge** option creates analytical surfaces (cones, cylinders, planes, and so on), which often result in split faces where the trim planes intersect the bodies. If cleared, results are spline-based, so surfaces and faces may appear smoother and original faces remain intact.

The **Flex Options** control surface quality. Increasing the quality also increases the success rate of the flex feature. For example, if the flex shows an error message, move the slider towards the right. Increase the quality only as needed because this may decrease the computer performance.

The preview graphic shows the result of the bending.
Position the triad and trim planes to control the degree, location, and extent of bending if needed.

Click **OK**.

3. Creating the second Flex-Bending feature:

Select **Insert / Features / Flex** once again.

Select the **model** from the graphics area.

Click the **Bending** option (arrow).

Enter **45deg** for Angle, the Radius is adjusted automatically.

Leave all other settings at their **default** values.

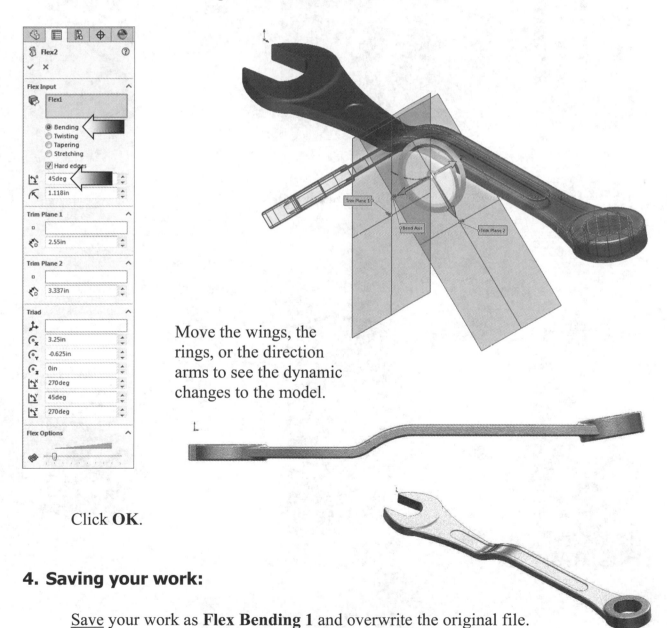

Move the wings, the rings, or the direction arms to see the dynamic changes to the model.

Click **OK**.

4. Saving your work:

Save your work as **Flex Bending 1** and overwrite the original file.

Flex examples

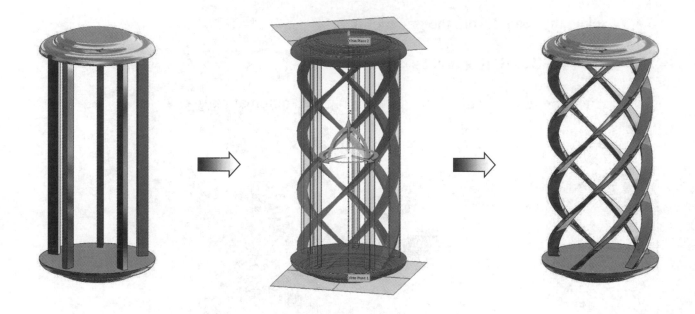

Flex Twisting

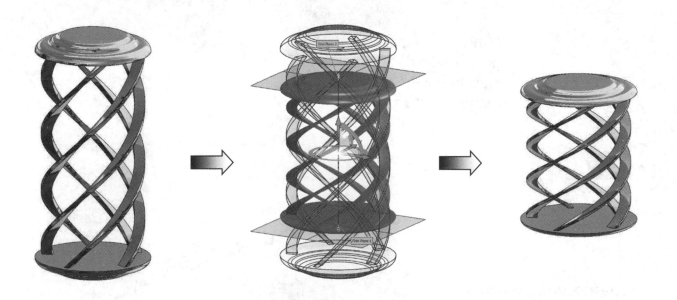

Flex Stretching

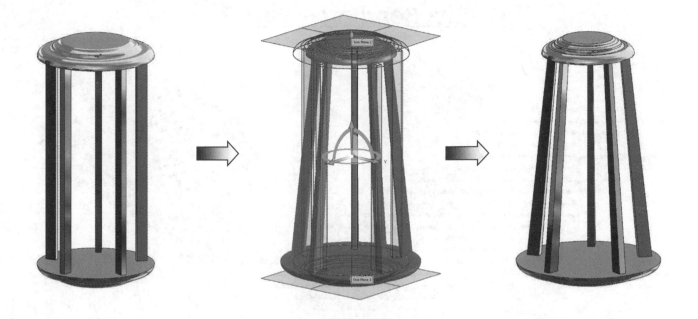

Flex Tapering

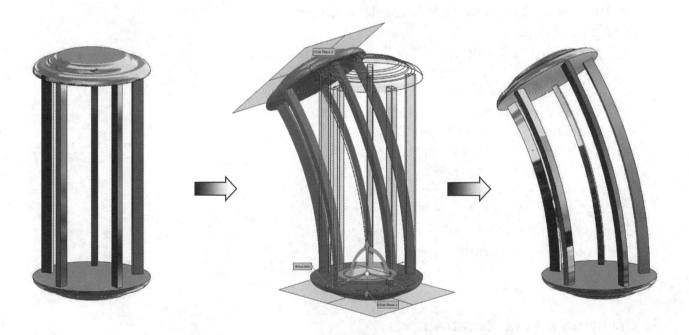

Flex Bending

Flex Bending – Part 2

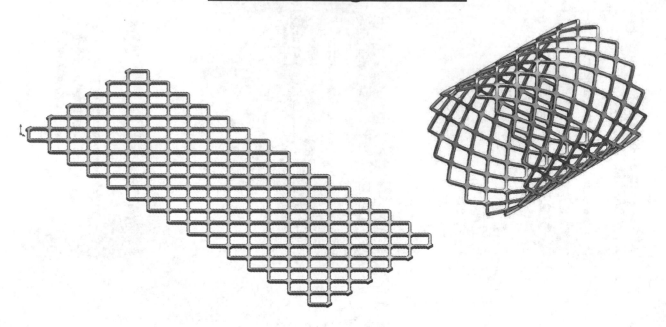

1. Opening a part document:

Select **File / Open**.

Browse to the **Training Files** folder and open a part document named:
Flex Bending 2.sldprt.

The Flex feature uses the Triad to position the center of the bend and the 2 Trim Planes to define the bend angle or the bend radius. The neutral plane of the bend passes through the triad origin and corresponds to the triad's X-Z plane. The arc length between the trim planes along the neutral plane stays constant throughout the bending operation.

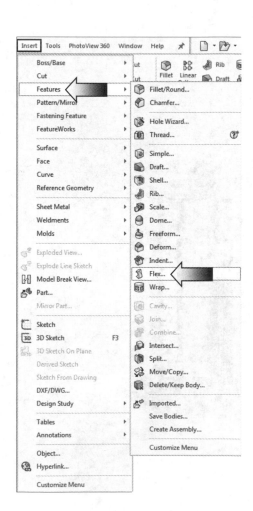

2. Creating a Flex-Bending:

Select **Insert / Features / Flex** (arrows).

Select **Insert / Features / Flex** once again.

Select the **model** from the graphics area.

Click the **Bending** option (arrow).

Enter **359.9deg** for Angle; the Radius is adjusted automatically.

Leave all other settings at their **default** values.

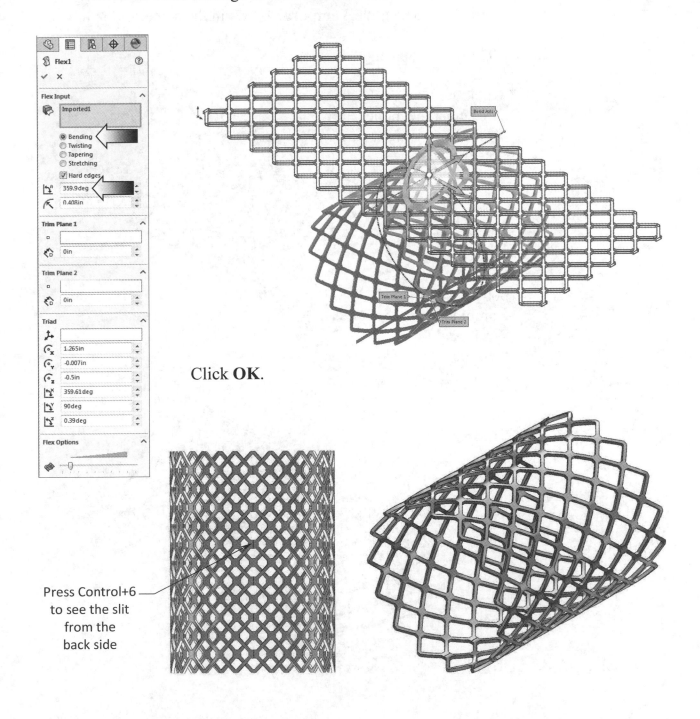

Click **OK**.

Press Control+6 to see the slit from the back side

3. Creating a Linear Pattern:

Switch to the **Features** tool tab.

Select **View / Hide/Show / Temporary Axis**. We will use the temporary axis in the center of the model as the pattern direction.

Click the **Linear Pattern** command and enter the following parameters:

* Pattern Direction: Select the **Temporary Axis** in the center of the model.

* Enter **1.0225in** for Spacing.

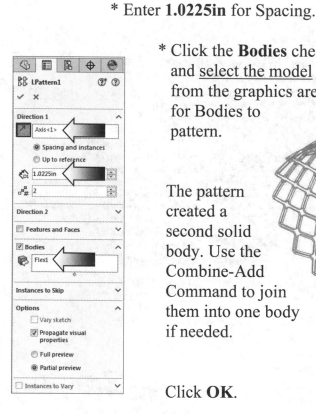

* Click the **Bodies** checkbox and <u>select the model</u> from the graphics area for Bodies to pattern.

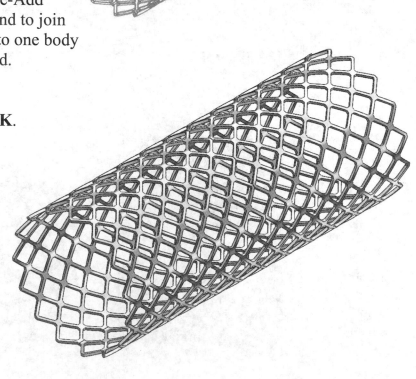

The pattern created a second solid body. Use the Combine-Add Command to join them into one body if needed.

Click **OK**.

4. Saving your work:

Save your work as **Flex Bending 2** and overwrite the original file when prompted.

Close all documents.

Exercise: Flex Bending a Tire

The Flex tool bends the model by moving the Trim Planes along the X,Y, or Z directions, or by entering the desired Bend Radius or Bend Angle.

1. Opening a part document:

Browse to the Training Folder and open the part document named: **Flex Bending_Tire**.

The fillets on the treads are skipped in this exercise to reduce the time it will take to create this complex bend.

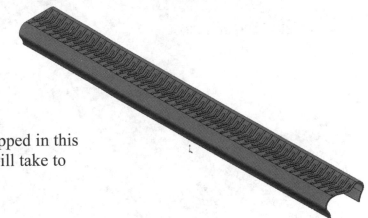

2. Creating a Flex Bending:

Select **Insert, Features, Flex**.

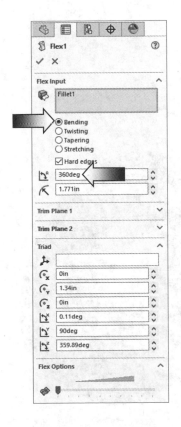

Click the **Bending** option.

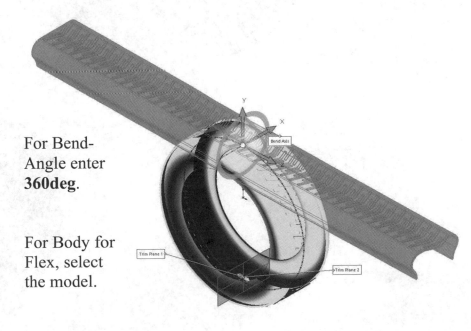

For Bend-Angle enter **360deg**.

For Body for Flex, select the model.

Click **OK**.

3. Saving your work:

Select **File, Save As**.

Enter **Flex Bending _Tire.completed** for file name.

Click **Save** and close the document.

Using the Deform feature

There are 3 options in deform: Deform with a Point, Curve-To-Curve deform, and Surface Push. Curve to curve deform is a more precise option to deform a complex shape. You deform the part by starting with an initial curve to a set of target curves.

We will take a look at the Curve-To-Curve option and learn to deform a solid and a surface body, using a 2D sketch curve to control the deform shape.

1. Opening a part document: (Disable 3D Interconnect under: **Tools, Options, Import**).

For the first half, we will learn how to deform a <u>Solid-Body</u>.
From the training folder, locate and open the document named **Deform Solid**.

Click **NO** to bypass the Feature Recognition options.

This part document comes with a 2D sketch curve. Change to the Front view orientation (Control+1).

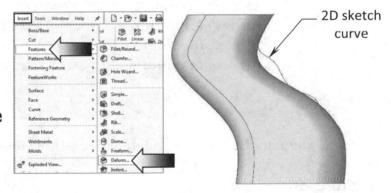

2D sketch curve

2. Creating a Curve-to-Curve deform:

Click **Insert / Features / Deform**.

Click the **Curve to Curve** option.

For Initial Curve, click the **right edge** of the part as shown.

For Target Curve, select the **2D curve sketch** as noted.

(more on next page...)

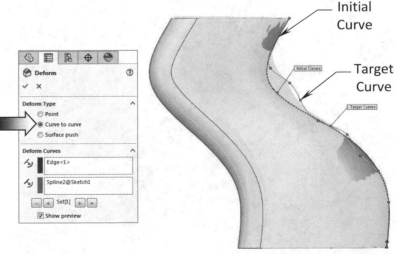

Initial Curve

Target Curve

Click in the **Anchor** section.

Select the **10 edges** as indicated. (The curved edges on top and bottom are touching the Target-Curve and cannot be used as the fixed edges.)

The 10 selected edges will remain fixed; they will not be affected by the deform feature.

Under the Shape Options* section, drag the **Accuracy** slider all the way to the right.

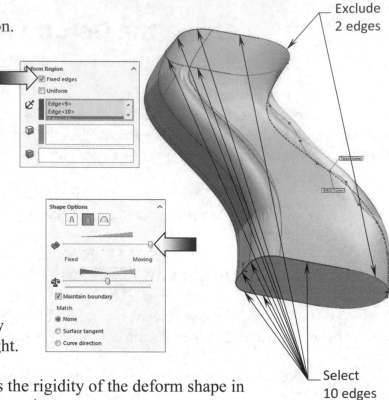

Exclude
2 edges

Select
10 edges

       Stiffness: Controls the rigidity of the deform shape in
               the deformation processes.
       Stiffness minimum: Least amount of rigidity.
       Stiffness medium: Medium amount of rigidity.
       Stiffness maximum: Largest amount of rigidity.

Click **OK**.

Change to the Front orientation to examine the results (Ctrl+1).

3. Save and close.

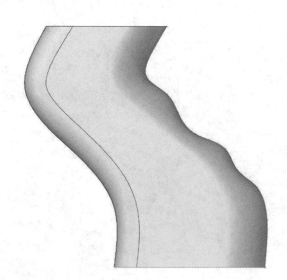

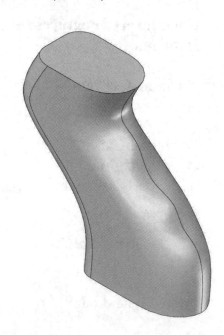

1. Opening a part document: (Disable 3D Interconnect under: **Tools, Options, Import**).

For the second half, we will try deforming a <u>Surface-Body</u>. Open the part document named **Deform Surface**.

Click **NO** to bypass the Feature Recognition options.

This document contains one half of the surface body part. After it is deformed, we will mirror it to make a complete surface body part.

2. Creating a Curve to Curve deform:

Select **Insert / Features / Deform**.

Click the **Curve to Curve** option, if not yet selected.

Change to the Front orientation (Ctrl+1).

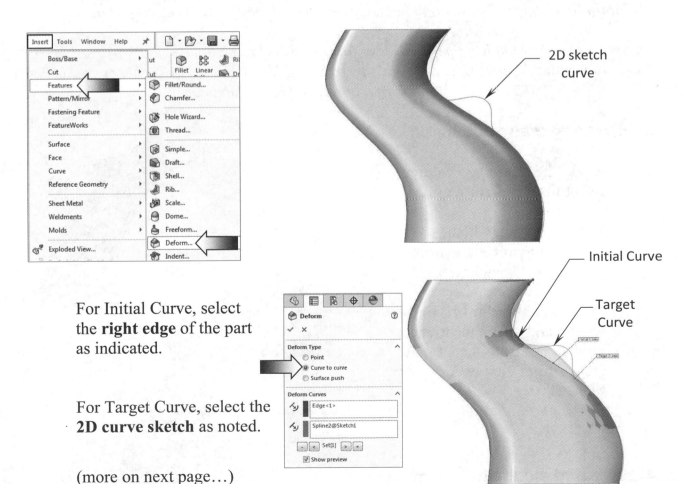

2D sketch curve

Initial Curve

Target Curve

For Initial Curve, select the **right edge** of the part as indicated.

For Target Curve, select the **2D curve sketch** as noted.

(more on next page…)

Click the **Anchor** section.

Select the **3 edges** as indicated to keep fixed.

Under the Shape Options drag the **Accuracy** slider all the way to the right.

(Note: Increasing the surface accuracy will decrease the computer performance.)

Click **OK**.

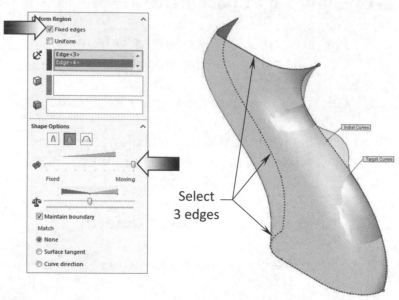

Select 3 edges

Shape accuracy: *Controls the surface quality.*
Weight: *Controls the degree of influence between two options: Fixed Edges and Moving Curves.*
Maintain boundary: *Ensures that boundaries selected for fixed curves/edges/faces are fixed.*
Surface tangent: *Matches target edges of faces and surfaces with a smooth transition.*
Curve direction: *Matches Target curves by mapping Initial curves to Target curves using the normal of the Target curves to shape the deform.*

3. Mirroring a surface body:

Click ⊞ from the Features toolbar or select **Insert / Features / Mirror**.

For Mirror Face / Plane, select the **Front** plane from the FeatureManager tree.

Expand the **Bodies to Mirror** section and click the **surface body** either from the graphics area or from the Feature-Manager tree.

Click **OK**.

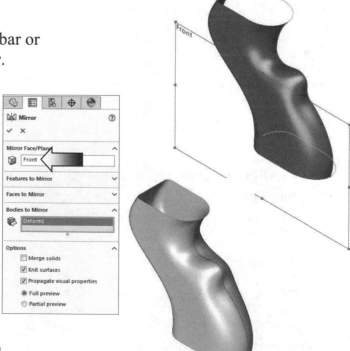

4. Save and close all documents.

Deform Using a 3D Sketch

1. Opening a part document:

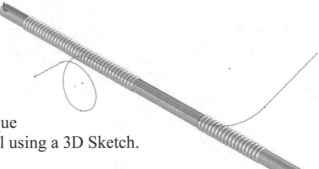

Browse the Training Files folder
and open the part document named:
Deform with 3D Sketch.sldprt

This exercise will demonstrate a unique
technique on deforming a solid model using a 3D Sketch.

2. Copying a solid body:

We will make a copy of the model and deform the original body so that we can
compare the results between the 2 solid bodies side by side.

Click 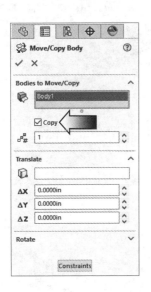 or select: **Insert, Features, Move/Copy**.

For Bodies to Move/Copy, select the **model** from the graphics area.

Enable the **Copy** checkbox (arrow).

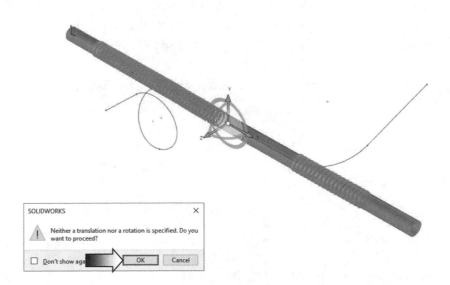

Click **OK**.

Click **OK** in the pop-up dialog to proceed. A copy of the solid body is created and
placed over the original body.

3. Deforming with a 3D Sketch:

Select **Insert, Features, Deform**.

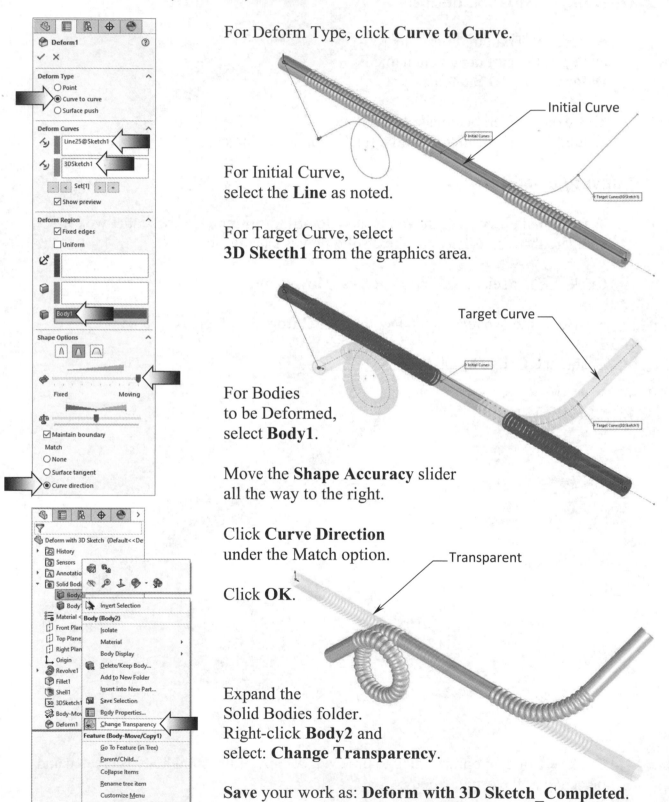

For Deform Type, click **Curve to Curve**.

For Initial Curve,
select the **Line** as noted.

For Target Curve, select
3D Skecth1 from the graphics area.

For Bodies
to be Deformed,
select **Body1**.

Move the **Shape Accuracy** slider
all the way to the right.

Click **Curve Direction**
under the Match option.

Click **OK**.

Expand the
Solid Bodies folder.
Right-click **Body2** and
select: **Change Transparency**.

Save your work as: **Deform with 3D Sketch_Completed**.

CHAPTER 7

Sweep with Guide Curves

Sweep with Guide Curves
Plastic Bottle

A sweep creates a base, boss, cut, or surface by moving a profile along a path, or by specifying a path and a diameter.

For a circular profile, you do not need to sketch the profile. You only need to select an existing path and specify a diameter for the profile.

If the path is a 3D curve, one or more guide curves are used to help control the twisting in the sweep. The profile must be constrained to the guide curves with either a coincident or a pierce relation.

There are three types of sweeps: Sweep Boss/Base, Cut Sweep, and Surface Sweep. The Swept Boss/Base tool is used when creating the external threads, and the Swept Cut is used to make the internal threads.

You can create helical threads on cylindrical faces using profile sketches, and you can store custom thread profiles as library features.

Using the Thread tool, you can define the start thread location, specify an offset, set end conditions, specify the type, size, diameter, pitch, and rotation angle, and choose options such as right-hand or left-hand thread.

You can create sweeps for a mid-path profile in either direction or the entire path using Direction 1, Direction 2, or Bidirectional. The bi-directional option is available for swept boss/base, swept surface, and swept cut in parts and assemblies.

Sweep with Guide Curves
Plastic Bottle

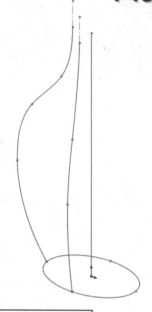

View Orientation Hot Keys:

Ctrl + 1 = Front View
Ctrl + 2 = Back View
Ctrl + 3 = Left View
Ctrl + 4 = Right View
Ctrl + 5 = Top View
Ctrl + 6 = Bottom View
Ctrl + 7 = Isometric View
Ctrl + 8 = Normal To Selection

Dimensioning Standards: **ANSI**

Units: **INCHES** – 3 Decimals

Tools Needed:

 Swept Boss/Base   Split Line   Fillet

 Ellipse   Offset Surface   Knit Surface

 Cut with Surface   Revolved Boss/Base   Shell

 Plane   Helix/Spiral

1. Opening a part document:

Select **File / Open**.

Browse to the Training Files location and open the part document named: **Sweep with Guide Curves.sldprt**.

Some of the sketches have been created to help focus on the Sweep feature.

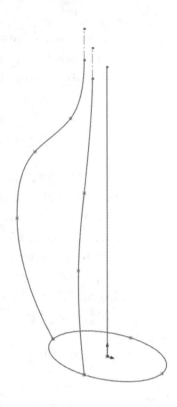

2. Creating the main body:

Switch to the **Features** tool tab.

Select the **Swept Boss/Base** command .

For Sweep Profile, select the **Ellipse** (or Sketch4).

For Sweep Path, select the **Vertical Line** (or Sketch3).

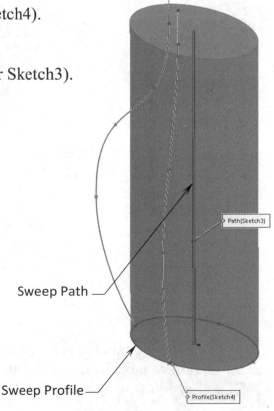

Sweep Path

Sweep Profile

(More settings on the next page…)

Expand the Guide Curves section and select **Sketch1** and **Sketch2** either from the graphics area or from the FeatureManager tree.

Click the **Preview Section**s button (arrow) and press the Up / Down arrow to see the preview graphics on how the profile is moved along the path.

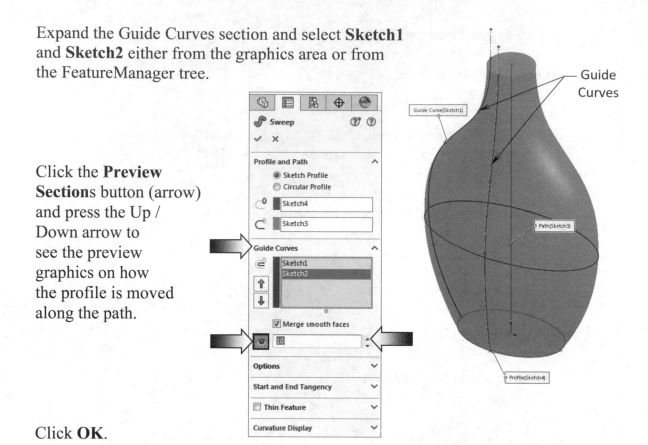

Guide Curves

Click **OK**.

Note: The Swept feature is shown with
         RealView Graphics enabled.

(RealView Graphics is hardware [graphics card] support of advanced shading in real time, including self-shadowing and scene reflections.)

3. Creating the first split line:

The Split Line command is used to split
a single face into 2 faces so that different
operations such as fillets, chamfers, or offsets
can be applied to the individual faces.

Select the <u>Front</u> plane and open a
new sketch.

Sketch a horizontal **Line** across
the bottom of the model.

Add the **Dimensions** shown to
fully define the sketch.

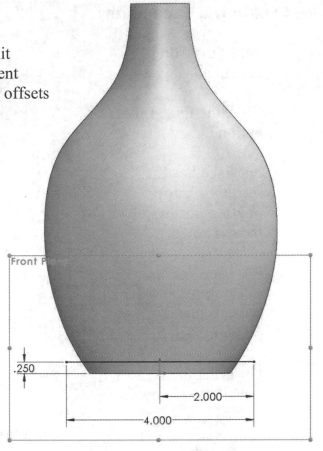

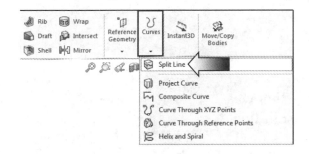

Switch to the **Features** tool tab.

Select the **Split Line**
command under the
Curves drop down.

For Type of Split, click
Projection type.

For Selections, use
the **Current Sketch**.

For faces to Split,
click **the body** of the
model as noted.

Click **OK**.

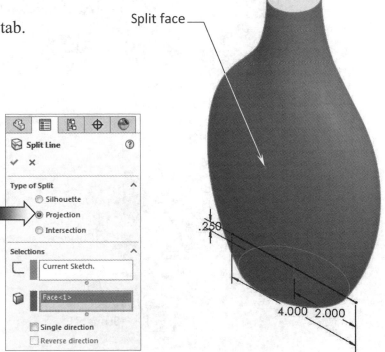

Split face

4. Creating a Face Fillet:

Select the **Fillet** command.

Click the **Face Fillet** option (arrow).

For Face Set 1, select the **upper portion** of the split face as noted.

For Face Set 2, select the **bottom face** of the model.

For Fillet Parameters, select **Hold Line**.

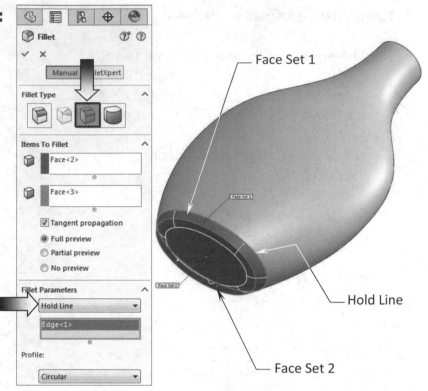

Face Set 1

Hold Line

Face Set 2

Select the **edge** that was created by the split line. (The Hold Line is used to determine the shape and size of the fillet.)

Click **OK**.

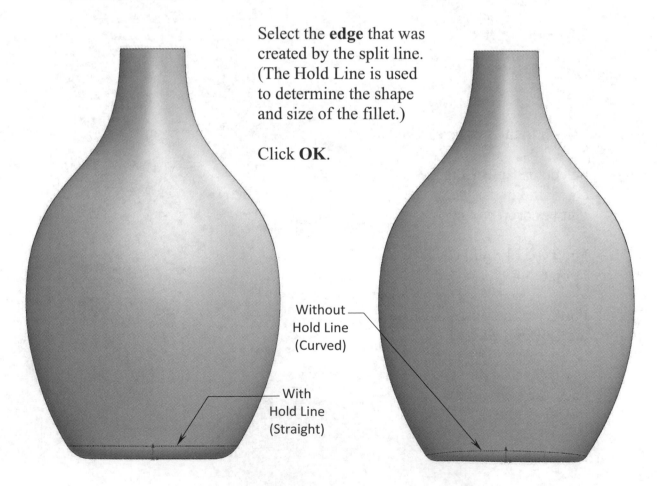

With Hold Line (Straight)

Without Hold Line (Curved)

5. Creating the second split line:

Select the **Front** plane and open a **new sketch**.

Sketch an **Ellipse** 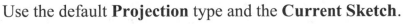 approximately as shown.

Add a **vertical relation** as noted to keep the ellipse from rotating.

Add the **dimensions** shown to fully define the sketch.

Switch to the **Features** tool tab.

Select the **Split Line** command below the Curves tool.

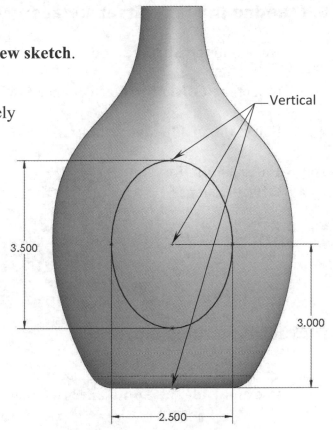

Use the default **Projection** type and the **Current Sketch**.

Select the **face** as indicated to split.

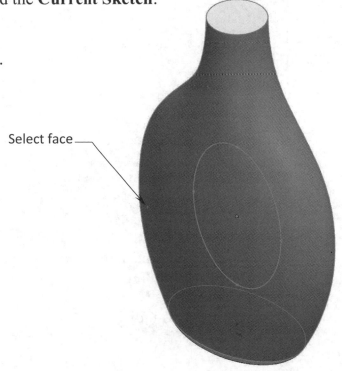

Click **OK**.

6. Creating the first offset surface:

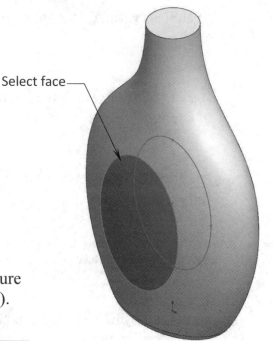

Switch to the **Surfaces** tool tab.

Select the **Offset Surface** command.

Select the **inside face** as noted.

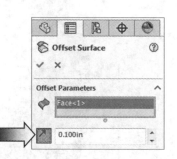

Enter **.040in**. for Offset Distance.

Click the **Reverse** button (arrow).

Click **OK**.

7. Creating a ruled surface:

Expand the Solid Bodies folder from the Feature tree and **Hide** the **Split Line2** (the main body).

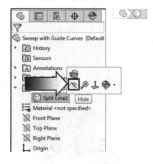

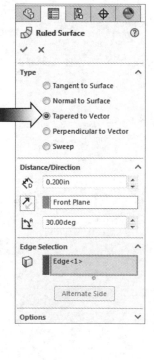

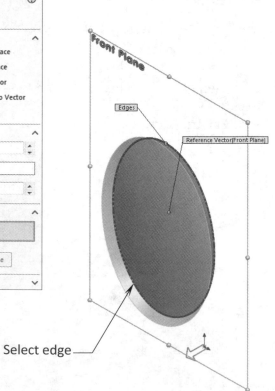

Select **Ruled Surface**.

Click **Taper to Vector**.

Enter **.200in**. for Distance.

Select the **Front** plane and enter **30.00deg** for Angle.

Select the **edge** as indicated.

Click **OK**.

8. Creating the first knit surface:

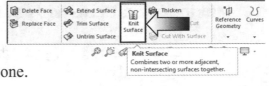

The Knit Surface command is used to combine two or more faces and surfaces into one.

Click the **Knit Surface** command.

Select 2 faces

Select the **2 surfaces** as noted.

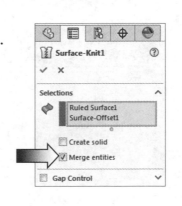

Enable the **Merge Entities** checkbox (arrow).

Click **OK**.

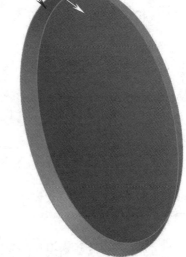

9. Adding the first .175" fillet:

Click the **Fillet** command and select the **Constant Radius Size** button (arrow).

Enter **.175in** for radius.

Select the **edge** in the middle of the 2 surfaces as indicated.

Enable the **Full Preview** option.

Click **OK**.

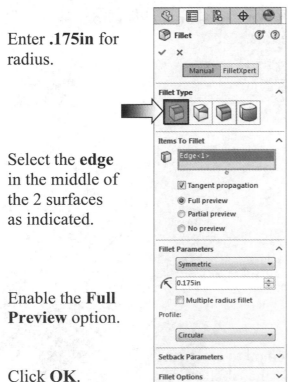

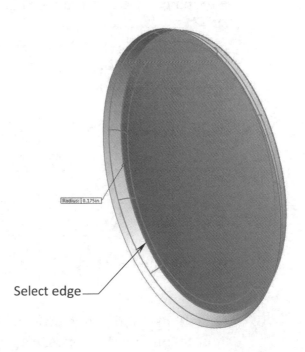

Radius: 0.175in

Select edge

10. Making a first cut-with-surface:

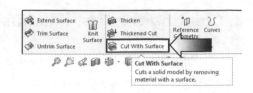

Using the Feature tree, <u>show</u> the **Split Line2** (the main body).

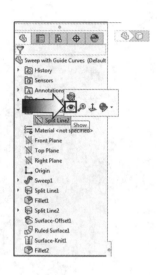

Arrow points to the side-to-remove

Click the **Cut with Surface** command from the **Surfaces** tool tab.

For Surface Cut Parameters, select the **Surface-Knit1** either from the graphics area or from the Feature tree.

Locate the Direction-Arrow and click to point it downward and remove the inner portion. Clicking Reverse also works.

Click **OK**.

Hide the Ruled Surface.

(We will take a look at a different method to create the cutout feature on the opposite side for the label area.)

11. Creating the first offset surface:

Repeat step number 6 and create a 2nd offset surface on the opposite side.

Use the same offset distance of **.100in**.

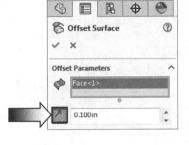

Click **Reverse** to place the new surface **inside** of the model.

Click **OK**.

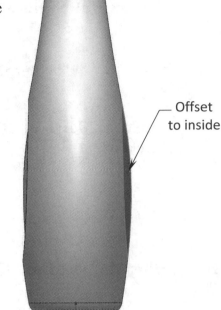

Offset to inside

12. Creating the second offset surface:

Hide the SurfaceCut1 (arrow).

Click **Ruled Surface**.

Use the default **Taper to Vector** option.

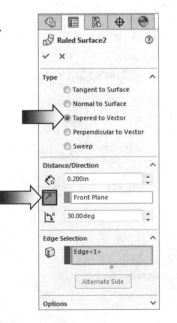

Enter **.200in** for distance.

Select the **Front** plane for Direction.

Enter **30deg** for Taper Angle.

Select the **edge** as noted.

Click **OK**.

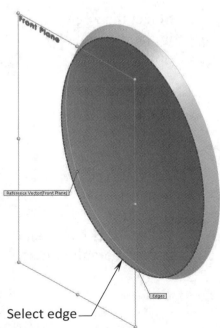

Select edge

13. Creating the second knit surface:

Click the **Knit Surface** command once again.

Select the **2 surfaces** as noted.

Enable the **Merge Entities** Checkbox.

Click **OK**.

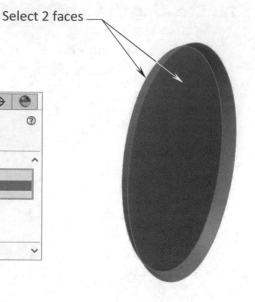

Select 2 faces

14. Adding the second .175" fillet:

Click the **Fillet** command

Select the **Constant Radius Size** button (arrow).

Enter **.175in** for radius.

Select the **edge** in the middle of the 2 surfaces as indicated.

Enable the **Full Preview** option.

Leave other parameters at their default values.

Click **OK**.

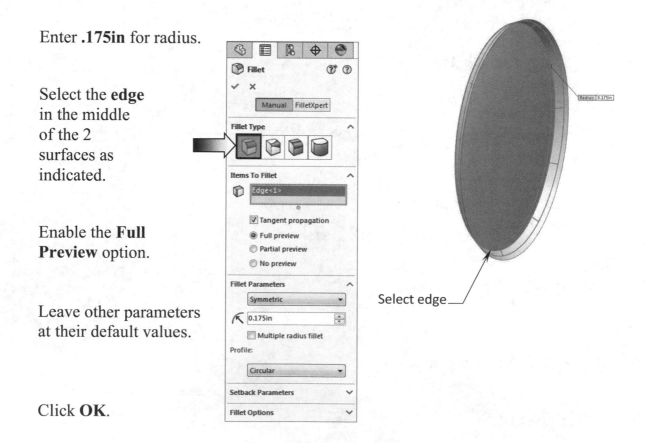

Select edge

15. Making a second cut-with-surface:

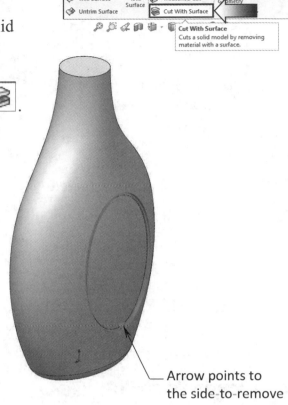

Using the Feature tree, **show** the main solid body.

Click the **Cut with Surface** command.

For Surface Cut Parameter, select the **Surface-Knit2** either from the graphics area or from the Feature tree.

Locate the Direction Arrow and click to point it upward and remove the inner portion. Click Reverse will also flip the direction arrow.

Arrow points to the side-to-remove

Click **OK**.

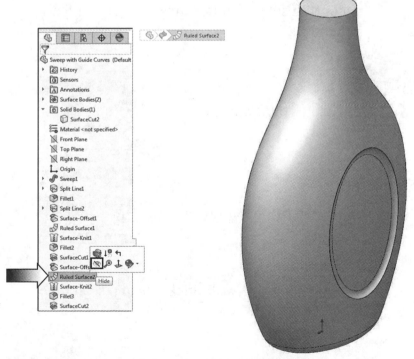

Hide the Ruled Surface (arrow).

Note: The recessed feature can be easily mirrored, but for the purpose of practicing, we went ahead and created both of them individually.

16. Adding the third .175" fillet:

Click the **Fillet** command.

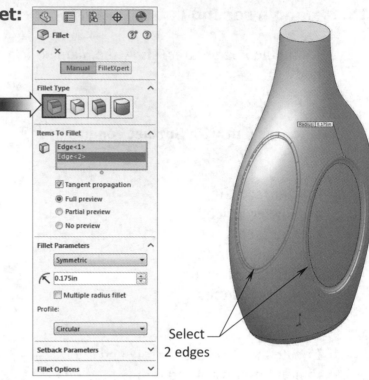

Use the default
Constant Size Radius
type (arrow).

Select the **2 edges** as noted.

Enter **.175in**. for radius.

Click **OK**.

Select 2 edges

17. Creating a shell:

The shell command hollows out a part, leaves open the faces you select, and creates thin-walled features on the remaining faces.

Face to remove

Switch to the **Features** tool tab.

Click the **Shell** command.

Enter **.060in**. for wall thickness and select the upper surface to remove.

Click **OK**.

18. Adding the neck feature:

The neck feature is added next so that the threads can be added after.

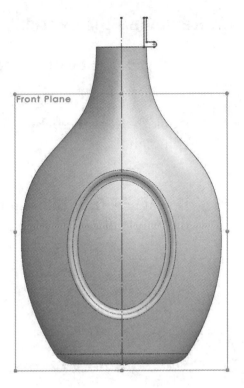

Select the <u>Front</u> plane and open a **new sketch**.

Sketch a **vertical centerline** that starts at the origin. This centerline will be used to add the dimensions and also to revolve the sketch.

Sketch the profile below using the **Line** tool.

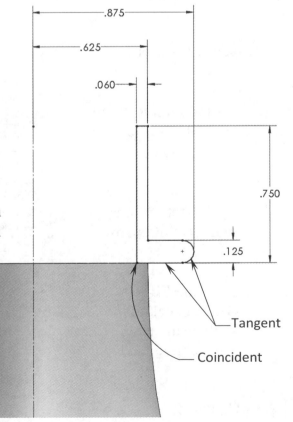

Add the dimensions shown. Hold the **Shift** key when adding the dimension .875. The shift key attaches the dimension to the tangent of the arc (called Maximum Condition).

Add the **Tangent** relation as noted to fully define the sketch.

19. Revolving the sketch:

Change to the **Features** tool tab.

Click the **Revolved Boss/Base** command.

Keep the default **Blind** type and the Angle at **360°**.

Click **OK**.

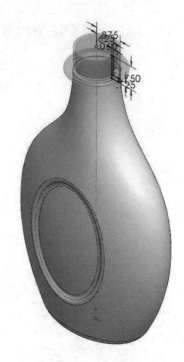

20. Creating a new plane:

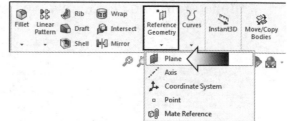

Using the "traditional method" a plane is used to define the offset distance of the threads.

Select **Reference Geometry / Plane** (arrow).

For the First Reference, select the **top face** of the neck feature.

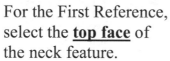

For Offset Distance, enter **.125in** (arrow).

Click the **Flip-Offset** checkbox to place the new plane <u>above</u> the face.

Click **OK**.

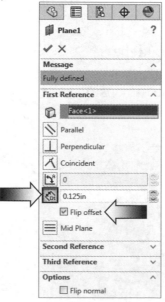

.125" below the top surface

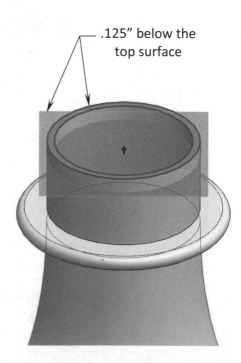

21. Creating a helix:

Select the <u>new plane</u> (Plane1) and open a **new sketch**.

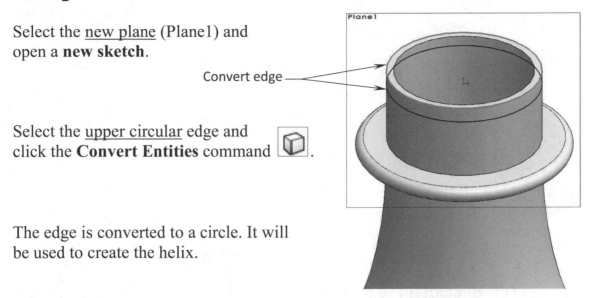

Convert edge

Select the <u>upper circular</u> edge and click the **Convert Entities** command .

The edge is converted to a circle. It will be used to create the helix.

Switch to the **Features** tool tab.

Select the **Helix/Spiral** command under the **Curves** drop down list.

For Defined by, select **Pitch and Revolution** (default).

For Parameters, set the following:

    * **Constant Pitch**

    * Pitch: **.100in**.

    * Reverse Direction: **Enabled**.

    * Revolution: **3.5**

Click **OK**.

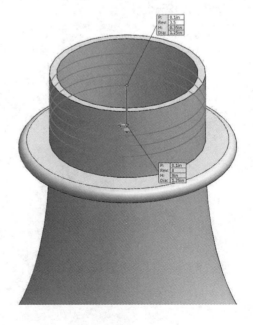

22. Sketching the thread profile:

Select the <u>Right</u> plane and open a **new sketch**.

Sketch the profile shown. Use the Mirror command to keep the entities Symmetric to each other.

The 2 centerlines at the left corner are there for clarity only. Do not create them.

Add the **dimensions** and the **Pierce** relation to fully define the sketch.

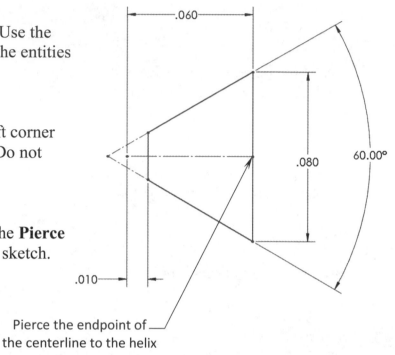

Pierce the endpoint of the centerline to the helix

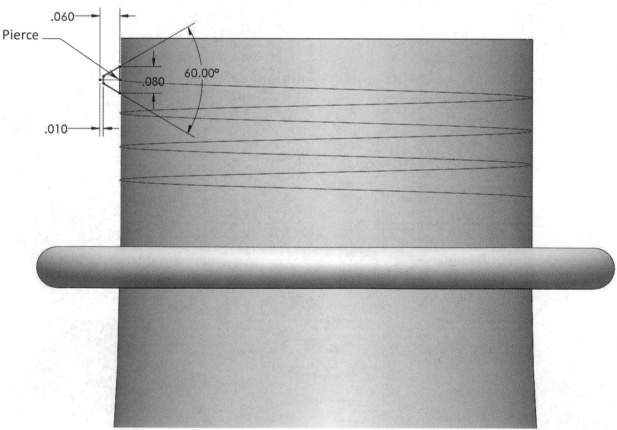

23. Making the Threads:

Switch to the **Features** tool tab.

Select the **Swept Boss/Base** command.

For Sweep Profile, select the sketch of the **Triangle**.

For Sweep Path, select the **Helix**.

Click **OK**.

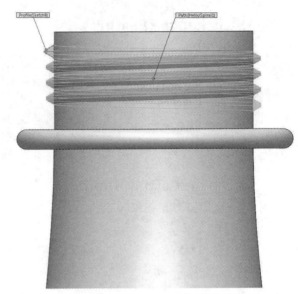

24. Rounding off the ends of the threads:

The ends of the threads need to be rounded off so that the cap can screw on more easily.

Convert face into sketch

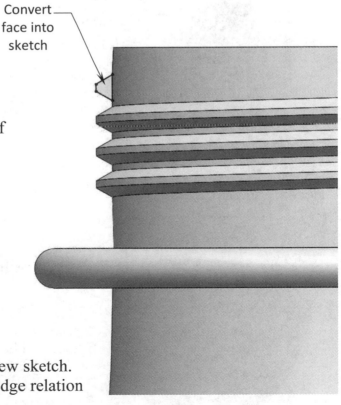

Select the **planar face** at the end of the thread and open a new sketch.

Highlight the planar face and press **Convert Entities** .

The planar face is converted to a new sketch. Each converted entity has an On-Edge relation added to them automatically.

25. Revolving the sketch:

Switch to the **Features** tool tab.

Press **Revolved Boss/Base**.

Select the **Vertical Line** for Axis of Revolution.

Keep the default **Blind** type and the angle of **360°**.

Click **OK**.

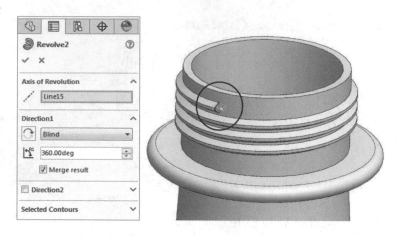

26. Rounding off the opposite Side:

Repeat steps 24 and 25 and close off the opposite side with the same revolved feature.

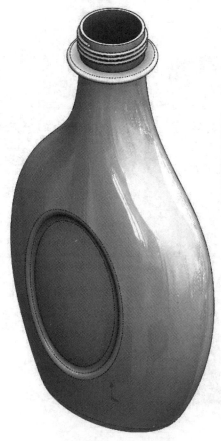

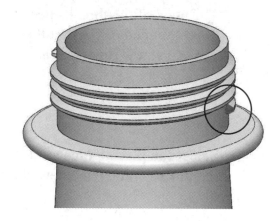

27. Saving your work:

Select **File / Save As**.

Use the same file name and overwrite the original document when prompted.

Click **Save**.

Sweep with Solid Body

Sweep with Solid Body
Creating a Cam Feature

Beside the traditional sweep using a sketch profile, we now have the option to sweep using a solid body. It is only available for the Cut Sweep command.

This feature is useful for end-mill simulation, where a solid body is swept along a path to create cuts around cylindrical bodies, or some other cam features.

Similar to the standard sweep that uses a profile and path, solid sweep requires the path to be tangent within itself and has no sharp corners.

The path can either be an open or closed contour and can be a set of 2D or 3D sketched lines, curves, or model edges.

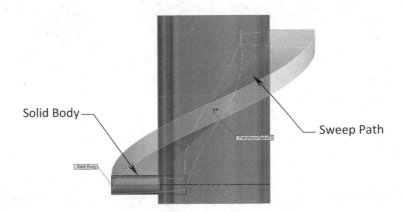

Solid Body

Sweep Path

Solid Body

The profile should be a revolved feature that consists of analytical geometry only, such as lines and arcs.

A plane perpendicular to the path is created, and the profile gets created on this plane and constrained to the sweep path.

Sweep with Solid Body
Creating a Cam Feature

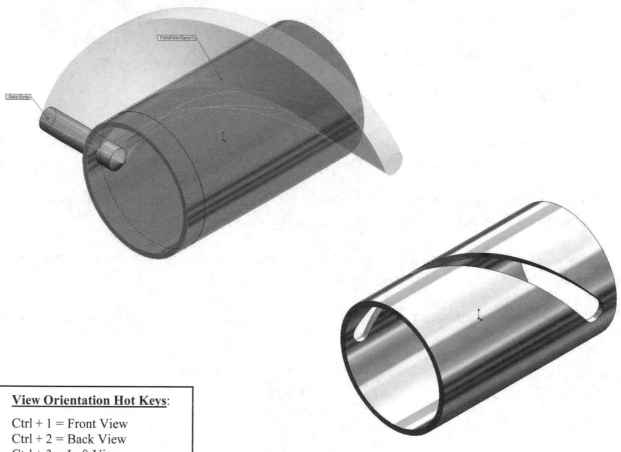

View Orientation Hot Keys:

Ctrl + 1 = Front View
Ctrl + 2 = Back View
Ctrl + 3 = Left View
Ctrl + 4 = Right View
Ctrl + 5 = Top View
Ctrl + 6 = Bottom View
Ctrl + 7 = Isometric View
Ctrl + 8 = Normal To
      Selection

Dimensioning Standards: **ANSI**

Units: **INCHES** – 3 Decimals

Tools Needed:

 Insert Sketch   Circle   Convert Entities

 Helix/Spiral   Plane   Swept Cut

1. Creating the main body:

Select the <u>Front</u> plane from the Feature tree and open a **new sketch** .

Sketch 2 **circles** centered on the origin as shown.

Add the OD and ID dimensions to fully define this sketch.

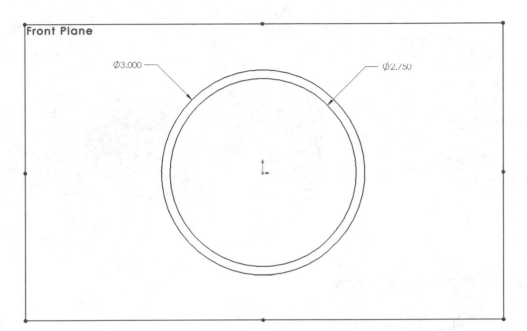

From the Features toolbar click **Extruded Boss-Base** .

Select the **Mid Plane** extrude option from the Direction 1 list.

Enter **5.000**in for extrude depth.

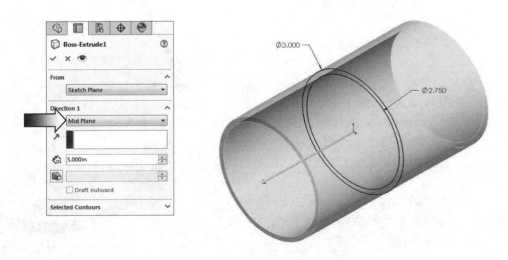

Click **OK**.

2. Creating an Offset-Distance plane :

Select the **Front** plane from the FeatureManager tree, hold the <u>Control</u> key and <u>drag</u> the edge of the front plane to the left, and then release the mouse button.

This technique makes a parallel copy of the selected plane.

The plane options appear on the tree and the Offset-Distance button is selected automatically.

Enter **2.000**in. for distance.

Click **OK**.

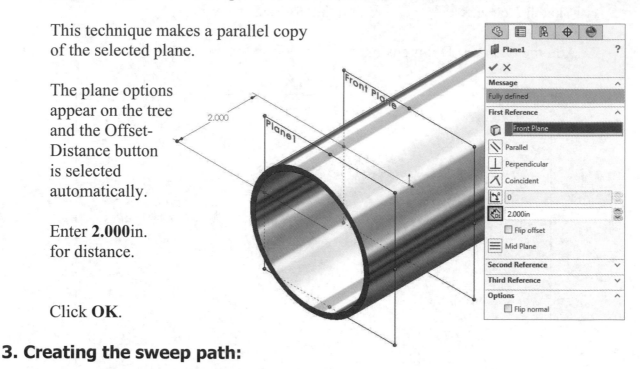

3. Creating the sweep path:

Select <u>Plane1</u> and open a new sketch 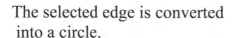.

Press **Control + 1** to change to the front orientation.

Select the <u>outer</u> circular edge (as noted) and click:

Convert Entities 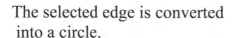 .

The selected edge is converted into a circle.

By converting the edge of the cylinder to a circle these 2 entities are now linked to one another.

Click the Helix 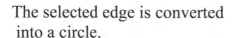 command from the **Features / Curves** toolbar.

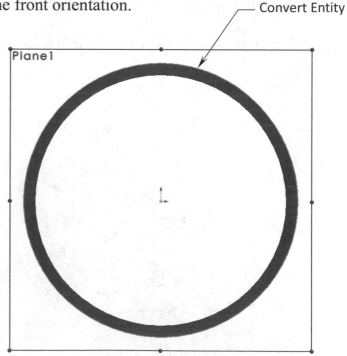

Convert Entity

For Defined By, use the **default** Pitch and Revolution option.

For Pitch, enter **7.750"**.

Click the **Reverse-Direction** checkbox.

Enter **0.5** (1/2) for revolutions.

For Start Angle, enter **180deg**.

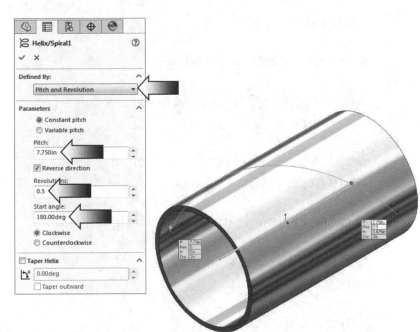

Use the default **Clockwise** direction.

Click **OK**.

4. Creating a plane Normal-To-Curve  :

Hold the <u>Control</u> key, select the Left endpoint of the helix <u>and</u> click the helix.

Click the **Plane** command (or select **Insert / Reference Geometry / Plane**).

The **Perpendicular** option should already be selected.

Click **OK**.

A new plane is created, normal to the helix and coincident to its left end.

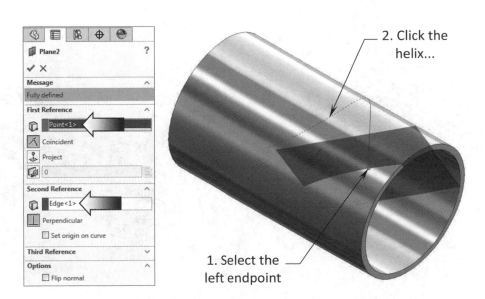

2. Click the helix...

1. Select the left endpoint

5. Creating the tool body:

Select the new <u>Plane2</u> and open a new sketch 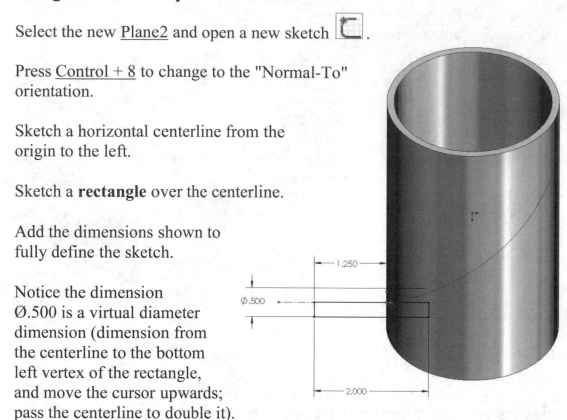.

Press <u>Control + 8</u> to change to the "Normal-To" orientation.

Sketch a horizontal centerline from the origin to the left.

Sketch a **rectangle** over the centerline.

Add the dimensions shown to fully define the sketch.

Notice the dimension Ø.500 is a virtual diameter dimension (dimension from the centerline to the bottom left vertex of the rectangle, and move the cursor upwards; pass the centerline to double it).

Press **Revolve Boss-Base** from the Features toolbar.

Use the default **Blind** and **360deg**. options.

<u>Clear</u> the **Merge Result** check box to make this feature a separate body from the first one.

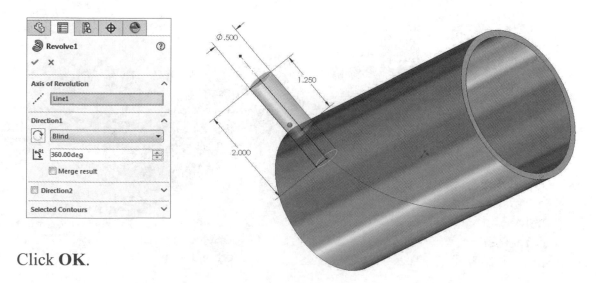

Click **OK**.

6. Creating the sweep cut with a solid body:

Click **Swept Cut** 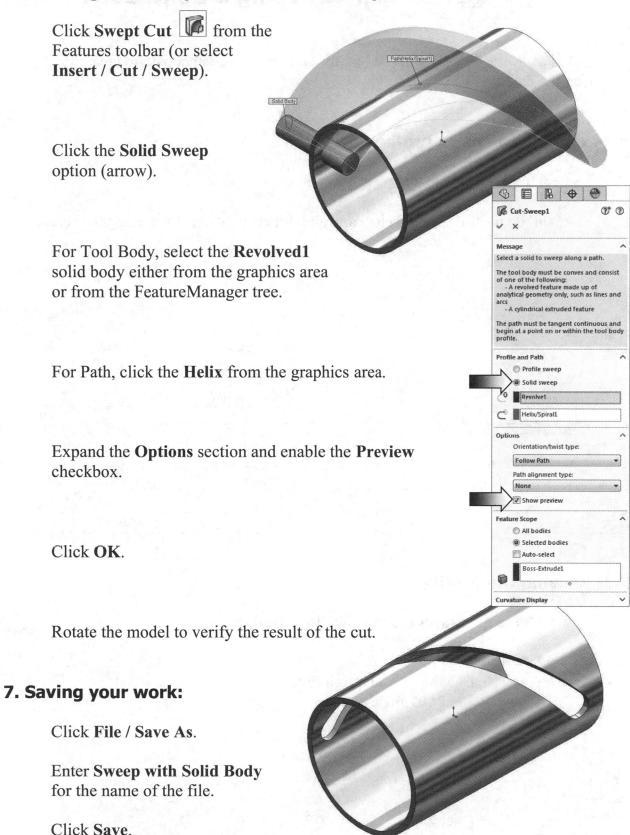 from the Features toolbar (or select **Insert / Cut / Sweep**).

Click the **Solid Sweep** option (arrow).

For Tool Body, select the **Revolved1** solid body either from the graphics area or from the FeatureManager tree.

For Path, click the **Helix** from the graphics area.

Expand the **Options** section and enable the **Preview** checkbox.

Click **OK**.

Rotate the model to verify the result of the cut.

7. Saving your work:

Click **File / Save As**.

Enter **Sweep with Solid Body** for the name of the file.

Click **Save**.

Exercise: Sweep Cut using a Solid Body

This exercise simulates a grinding tool that moves along a path to create the flutes on the drill bit.

1. Sketching the main profile:

Select the Right plane and open a **new sketch**.

Sketch the profile shown below and add dimensions to fully define the sketch.

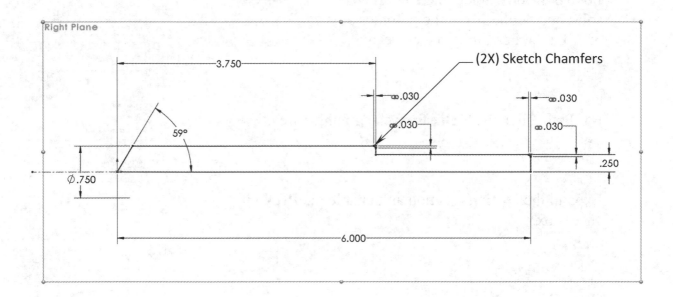

2. Revolving the profile:

Switch to the **Features** tab and click **Revolved Boss-Base**.

Use the default **Blind** type and **360°** angle.

Click **OK**.

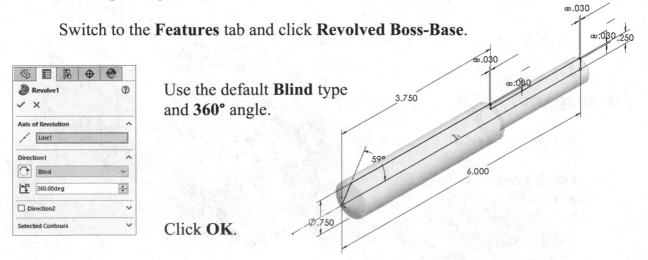

3. Creating the sweep path:

Select the <u>Front</u> plane and open a **new sketch**.

Select the <u>circular edge</u> on the left end of the model and click: **Convert Entities**.

The circular edge is converted into a sketch circle. It has an On-Edge relation added to it automatically.

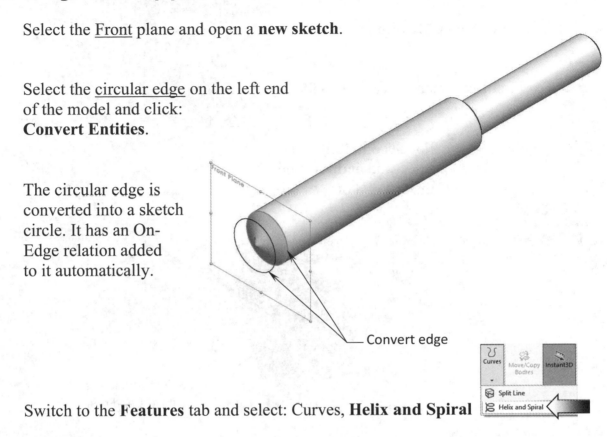

Convert edge

Switch to the **Features** tab and select: Curves, **Helix and Spiral**

For Define By, select **Height and Revolution**.

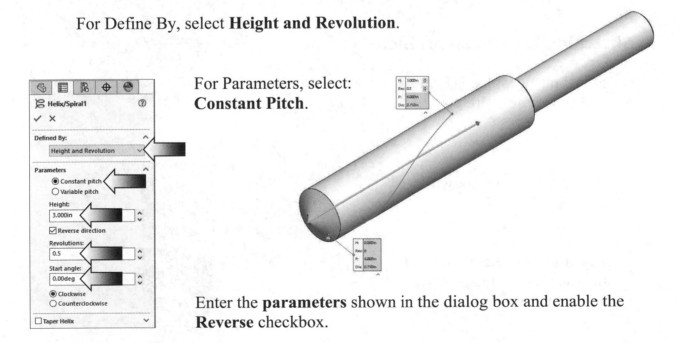

For Parameters, select: **Constant Pitch**.

Enter the **parameters** shown in the dialog box and enable the **Reverse** checkbox.

Click **OK**.

4. Creating a plane Normal to Curve:

Click **Reference Geometry**, **Plane**.

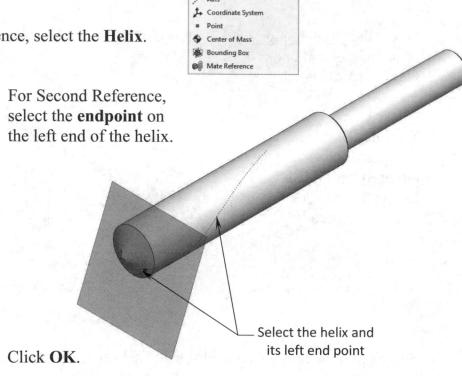

For First Reference, select the **Helix**.

For Second Reference, select the **endpoint** on the left end of the helix.

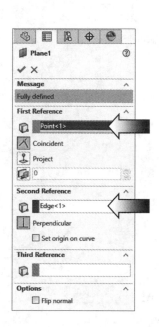

Select the helix and its left end point

Click **OK**.

5. Sketching the sweep profile:

Select the <u>new plane</u> (Plane1) and open a **new sketch**.

Sketch the profile shown.

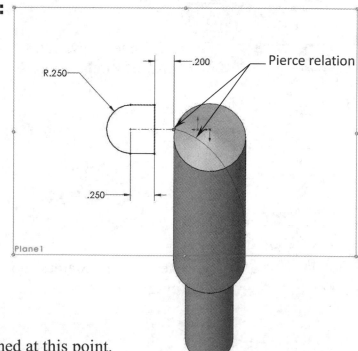

Add a **Pierce** relation between the right end of the centerline and the helix as noted.

The sketch should be fully defined at this point.

Switch to the **Features** tab and click **Revolved Boss-Base**.

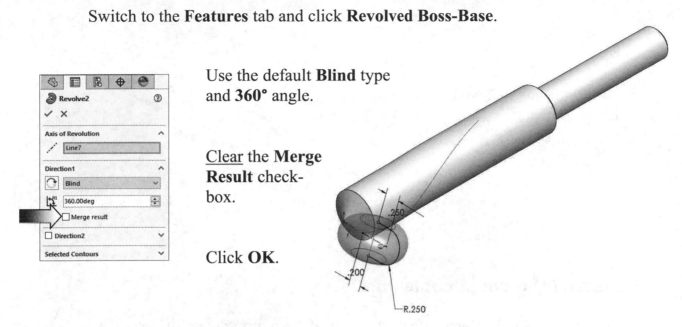

Use the default **Blind** type and **360°** angle.

Clear the **Merge Result** check-box.

Click **OK**.

6. Making a sweep cut:

Click **Swept Cut** on the Features tab.

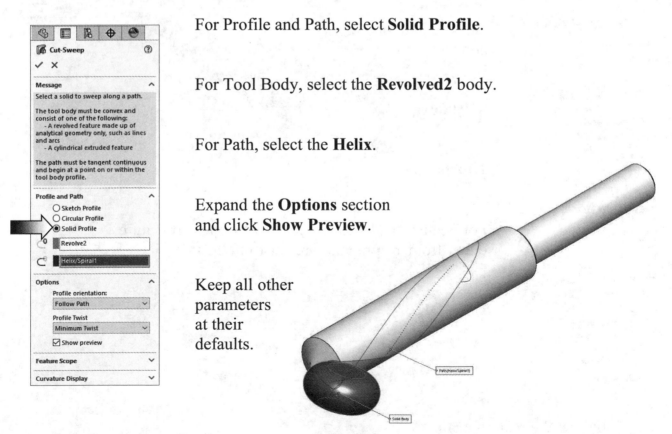

For Profile and Path, select **Solid Profile**.

For Tool Body, select the **Revolved2** body.

For Path, select the **Helix**.

Expand the **Options** section and click **Show Preview**.

Keep all other parameters at their defaults.

Click **OK**.

Verify the cut against the image shown here.

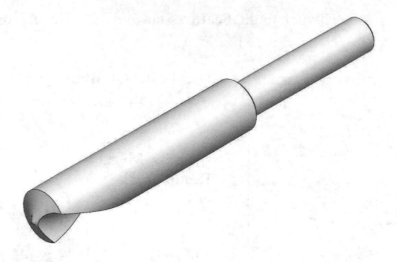

7. Patterning the swept cut feature:

Select **Circular Pattern** below the Linear Pattern drop-down list.

Pattern direction

For Pattern Direction, select the **circular edge** as noted.

Click **Equal Spacing** and enter:

360.00deg

2 Instances

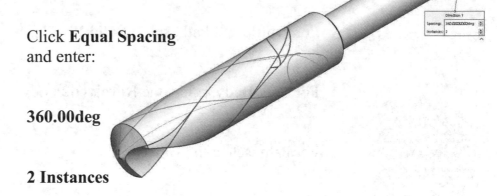

For Features and Faces, select the **Swept Cut** feature either from the graphics area or from the tree.

Click **OK**.

Inspect the pattern against the image shown on the right.

8. Assigning material:

Right-click the Material option and select:
Plain Carbon Steel.

Enable Realview Graphics for a more
realistic model display (if applicable).

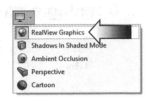

Click **View Settings**, **Realview Graphics**.

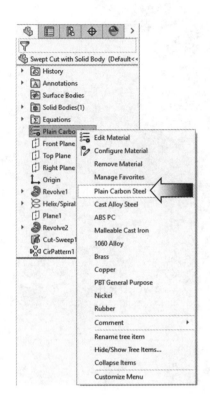

*Note: Realview Graphics is
available with supported
graphics cards only.*

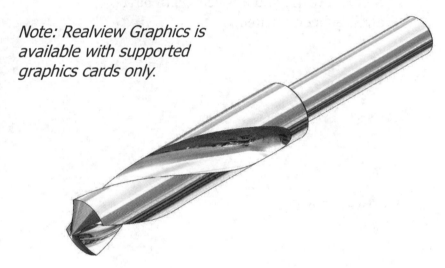

9. Saving your work:

Select **File, Save As**.

Enter **Swept Cut using a Solid Body** for the files name.

Click **Save**.

Close all documents.

Exercise: Sweep Using a Sketch Profile

1. Opening a part document:

Open a part document named:
Swept Cut with Sketch.sldprt

To help focus on the Sweep feature, a
Sweep Profile and a Sweep Path have
already been created.

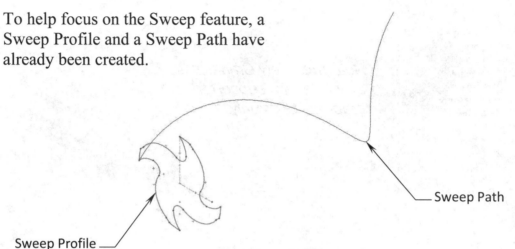

Sweep Path

Sweep Profile

2. Creating the main body:

Switch to the **Features** tab
and click **Swept Boss-Base**.

For Sweep Profile, select
Sketch2 in the graphics area.

For Sweep Path, select the
Helix also from the graphics
area.

Click **OK**.

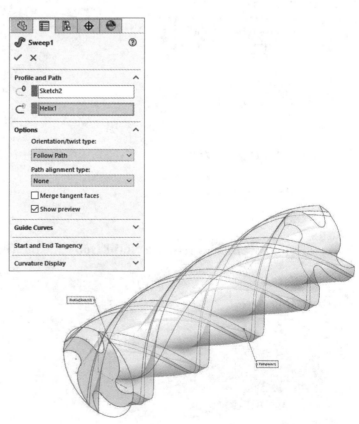

3. Adding the shank body:

Select the <u>planar face</u> on the right side and open a **new sketch**.

Sketch a **Circle** that centered on the Origin.

Add a **Coradial** relation between the <u>Circle</u> and the <u>circular edge</u> of the model.

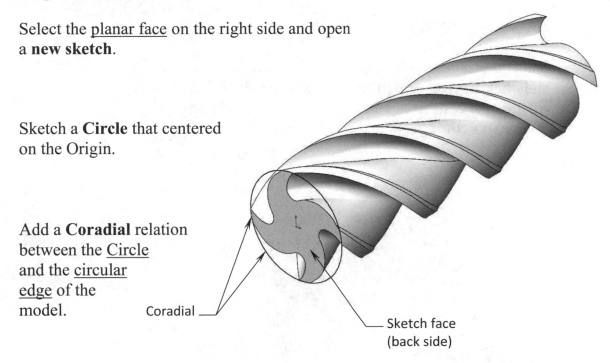

Coradial

Sketch face (back side)

The sketch should be fully defined at this point.

Switch to the **Features** tab and click: **Extruded Boss-Base**.

For Direction 1, use the default **Blind** type.

For Depth, enter **2.500in**.

Click **OK**.

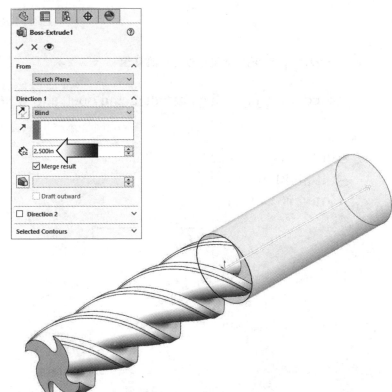

Change to the Isometric view.

4. Creating the sweep path:

Open a new **3D Sketch**.

Select the <u>edge</u> as indicated and click **Convert Entities**. Change the converted entity to **Construction** geometry.

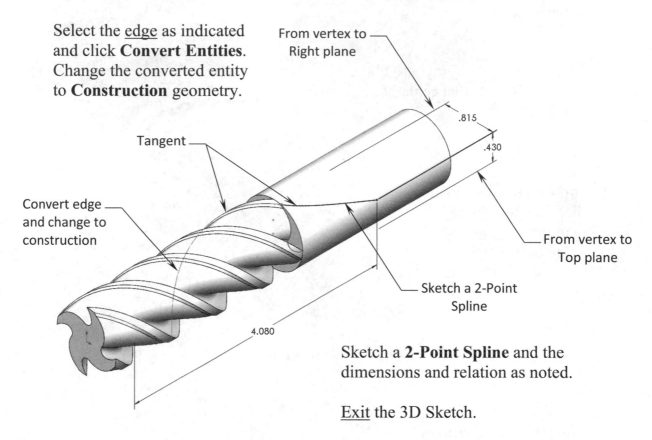

From vertex to Right plane

.815

.430

Tangent

Convert edge and change to construction

From vertex to Top plane

Sketch a 2-Point Spline

4.080

Sketch a **2-Point Spline** and the dimensions and relation as noted.

<u>Exit</u> the 3D Sketch.

5. Sketching the sweep profile:

Select the <u>planar face</u> as noted and open a **new sketch**.

While the selected face is still highlighted, click: **Convert Entities**.

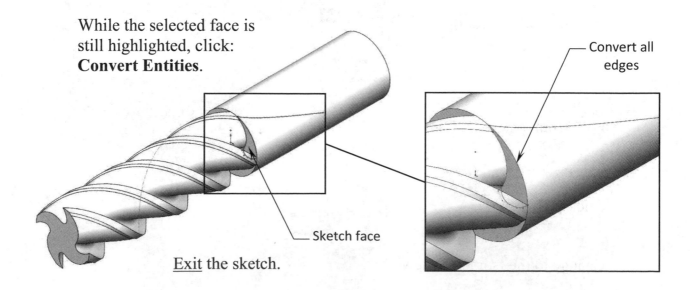

Convert all edges

Sketch face

<u>Exit</u> the sketch.

6. Removing the extra material:

Switch to the **Features** tab and click **Swept Cut**.

For Sweep Profile, select the **Sketch4** either from the graphics area or from the FeatureManager tree.

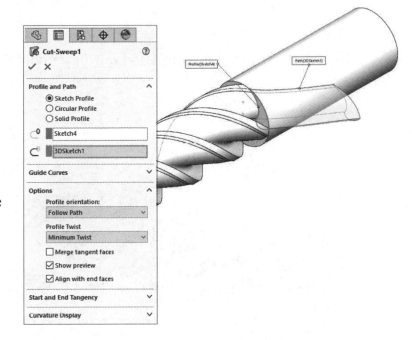

For Sweep Path, select the **3D Sketch**.

Click **OK**.

7. Circular Patterning the cut feature:

Click **Circular Pattern** below the Linear Pattern drop-down list.

For Pattern Direction, select the **circular edge** on the right side.

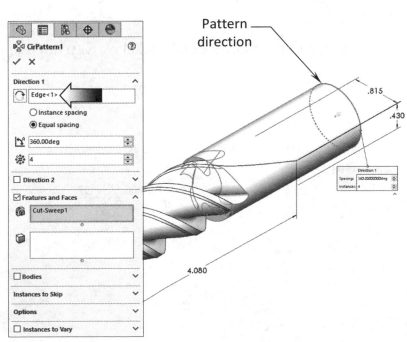

Use **Equal Spacing** and **4 Instances** of the Swept-Cut feature.

Click **OK**.

8. Adding the holding flats:

Select the <u>Right</u> plane and open a **new sketch**.

Sketch a **Corner Rectangle** and <u>mirror</u> it about the horizontal centerline.

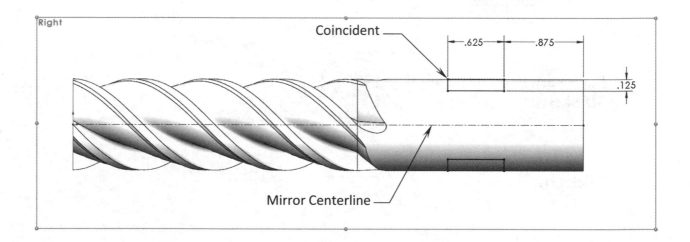

Add the dimensions shown to fully define the sketch.

Switch to the **Features** tab and click **Extruded Cut**.

For Direction 1, select **Through All-Both**.

Click **OK**.

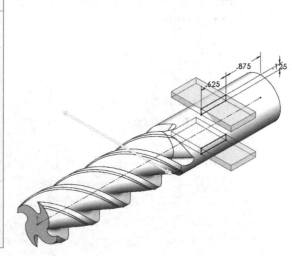

Press **Control+7**.

9. Adding the center hole:

Select the <u>planar face</u> as indicated and open a **new sketch**.

Sketch a **Circle** centered on the origin.

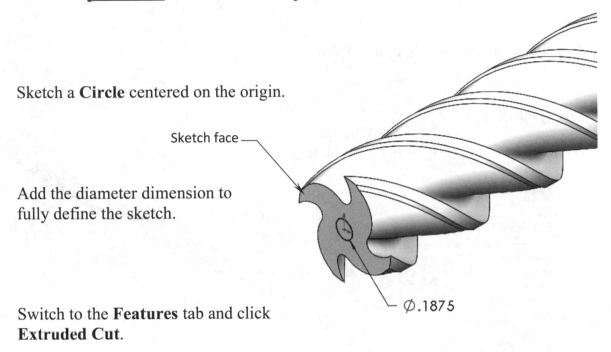

Sketch face

Add the diameter dimension to fully define the sketch.

Ø.1875

Switch to the **Features** tab and click **Extruded Cut**.

For Direction 1, use the default **Blind** type.

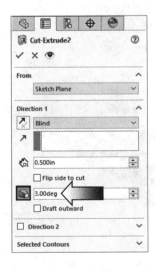

For Depth, enter **.500in**.

Enable the Draft option and enter **3.00deg**.

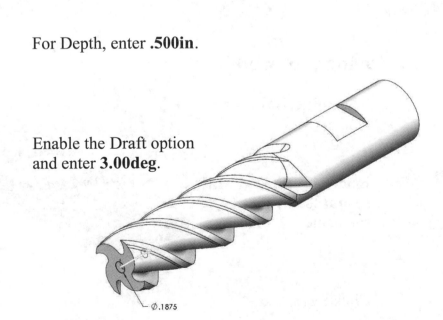

Ø.1875

Click **OK**.

10. Adding Chamfers:

Click **Chamfer** (below the Fillet drop-down list).

For Chamfer Depth, enter **.050in**.

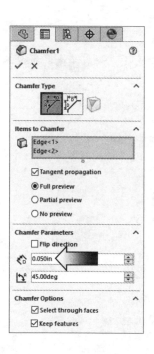

For Chamfer Angle, enter **45.00deg**.

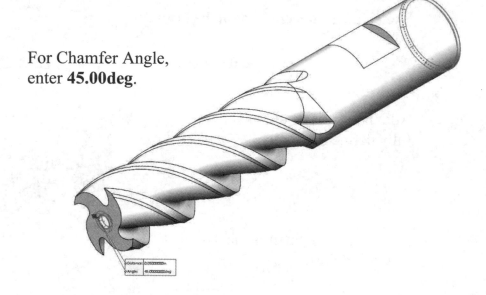

For Items to Chamfer, select the <u>edge of the shank</u> on the right end and the <u>edge of the center hole</u>.

Click **OK**.

11. Saving your work:

Select **File, Save As**.

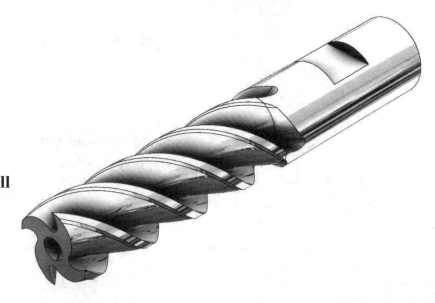

Enter **4-Flute End Mill (Completed)** for the file name.

Click **Save**.

CHAPTER 8

Lofts and Boundaries

Lofts and Boundaries
Solar Boat

A loft feature is created by making transitions between two or more profiles. Only the first and last profiles can be points; the others must be closed sketches to create a solid feature or opened sketches to create a surface feature.

For a solid loft, the first and last profiles must be model faces or faces created by split lines, planar profiles, or surfaces. A loft can be a base, boss, cut, or surface.

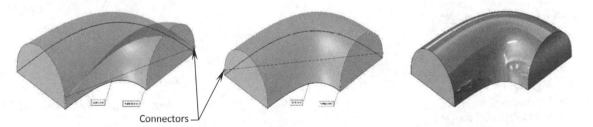

Connectors

The boundary command is similar to the loft, but it produces higher quality and accurate features useful for creating complex shapes. It uses 2 or more profiles and curves in two directions to define the boundary.

The Loft and Boundary commands use connectors to manipulate the twisting between the profiles. Right-click a connector to get to the connectors options to reset them.

Connectors

Lofts and Boundaries
Solar Boat

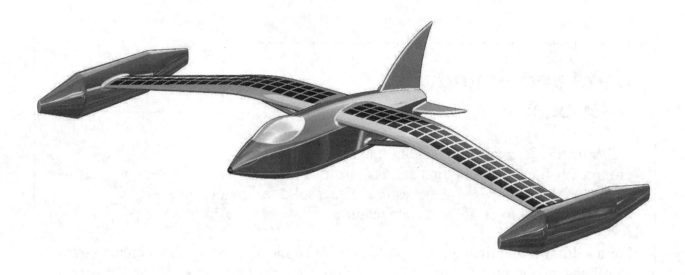

<u>**View Orientation Hot Keys**</u>:

Ctrl + 1 = Front View
Ctrl + 2 = Back View
Ctrl + 3 = Left View
Ctrl + 4 = Right View
Ctrl + 5 = Top View
Ctrl + 6 = Bottom View
Ctrl + 7 = Isometric View
Ctrl + 8 = Normal to Selection

Dimensioning Standards: **ANSI**

Units: **INCHES** – 3 Decimals

Tools Needed:

Insert Sketch	Split Lines	Extruded Boss/Base
Dimension	Add Relations	Plane
Mirror	Lofted Boss/Base	Boundary Boss/Base

1. Opening a part document:

Select **File / Open**.

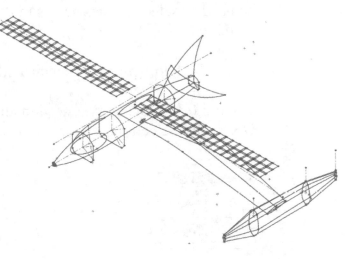

Browse to the training files folder and open a part document named **Loft and Boundary.sldprt**.

Most of the sketches have been created ahead of time to help focus on the main feature Loft and Boundary.

2. Creating a lofted feature:

A loft feature is created by making transitions between two or more profiles. Only the first, last, or first and last profiles can be points. The others must be closed sketches to create a solid feature, or opened sketches to create a surface feature.

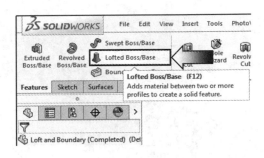

We will create the main body using the loft command. Switch to the **Features** tool tab and select the **Lofted Boss/Base** command.

Select the <u>profiles</u> by clicking their common points to prevent the transition from twisting as noted below.

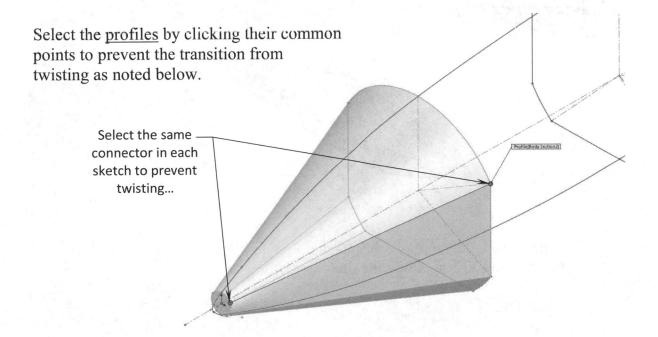

Select the same connector in each sketch to prevent twisting...

When creating a loft, a sweep, or a boundary feature, the SelectionManager can be used to select entities across multiple sketches as well as in combination with model edges.

Select the 5 loft-profiles as indicated.

Expand the Guide Curves section and select the 2 guide curves as noted; click **OK** in the SelectionManager after selecting each guide curve.

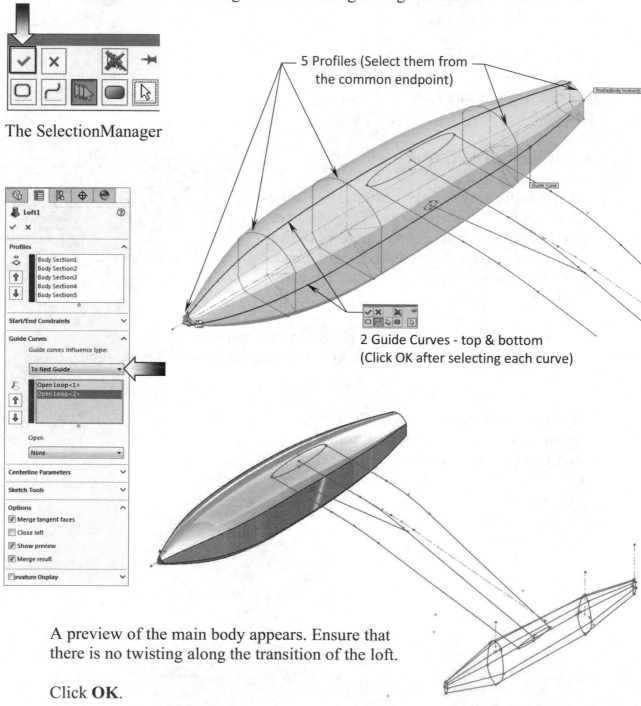

The SelectionManager

5 Profiles (Select them from the common endpoint)

2 Guide Curves - top & bottom
(Click OK after selecting each curve)

A preview of the main body appears. Ensure that there is no twisting along the transition of the loft.

Click **OK**.

1. Creating the boundary feature:

The boundary command produces very high quality, accurate features useful for creating complex shapes. It uses two or more profiles and curves in two directions to define the boundary.

We will create the float body using the boundary command. Click the **Boundary Boss/Base** command from the Features tool tab.

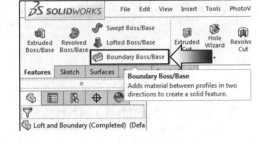

For **Direction 1**, select the 4 profiles at their common points as noted.

For **Direction 2**, select the 6 guides as indicated; click **OK** in the SelectionManager after selecting each guide.

Inspect the preview graphics to ensure that there is no twist issue with the boundary.

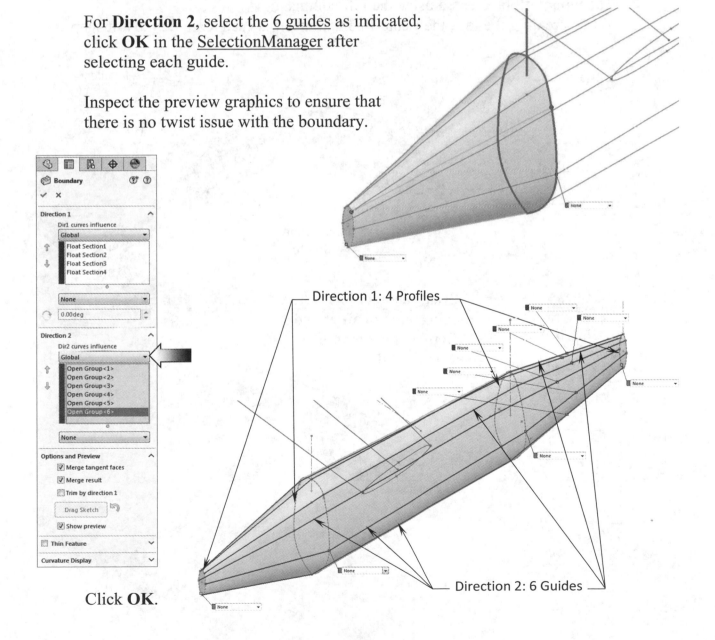

Direction 1: 4 Profiles

Direction 2: 6 Guides

Click **OK**.

The resulted Boundary feature.

Press **Control+7** to change to the Isometric view.

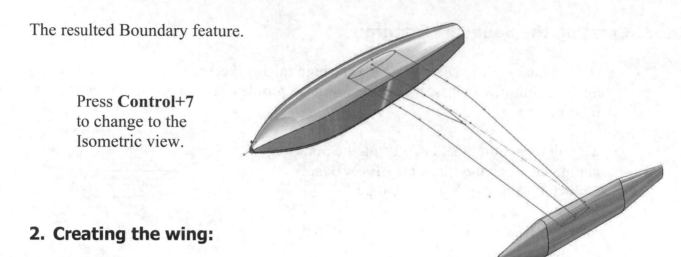

2. Creating the wing:

The wing will be created using the loft command.
While creating the loft, the Centerline Parameter is used to guide the loft shape.

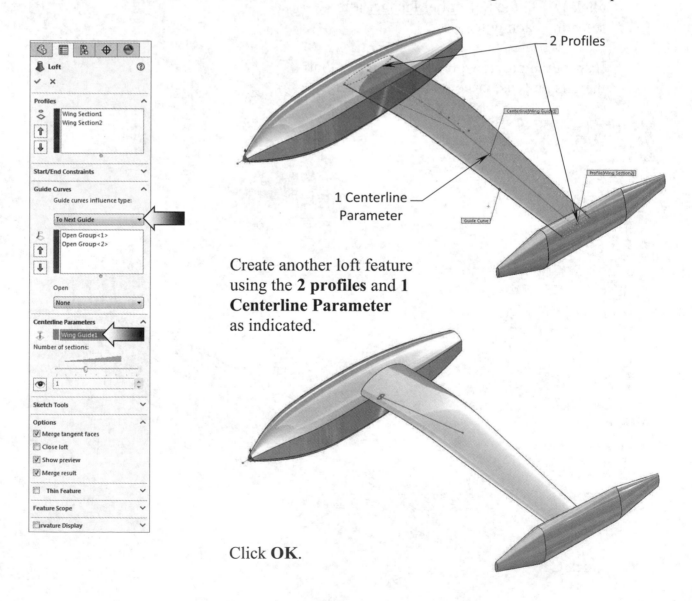

2 Profiles

1 Centerline Parameter

Create another loft feature using the **2 profiles** and **1 Centerline Parameter** as indicated.

Click **OK**.

3. Creating the wing-support:

The wing support is much simpler than the wing itself. It can be created using a simple sweep operation.

The sweep command creates a feature by moving a profile along a path.

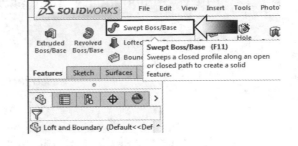

Click **Swept Boss/Base**.

Select the **Ellipse** for profile.

Select the **Line** for sweep path.

Leave all other settings at their default values.

Click **OK**.

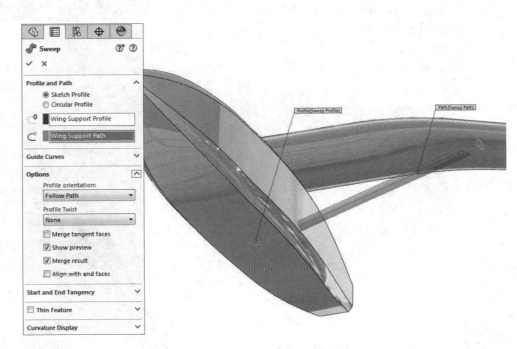

Press Control+1 to change to the Front orientation.

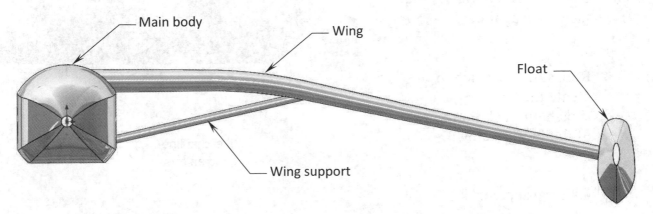

Main body
Wing
Float
Wing support

4. Constructing the nose feature:

The following examples will demonstrate the slightly different results created by the loft vs. the boundary commands.

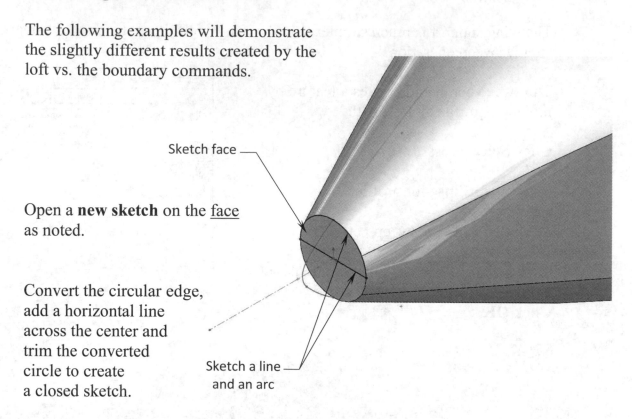

Open a **new sketch** on the <u>face</u> as noted.

Convert the circular edge, add a horizontal line across the center and trim the converted circle to create a closed sketch.

<u>Exit</u> the sketch.

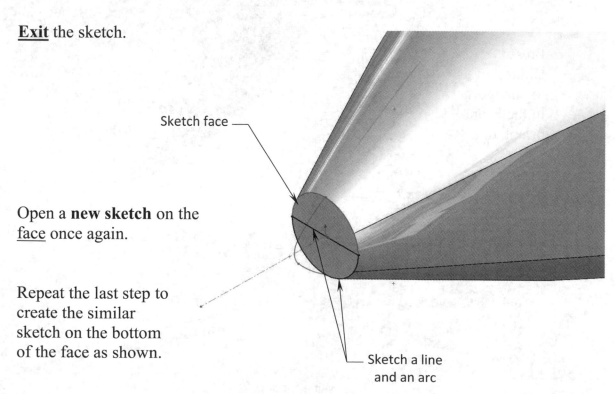

Open a **new sketch** on the <u>face</u> once again.

Repeat the last step to create the similar sketch on the bottom of the face as shown.

<u>Exit</u> the sketch.

Click the **Boundary Boss/Base** command.

For **Direction 1**, select the 2 nose profiles that were created in the last step.

For **Direction 2**, select the Nose Guide sketch from the Feature tree.

The preview graphic shows the result of the nose tip that may need further adjustments.

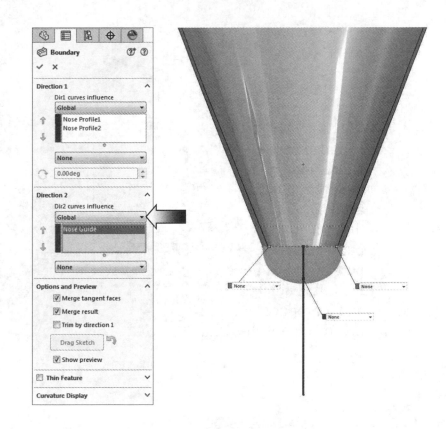

Click **Cancel**. We will try creating the nose-tip as a lofted feature instead.

Click **Lofted Boss/Base**.

For **Loft Profile**, select the 2 Nose Profiles that were created in the last step.

For **Guide Curve**, select the Nose Guide sketch from the Feature tree.

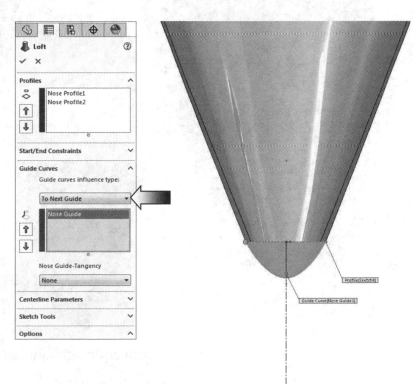

The preview graphic shows a much better result of the nose tip than the one created with the boundary command.

Click **OK**.

The Boundary command can create higher quality and more accurate features than the loft can, but for this particular feature the loft is quicker and it produces an acceptable result.

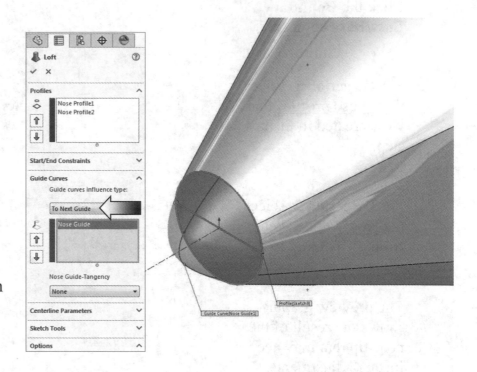

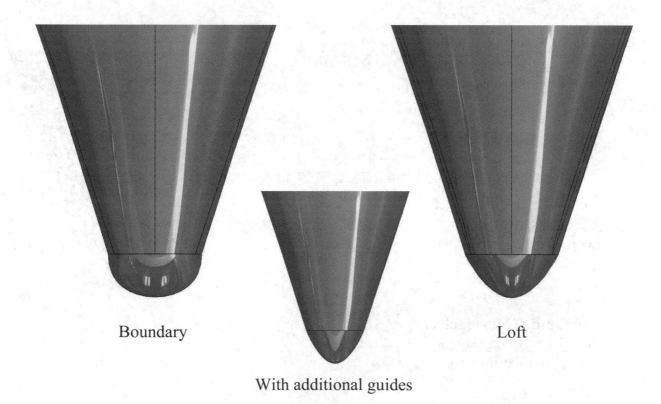

Boundary   Loft

With additional guides

Both the Loft and Boundary require additional guide curves to better control the transition between the nose tip to the body.

5. Creating the horizontal tail fins:

Open a **new sketch** on the Top plane. Press Control+5 to change to the top orientation.

Sketch the profile shown.

Use the mirror option to keep the sketch entities symmetrical with each other.

Add the dimensions and relations shown to fully define the sketch.

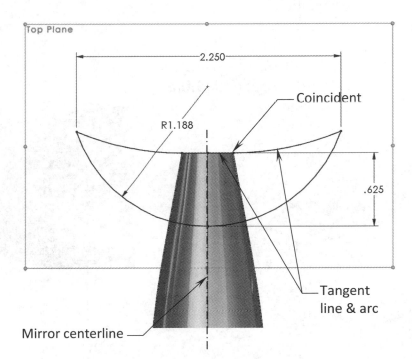

Switch to the **Features** tool tab.

Select **Extruded Boss/Base**.

Select the **Mid Plane** condition.

Enter **.060in** for extrude depth.

Enable the **Merge Result** checkbox.

Click **OK**.

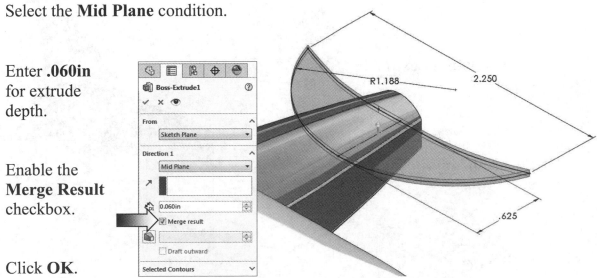

6. Creating the vertical tail fin:

Open a **new sketch** on the Right plane. Press Control+4 to change to the right orientation.

Sketch the profile using the **3-Point Arc** and the **Line** commands.

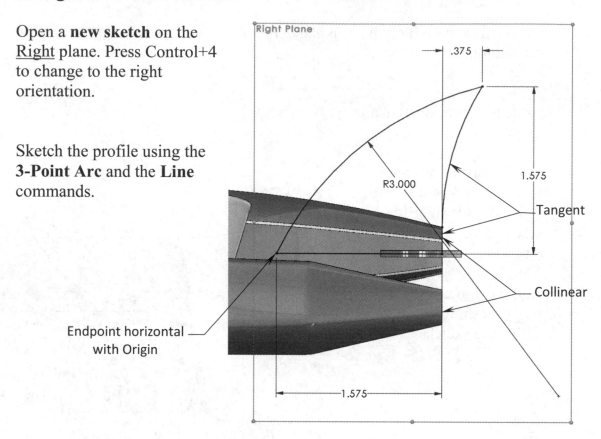

Add the dimensions and relations indicated to fully define the sketch.

Extrude the sketch with the **Mid Plane** condition.

Enter **.060in** for extrude depth.

Enable the **Merge Result** checkbox.

Click **OK**.

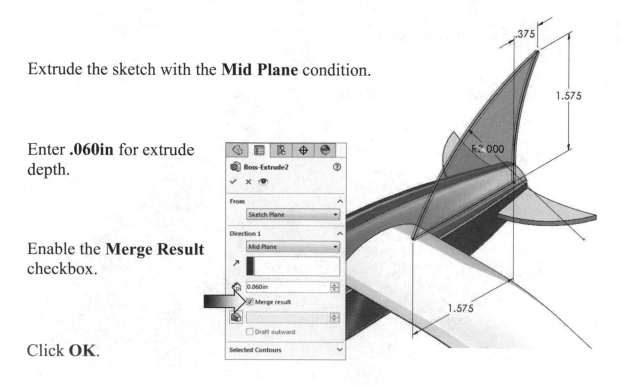

7. Creating a split line:

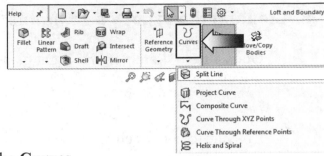

The split line is used to define the shape, size, and location of the canopy feature.

Select the **Split Line** command from the **Curves** pull-down menu.

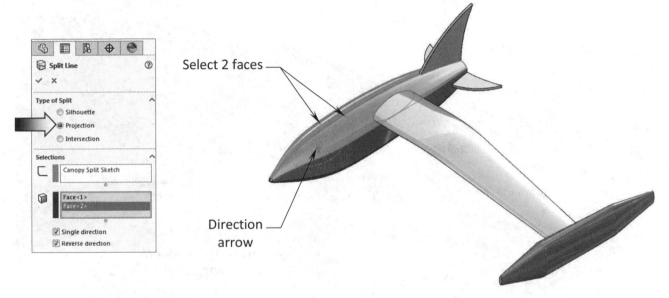

For Type of Split, select the **Projection** type.

For Sketch Selection, select the **Canopy Split Sketch** from the Feature tree.

For Faces to Split, select the **2 upper Faces** of the main body as noted.

Enable the **Single Direction** checkbox and click the direction arrow to reverse it, also clear the **Merge Tangent Faces** option.

Click **OK**.

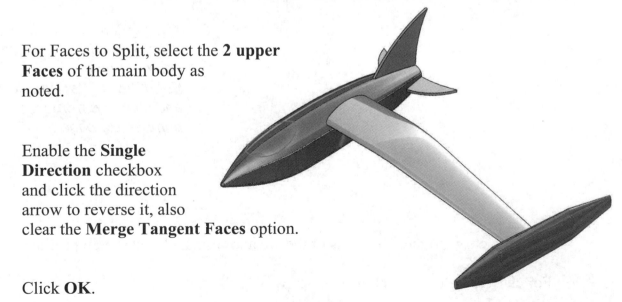

8. Creating the canopy's 1st profile*:

Open a **new sketch** on the <u>Right</u> plane. Press Control+4 to change to the right orientation.

Convert the <u>upper edge</u> of the body to a curve.

Add a **3-Point-Arc** and a **Centerline** shown.

Add the dimensions indicated to fully define the sketch.

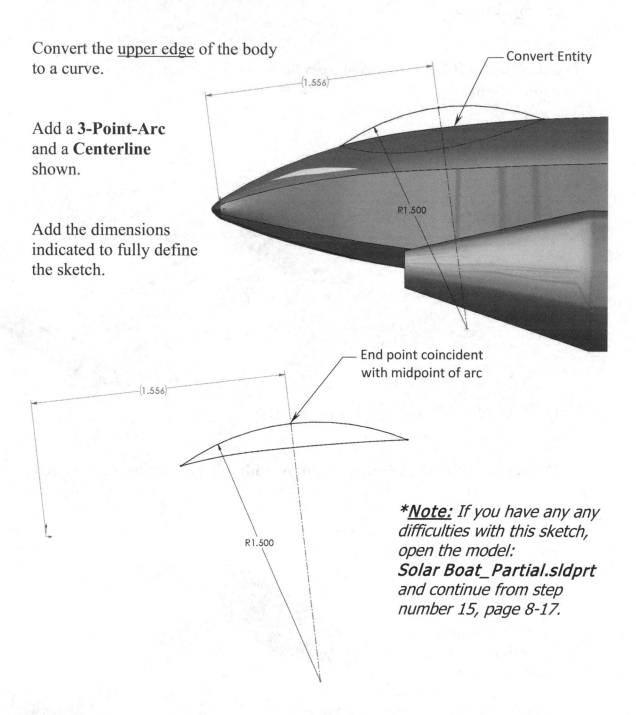

Note: If you have any any difficulties with this sketch, open the model: Solar Boat_Partial.sldprt and continue from step number 15, page 8-17.

Trim the curve to meet the ends of the arc and form a closed sketch profile.

For clarity, the model is hidden to show only the sketch.

9. Creating a new plane:

A new plane is needed to sketch the guide curve.

Select the **Plane** command from the Reference Geometry drop down menu, or select **Insert / Reference Geometry / Plane**.

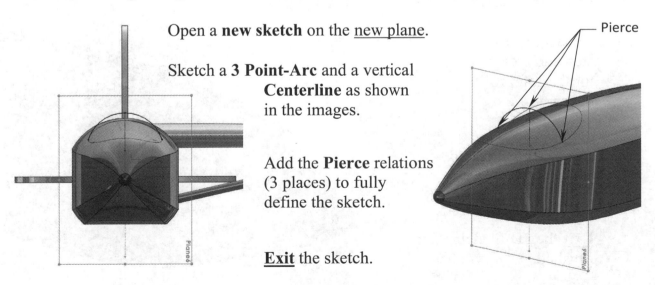

Select centerline

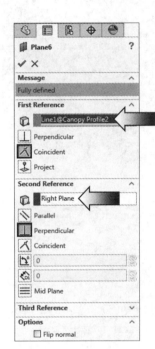

For First Reference, select the **Vertical Centerline** from the previous sketch.

For Second Reference, select the **Right** plane from the Feature tree.

SOLIDWORKS automatically places the new plane on the vertical centerline and rotates it to the perpendicular position to the right reference plane.

10. Sketching the guide curve:

Open a **new sketch** on the new plane.

Sketch a **3 Point-Arc** and a vertical **Centerline** as shown in the images.

Add the **Pierce** relations (3 places) to fully define the sketch.

Pierce

Exit the sketch.

11. Creating the canopy's 2nd loft profile:

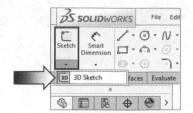

When working on non-planar surfaces, sometimes it is quicker to create the geometry in 3D sketch rather than creating them on a plane and then later having to projecting it onto the surfaces.

We will create the 2 loft profiles as 3D sketch profiles.

Select the **3D sketch** 3D command from the Sketch drop down.

Select the 2 edges shown and click **Convert Entities**.

The 2 selected edges are converted to 2 splines. (Make any adjustments needed to ensure the corners are coincident with one another.)

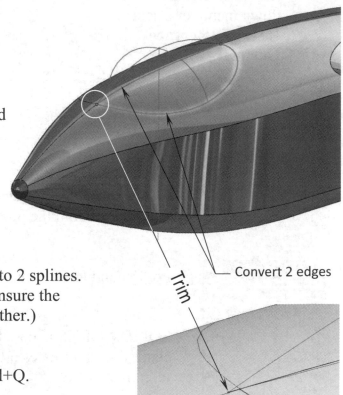

Convert 2 edges

Trim

Exit the 3D-Sketch or press Control+Q.

12. Creating the canopy's 3rd loft profile:

Repeat the last step and create the 2nd loft profile the same way.

Convert the opposite 2 edges to 2 splines.

Exit the 3D-Sketch when finished.

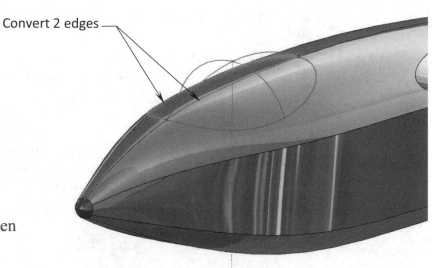

Convert 2 edges

13. Creating the canopy feature:

Select the **Lofted Boss/Base** command from the Features tool tab.

For <u>Profiles</u>, select the sketches of the **3 loft profiles**.

For <u>Guide Curves</u>, select the **arc** as noted.

Click **OK**.

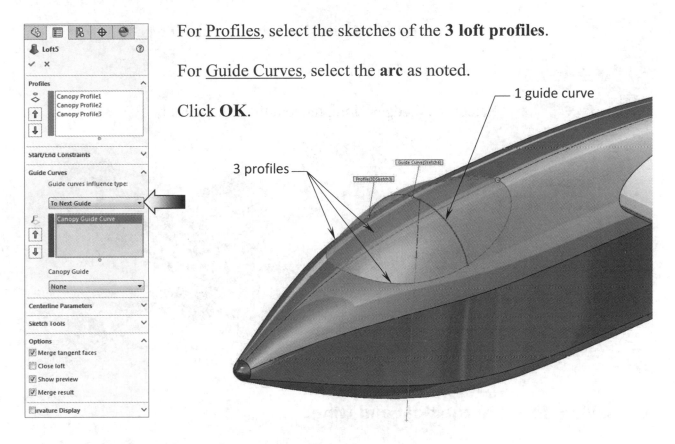

14. Creating a mirror feature:

Select the **Mirror** command from the Features tool tab.

For <u>Mirror Face/Plane</u>, select the **Right** plane from the Features tree.

For <u>Features to Mirror</u>, select from the graphics area the **Wing**, the **Float**, and the **Wing Support** features.

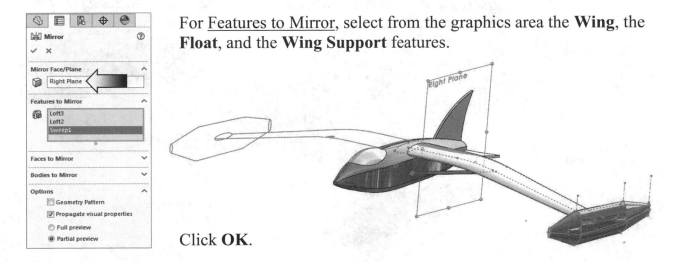

Click **OK**.

15. Adding fillets to the main body:

Select the **Fillet** command from the **Features** tool tab.

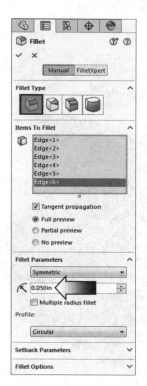

The **Constant Size Fillet** should be the default type.

Enter **.050in** for radius.

Select the **6 edges** along the length of the main body.

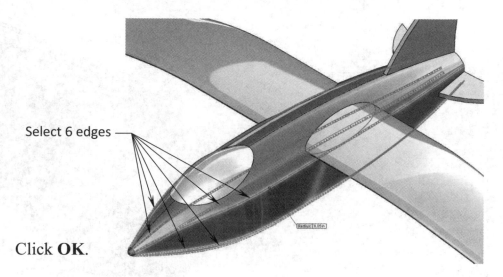

Select 6 edges

Click **OK**.

16. Adding fillets to the float and wing:

Click the **Fillet** command and Enter **.050in** for radius.

Select the **surfaces** of the **wing**, the **float**, and the **wing** support to add the fillets.

NOTE: It is quicker to select a face of a feature than selecting one edge at a time.

Click **OK**.

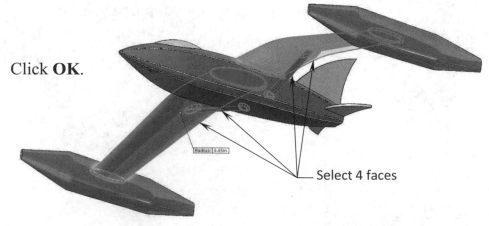

Select 4 faces

17. Adding fillets to the ends of the floats:

Add a fillet of **.008in** to the left and right ends of both floats.

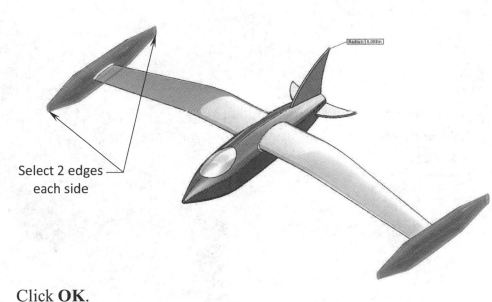

Select 2 edges
each side

Click **OK**.

18. Adding fillets to the tail fins:

Add a fillet of **.030in** to <u>all edges</u>
of the 3 fins.

(2X) Do not select

There should be a total
of 9 edges.

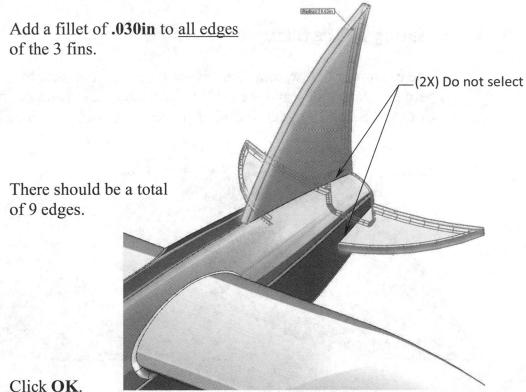

Click **OK**.

19. Adding fillets to the ends of the fins:

Add another fillet of **.030in** to the ends of the 3 fins.

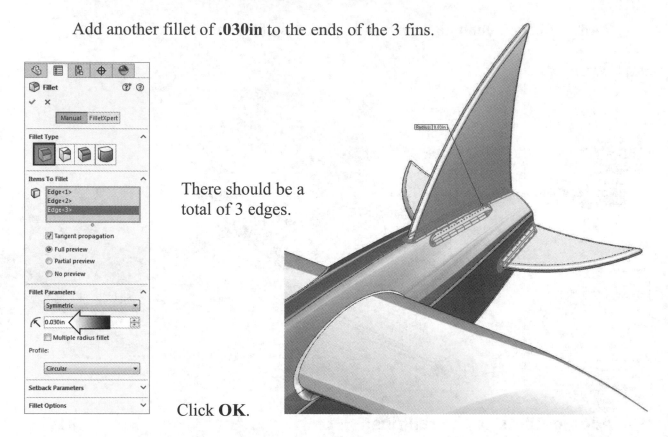

There should be a total of 3 edges.

Click **OK**.

20. Creating a face fillet:

The examples below show 2 different types of fillet used to create the blend between the canopy and the main body. One of them was created using the Constant Size Fillet, and the other was created using the Face Fillet.

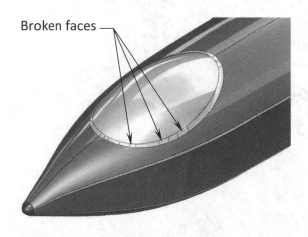

Broken faces

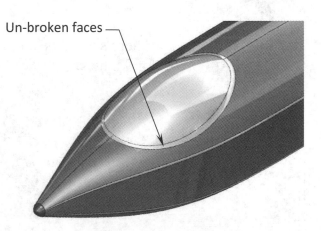

Un-broken faces

<u>Constant Size Fillet</u>

<u>Face Fillet</u>

We will use the Face Fillet option to create a smoother blend between the canopy and the main body.

Select the **Fillet** command once again.

Click the **Face Fillet** option (arrow).

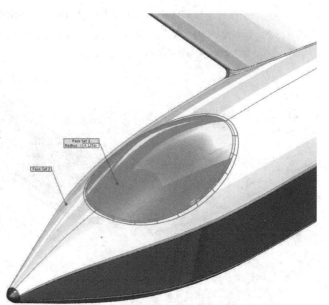

Enter **.125in** for radius.

For <u>Face Group 1</u>, select the **upper 2 faces** of the main body.

For <u>Face Group 2</u>, select the face of the canopy.

Click **OK**.

21. Creating a split line for the solar panels:

From the Features tool tab, select the **Split Line** command from the Curves pull-down menu.

Use the default **Projection** type.

For Sketch to Split, select the **Sketch7** from the Feature tree.

For Faces to Split, select the **2 wings**. Click Single Direction and Reverse to split only the top halves of the 2 wings.

Click **OK**.

22. Changing the color of the solar squares:

The color of any faces, features, or bodies can be changed at any time. To further enhance the appearance of the solar squares, we will change them to the black color.

Click on the inside of one of the squares and select the **Appearance** button, then click the **Face** option (arrows).

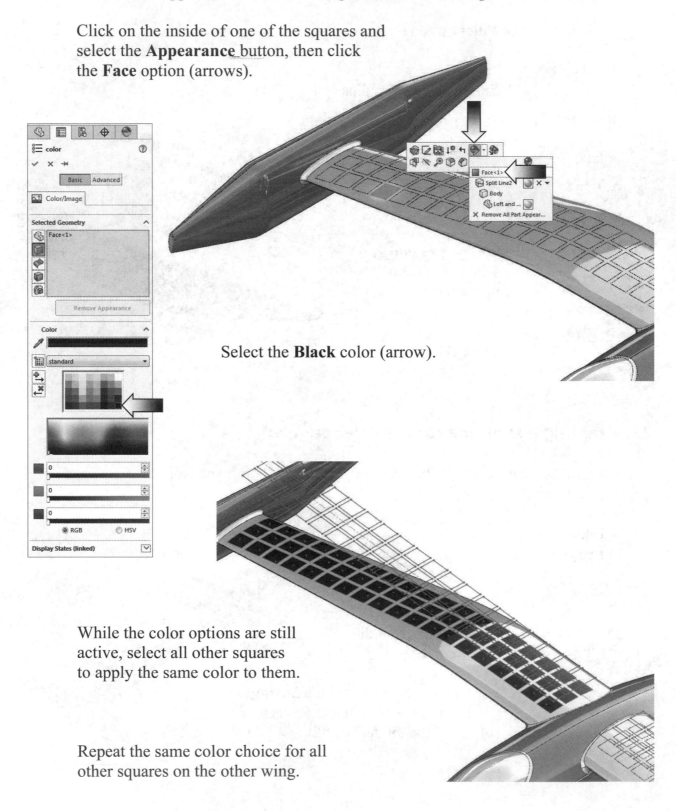

Select the **Black** color (arrow).

While the color options are still active, select all other squares to apply the same color to them.

Repeat the same color choice for all other squares on the other wing.

23. Changing the edge display:

While creating the model we often use
the option Shaded with Edges for
convenience, but when the model is
completed we should change the display
to Tangent Edges Remove. Not only will this
shading option make the model look a little
more realistic but it also helps improve the computer performance.

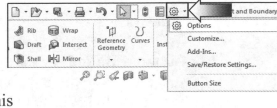

Click **Options**
(arrow).

Select the option
Display Selection.

Click the **Remove**
option under the
Part/Assembly
Tangent Edge
Display.

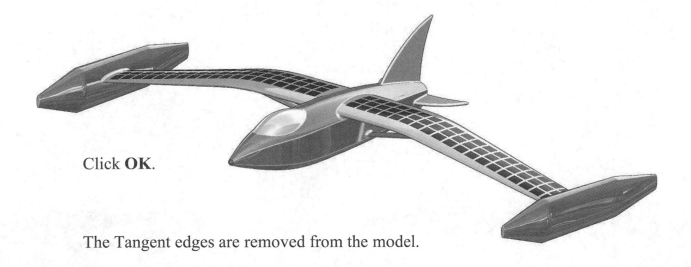

Click **OK**.

The Tangent edges are removed from the model.

24. Saving your work:

Select **File/ Save As**.

Enter **Loft and Boundary** for the name of the file.

Press **Save** and click **Yes** to overwrite the old file if prompted.

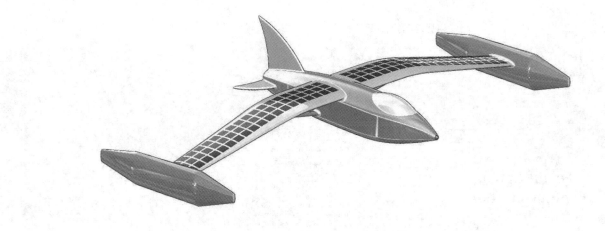

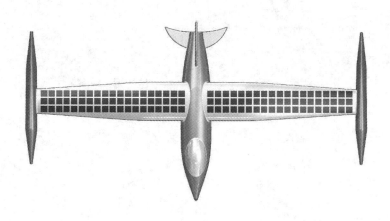

Exercise: Surface & Solid Modeling

1. Starting a new part document:

Click **File / New**.

Select the **Part Template** and click **OK**.

Set the System of Units to **ANSI, Inches** (or IPS), **3 decimal** places.

Select the <u>Front</u> plane and open a **new sketch**.

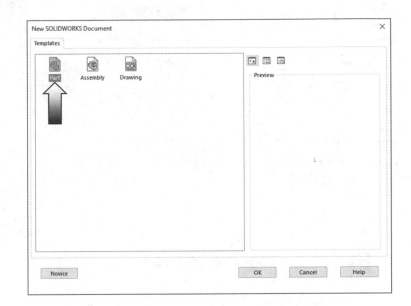

Sketch the profile shown below. Add the dimensions and relations as indicated to fully define the sketch.

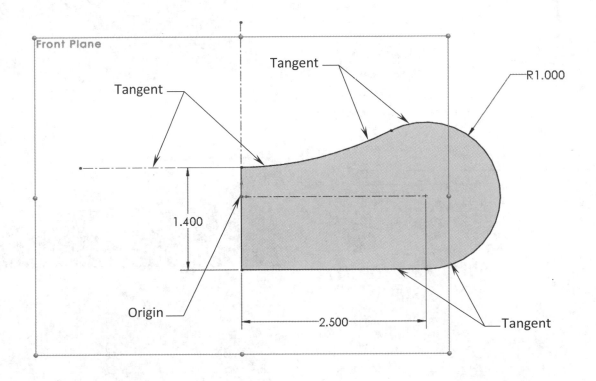

2. Revolving the sketch:

Change to the **Features** tool tab.

Click **Revolved Boss Base**.

Select the <u>vertical centerline</u> for Axis of Revolution.

Use the default **Blind** type and **360deg**.

Click **OK**.

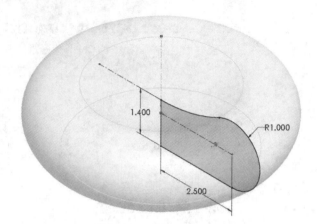

3. Making the sketch of the tip:

Open a **new sketch** on the <u>Top</u> plane.

Sketch the profile below. Also add 2 additional horizontal centerlines.

Add dimensions and relations shown to fully define the sketch.

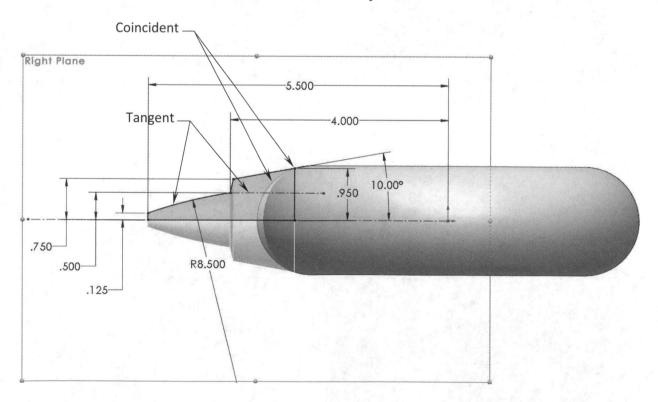

4. Revolving the sketch:

Switch back to the **Features** tab.

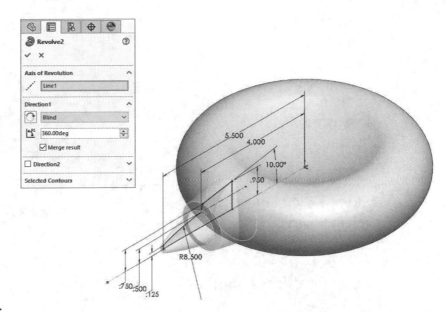

Click **Revolved Boss Base**.

Revolve about the centerline that is attached to the origin.

Use **Blind** and **360deg**.

Click **OK**.

5. Creating a Split Line:

The split line command is used to split a face into two faces. The split faces form a closed boundary which will be used to create the raised features for gripping in the next couple of steps.

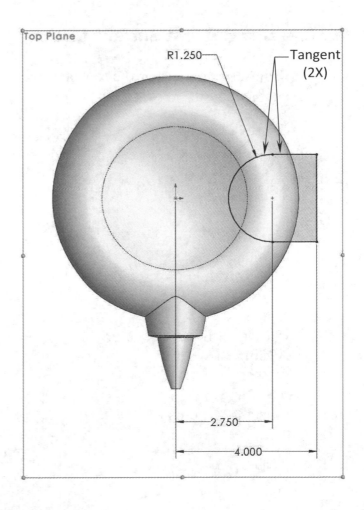

Open a **new sketch** on the Top plane.

Sketch the profile using the **Line** and the **Tangent Arc** commands.

Add the dimensions and relations needed to fully define the sketch.

Switch to the **Features** tab.

Select **Curves /
Split Line**.

For Type of Split,
click the
Projection
button.

The Split Sketch,
Sketch3, should be
selected already.

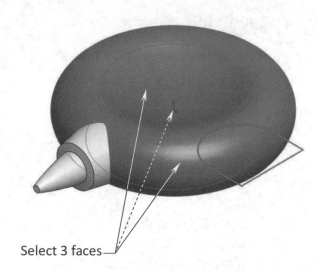

For Faces to Split,
select the **3 faces**: top, bottom, and side as indicated.

Click **OK**.

6. Creating the first Surface Offset:

Right-click on any of the tool tabs and
enable the **Surfaces** tab.

Select the **Surface
Offset** command.

For Offset-
Parameters, select
the **3 split faces** as noted.

For Offset Distance, enter
.125in.

Offset to the <u>outside</u>.

Click **OK**.

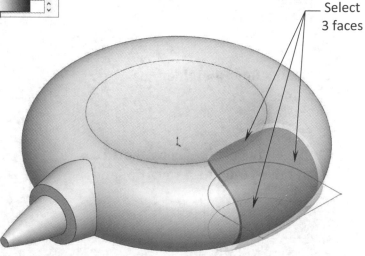

Select
3 faces

7. Creating the second Surface Offset:

Click **Surface Offset** once again.

For Offset Parameters, select the same 3 faces as last time.

For Offset Distance, enter **0.00in** (zero).

Click **OK**.

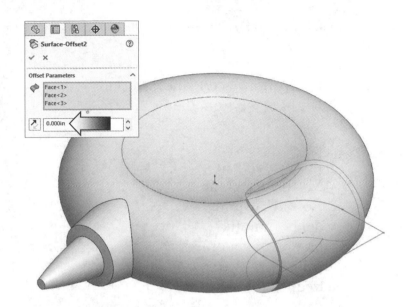

8. Creating a lofted feature:

Change to the **Features** tool tab.

Click **Lofted Boss Base**.

For Loft Profiles, select the **2 Surface Offsets** either from the graphics area or from the Surface Bodies folder.

The 2 connectors should line up with one another. If not, drag one of the connectors to reposition it right next to the other, to prevent the twisting.

Click **OK**.

Right click the **Surface Bodies** folder and select **Hide**.

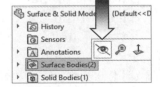

9. Mirroring the lofted feature:

Click **Mirror** on the Features tab.

For Mirror Face/Plane, select the **Right** plane from the Feature tree.

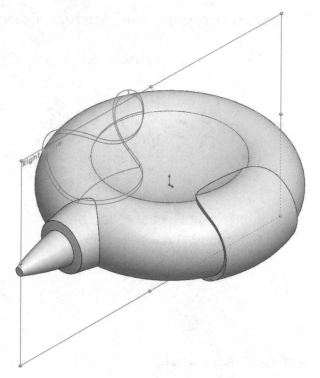

For Features to Mirror, select the **Lofted1** feature either from the graphics area or from the Feature tree.

Click **OK**.

10. Adding the .060" fillets:

Click **Fillet** on the Features tab.

For Fillet Type, select the **Constant Size Radius** button.

For Radius, enter **.060in**.

Select the <u>upper edges</u> of the lofted features and the 2 edges on the left end.

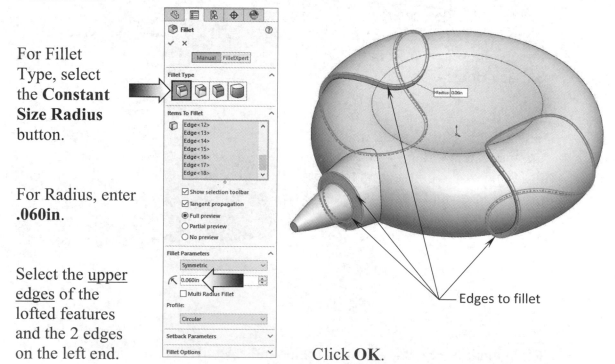

Edges to fillet

Click **OK**.

11. Adding the .090" fillets:

Click **Fillet** again.

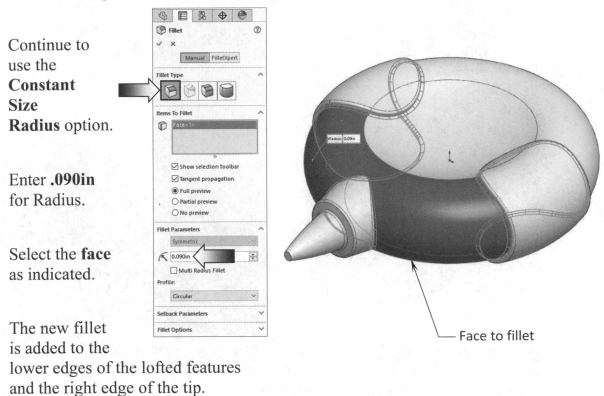

Continue to use the **Constant Size Radius** option.

Enter **.090in** for Radius.

Select the **face** as indicated.

The new fillet is added to the lower edges of the lofted features and the right edge of the tip.

Face to fillet

12. Shelling the model:

The Shell feature usually gets added towards the end of the model creation, so that the wall thickness remains constant throughout the part.

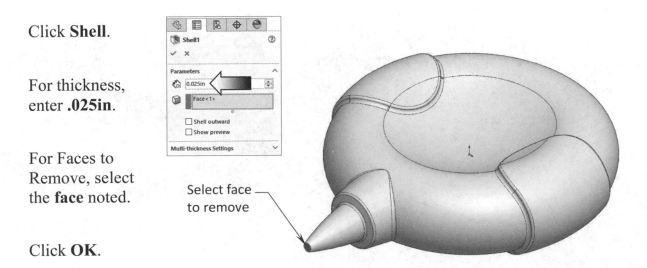

Click **Shell**.

For thickness, enter **.025in**.

For Faces to Remove, select the **face** noted.

Select face to remove

Click **OK**.

13. Creating a section view:

Click **Section View** on the View-Heads-Up Toolbar.

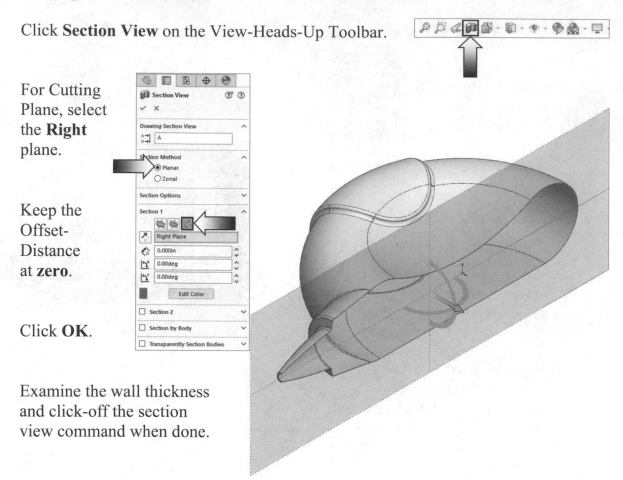

For Cutting Plane, select the **Right** plane.

Keep the Offset-Distance at **zero**.

Click **OK**.

Examine the wall thickness and click-off the section view command when done.

14. Saving your work:

Select **File / Save As**.

For the file name, enter **Surface & Solid Modeling.sldprt**

Click **Save**.

Close all documents.

Exercise: Using Split Lines

The Split Line tool uses a sketch to split a face or a surface and divides a selected face into multiple separate faces.

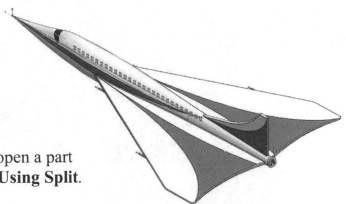

1. Opening a part document:

Browse to the Training Folder and open a part document named: **Supersonic Jet_Using Split**.

2. Creating the 1st split:

Switch to the **Features** tool tab and select: **Curves, Split Line**.

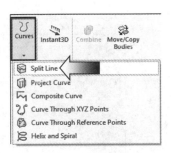

For Type of Split, select **Projection**.

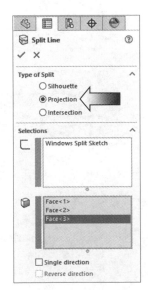

For Selections, select the sketch named: **Windows Split Sketch** from the FeatureManager tree.

For Faces to Split, select **3 faces**, the body, and the front and back of the center-stripe.

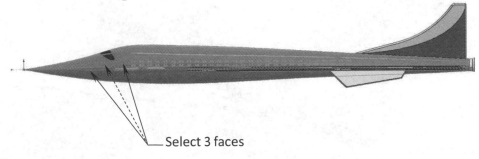

Select 3 faces

3. Changing the face color:

Click the <u>face</u> of the center-stripe and select: **Appearances, Face <1>**.

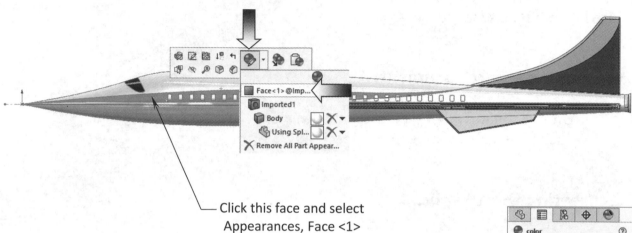

Click this face and select
Appearances, Face <1>

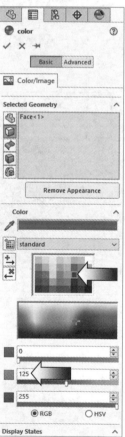

Select the **Navy Blue** color from the color palette.

Change the values of the colors to:

Red:	**0**
Green:	**125**
Blue:	**225**

Apply the same color to the other side of the center-stripe.

Click **OK**.

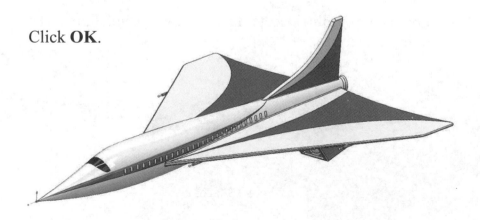

4. Creating the 2nd split:

Switch to the **Features** tool tab and select **Curves, Split Line**.

For Type of Split, select **Projection**.

For Selections, select the sketch named: **Top & Bottom Split Sketch** from the Feature tree.

For Faces to Split, select the **face of the main body**.

Click **OK**.

5. Changing the face color:

Click the <u>face</u> of the top-stripe and select: **Appearances, Face <1>**.

Select the **Navy Blue** color again.

Change the values of the colors to:

Red: 0
Green: 125
Blue: 225

Apply the same color to the bottom face and click **OK**.

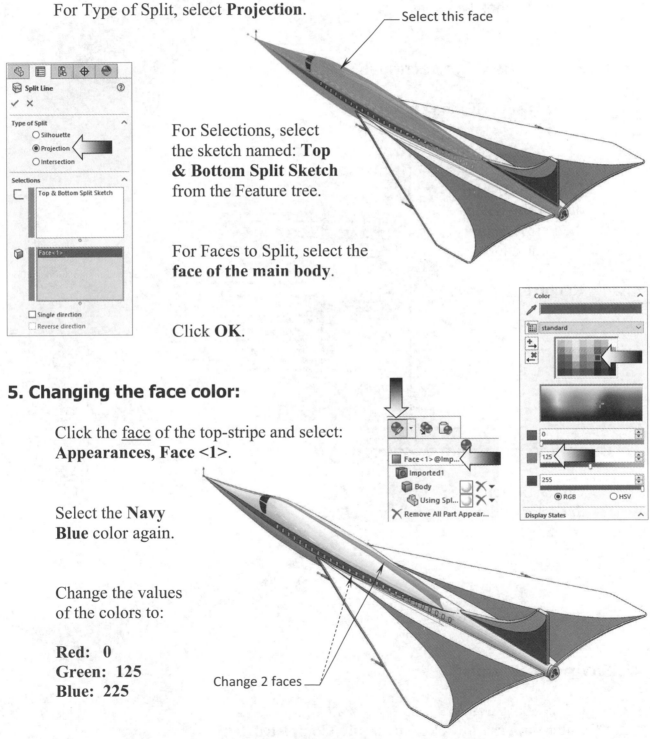

Select this face

Change 2 faces

The tangent edges are visible by default.

Changing the edge display from **Visible** to **Remove** would make the model look better.

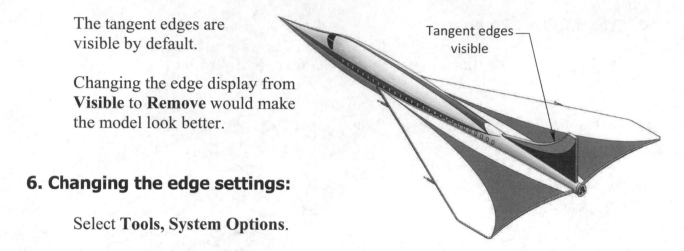

Tangent edges visible

6. Changing the edge settings:

Select **Tools, System Options**.

Click the **Display** option and change the Part/Assembly Tangent Edge Display to: **Remove**.

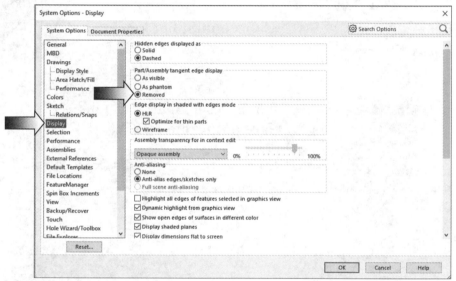

Click **OK**.

7. Changing the window color:

Optionally, change the color of the windows to **Black**.

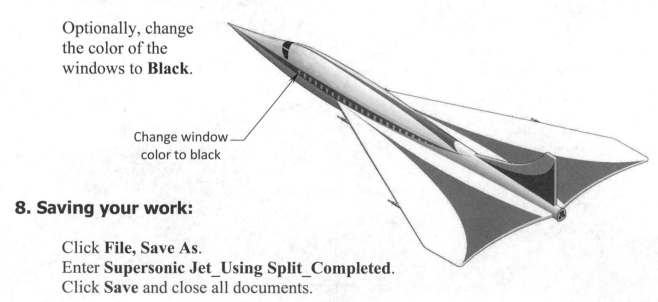

Change window color to black

8. Saving your work:

Click **File, Save As**.
Enter **Supersonic Jet_Using Split_Completed**.
Click **Save** and close all documents.

CHAPTER 9

Surfaces and Patches

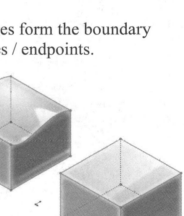

Surfaces and Patches
Welding Fixture

When creating models in SOLIDWORKS, it is best to
combine both Solid and Surfaces tools to develop the geometry.

Solid and Surface bodies are made up of two types of entities: Geometry and
Topology.

* Geometry describes shape, flat or curved.

* Topology describes relationships such as which edges form the boundary
  of which face, or which edges meet at which vertices / endpoints.

Look at the two images on the right. Geometrically
they have different shapes, but topologically
they are identical. Both are composed of six
faces, 12 edges, and eight vertices.

Creating the surface bodies is very similar to
creating the solid bodies but there are some exceptions:

* If you wish to add a cut in a **solid body**, simply sketch the
  geometry and extrude-cut it.

* If you wish to cut a **surface body**, you will need to sketch the geometry and
  either use the sketch to trim the surface body – or – extrude the sketch as a
  new surface, and then use it to trim the surface body.

This lesson discusses the use of the surface tools to create the surface bodies, trim,
knit and thicken them into a solid body.

Surfaces and Patches
Welding Fixture

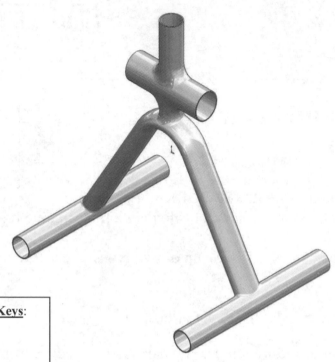

View Orientation Hot Keys:

Ctrl + 1 = Front View
Ctrl + 2 = Back View
Ctrl + 3 = Left View
Ctrl + 4 = Right View
Ctrl + 5 = Top View
Ctrl + 6 = Bottom View
Ctrl + 7 = Isometric View
Ctrl + 8 = Normal to Selection

Dimensioning Standards: **ANSI**

Units: **INCHES** – 3 Decimals

Tools Needed:

	Extruded Surface		Swept Surface		Lofted Surface
	Boundary Surface		Trim Surface		Filled Surface
	Fillet		Thicken Surface		Section View

1. Opening a part document:

Select **File / Open**.

Browse to the Training Files location and open a part document named **Surfaces & Patches.sldprt**.

The sketches have been previously created to help focus on the main topics: Surfaces and Patches.

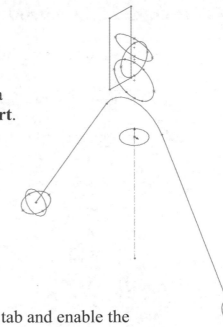

2. Creating a swept surface:

Right-click either the Sketch tab or the Features tab and enable the **Surfaces** tool tab.

Select the **Swept Surface** 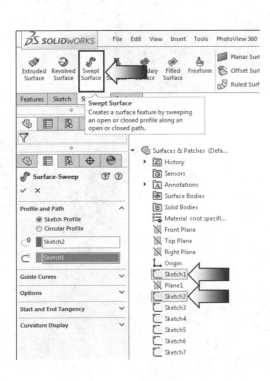 from the Surfaces tool tab (arrow).

For Sweep Profile, select the **Sketch2** from the Feature tree (arrow).

For Sweep Path, select the **Sketch1** also from the Feature tree (arrow).

Click **OK**.

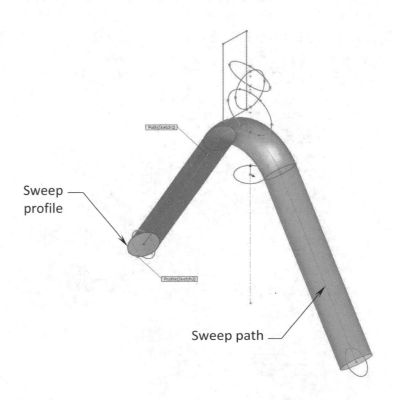

Sweep profile

Sweep path

3. Creating the first extruded surface:

Select the **Sketch3** from the FeatureManager tree; click **Extruded Surface** .

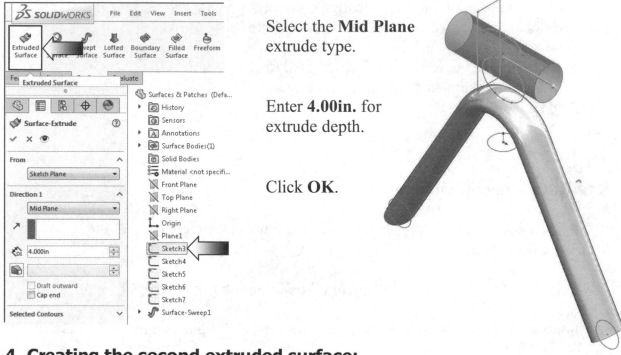

Select the **Mid Plane** extrude type.

Enter **4.00in.** for extrude depth.

Click **OK**.

4. Creating the second extruded surface:

Select the **Sketch4** and click the **Extruded Surface** command once again.

Select the **Blind** type and enter the depth of **6.500in**.

Click **OK**.

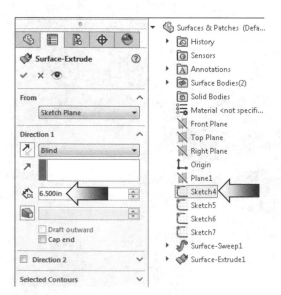

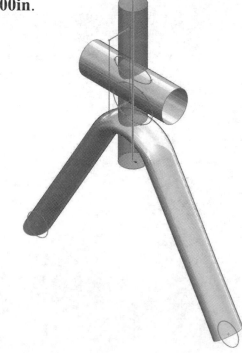

5. Creating the third extruded surface:

Select the **Sketch5** from the Feature tree and click **Extruded Surface** .

Select the **Mid Plane** type and a depth of **8.00in**.

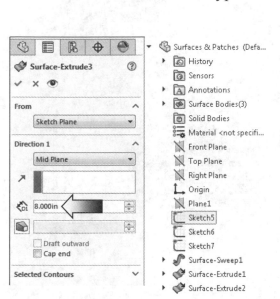

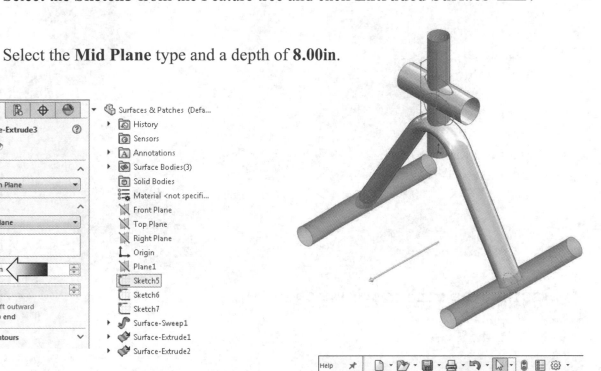

Click **OK**.

6. Creating the first trimmed surface:

Select the **Trim-Surface** command from the Surfaces tool tab (arrow).

For Trim Type, select **Standard Trim**.

For Trim Tool, select the **Sketch6** from the Feature tree.

(more settings on next page...)

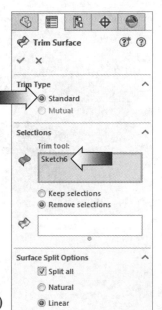

Trim sketch

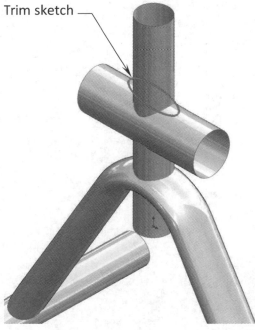

Click the **Keep Selections** option (arrow).

Select the **2 Extruded Surfaces** to keep (arrow).

Click the **Natural** option. (Boundaries extend tangent from the ends of the Trim tool.)

Click **OK**.

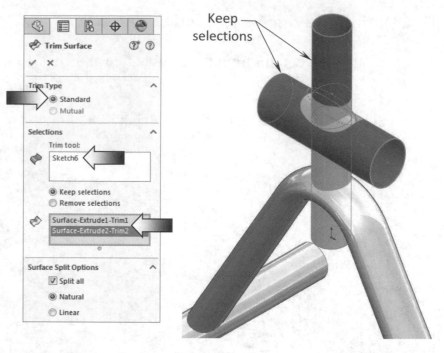

Keep selections

7. Creating the second trimmed surface:

Click the **Trim Surface** command once again.

For Trim Type, use the default **Standard Trim** option.

For Trim Tool, select the **Sketch7** from the Feature tree.

(more settings on next page…)

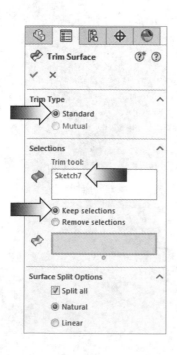

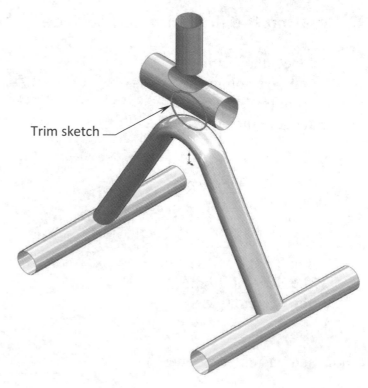

Trim sketch

Click the **Keep Selections** option (arrow).

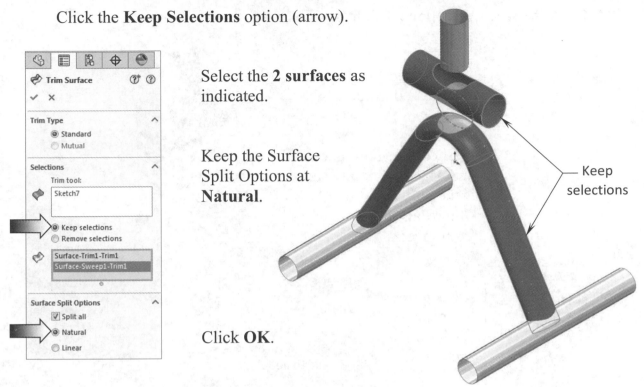

Select the **2 surfaces** as indicated.

Keep the Surface Split Options at **Natural**.

Click **OK**.

Keep selections

8. Creating the third trimmed surface:

Click the **Trim Surface** command again.

For Trim Type, select the **Mutual** option (arrow).

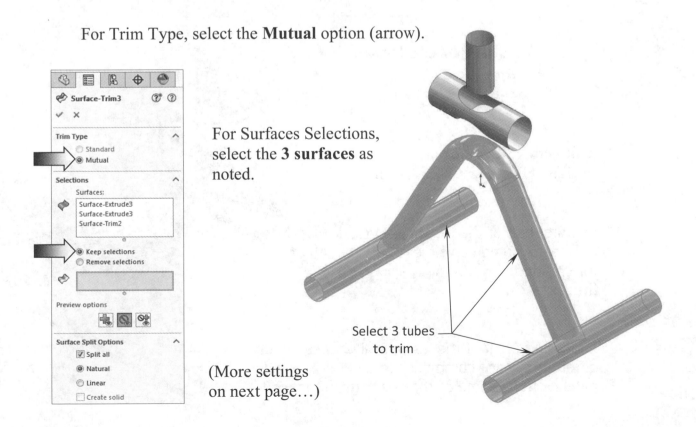

For Surfaces Selections, select the **3 surfaces** as noted.

Select 3 tubes to trim

(More settings on next page...)

Click the **Keep Selections** option (arrow).

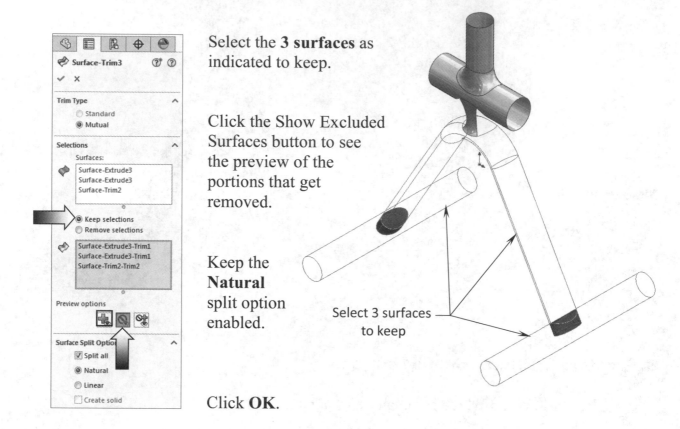

Select the **3 surfaces** as indicated to keep.

Click the Show Excluded Surfaces button to see the preview of the portions that get removed.

Select 3 surfaces to keep

Keep the **Natural** split option enabled.

Click **OK**.

9. Saving the first half of the lesson:

Click **File / Save As**.

Use the **same file name** and overwrite the original document when prompted.

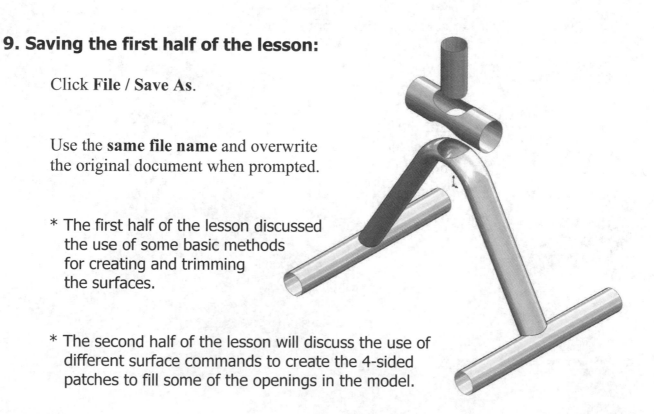

* The first half of the lesson discussed the use of some basic methods for creating and trimming the surfaces.

* The second half of the lesson will discuss the use of different surface commands to create the 4-sided patches to fill some of the openings in the model.

10. Creating a lofted surface:

Click the **Lofted Surface** command 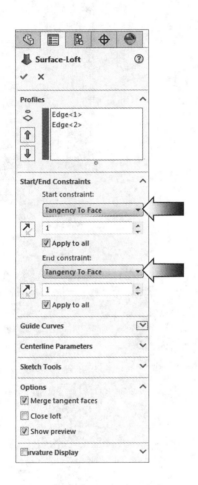 from the surfaces tool tab.

For Loft Profiles, select the **2 edges** near the centers of the 2 surfaces as noted.

> **Lofted Surface**
>
> A Lofted Surface is created by making transitions between two or more profiles and guided by guide curves.

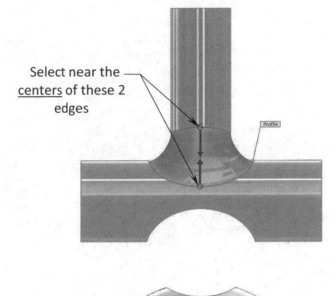

Select near the centers of these 2 edges

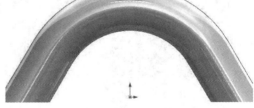

Expand the Start/End Constraints section and select **Tangent to Face** for both directions (arrow).

Keep the **Tangent Weight** at their default value of **1**. Enable the **Apply to All** check boxes for both directions.

Click **OK**.

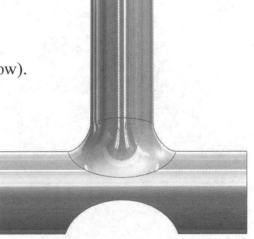

11. Creating the first boundary surface:

Click the Boundary **Surface command** from the Surfaces tool tab, and change to the Front orientation (Control+1).

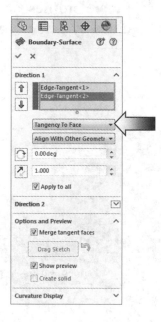

For Direction 1, select the **2 edges** near the ends to prevent twisting.

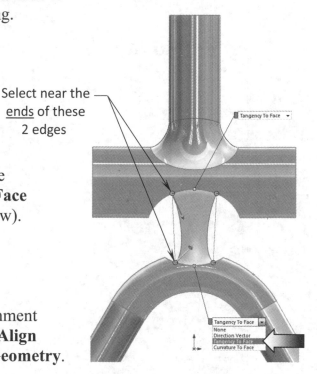

Select near the ends of these 2 edges

Select the same **Tangency to Face** for edges (arrow).

Keep the Alignment at the default: **Align With Other Geometry**.

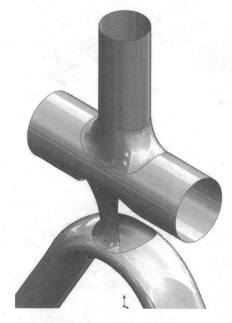

Keep the **Tangent Weight** at the default value of **1** for both edges.

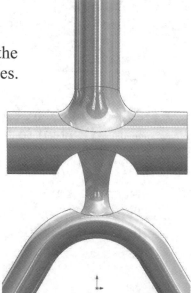

Enable **Apply to All** for both edges.

Click **OK**.

12. Creating the second boundary surface:

Click the **Boundary Surface** command 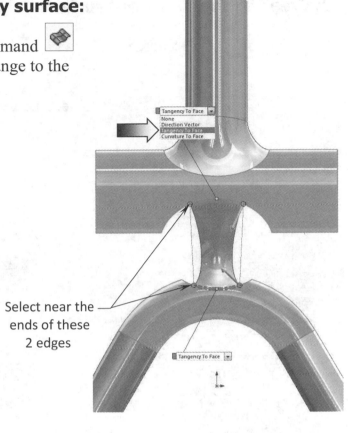 from the Surfaces tool tab and change to the Back orientation (Control+2).

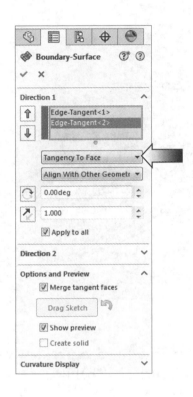

Select near the ends of these 2 edges

For Direction 1, select the **2 edges** near the ends to prevent twisting.

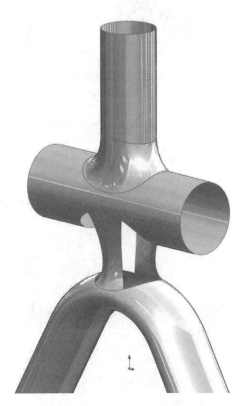

Select the same **Tangency to Face** for both edges (arrow).

Keep all other settings at their default values.

Click **OK**.

13. Creating the first filled surface:

Click the **Filled Surface** command 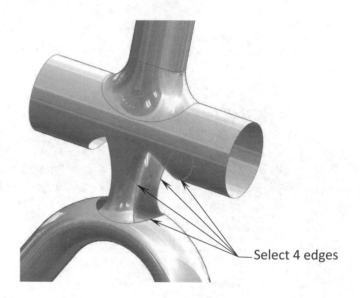 from the Surfaces tool tab.

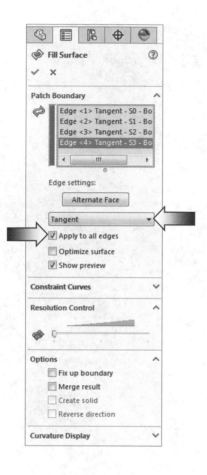

Select the **4 edges** on the right side of the opening.

Select 4 edges

When a closed boundary is defined, a **preview mesh** appears showing a patch surface is being generated.

Select **Tangent** from the list (arrow).

Enable the **Apply to All Edges** checkbox.

Clear the **Optimize surface** checkbox.

Click **OK**.

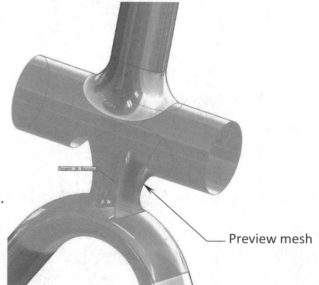

Preview mesh

14. Creating the second filled surface:

Click the **Filled Surface** command from the Surfaces tool tab.

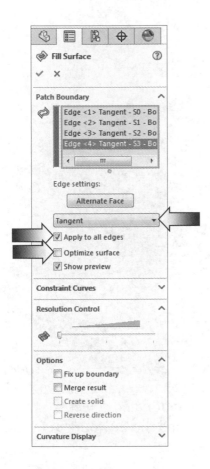

Select the **4 edges** on the left side of the opening.

Select **Tangent** from the list (arrow).

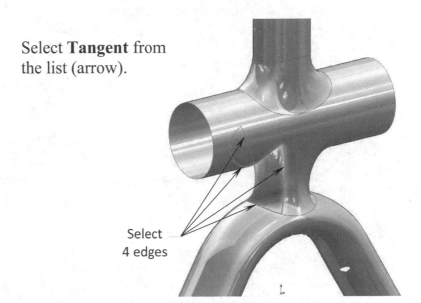

Select
4 edges

Enable the **Apply to All Edges** checkbox.

<u>Clear</u> the **Optimize surface** checkbox.

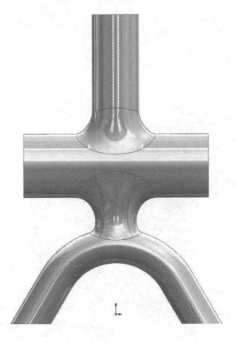

Click **OK**.

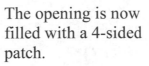

The opening is now filled with a 4-sided patch.

15. Creating a face fillet:

Switch to the Features tool tab and click the **Fillet** command .

Add a **Constant Radius Size** of **.125"** to the <u>edge</u> shown.

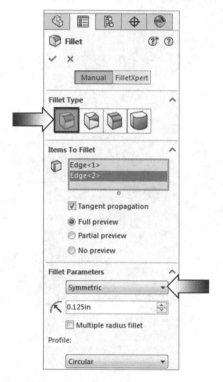

Notice the "uneven width" from side to side?

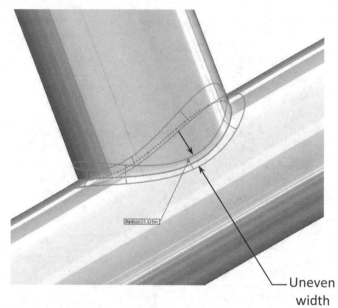

Uneven width

Change to the Face Fillet option and select the **top surface** for <u>Face1</u> and the **bottom surface** for <u>Face2</u>.

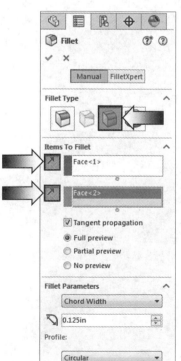

Set all the parameters as shown.

Notice the "even width" fillet all around?

Click **OK**.

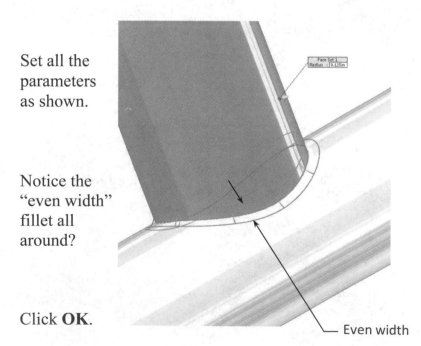

Even width

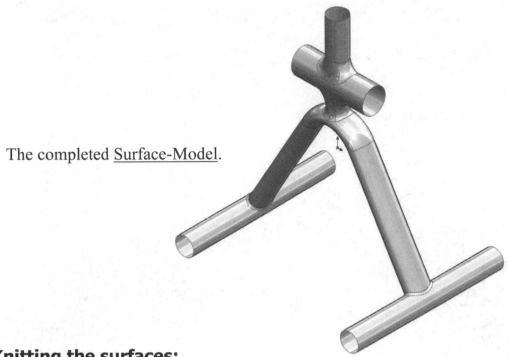

The completed <u>Surface-Model</u>.

16. Knitting the surfaces:

Select the **Knit Surface** command .

Switch to the **Surfaces** tool tab.

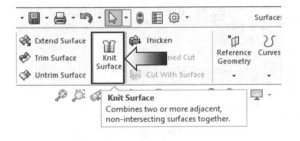

Select **all surfaces** from the graphics area.
(Ensure all surfaces are highlighted.)

Enable the **Merge Entities** checkbox (arrow).

Click **OK**.

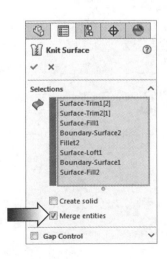

17. Thickening the surface model:

Click the **Thicken** command .

Select the **model** from the graphics area.

Enter **.040in**. for thickness.

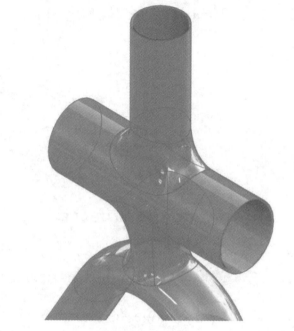

Select the **Material Inside** option (arrow).

Click **OK**.

The completed <u>Solid-Model</u>.

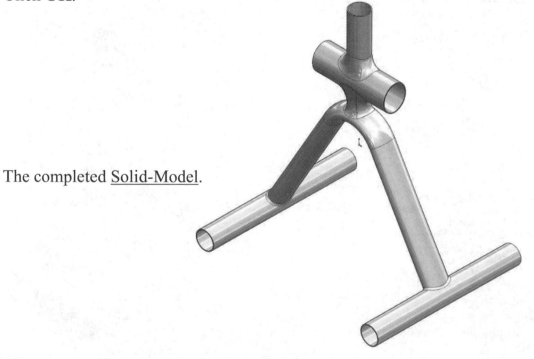

18. Creating a section view:

Click the **Section View** command from the View Heads-Up toolbar (arrow).

Select the **Front** plane as the cutting plane.

Keep the section location at **Zero** (default).

The preview graphic shows a half section of the model.

The red (or blue) color is the wall thickness of the solid model.

Click-off the section view command to return to the full view.

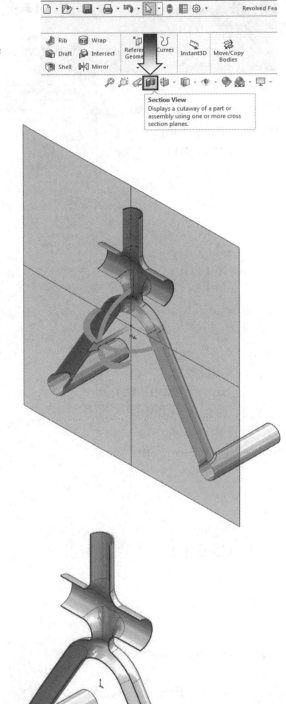

19. Changing the edge display:

The model can look more realistic simply by changing the edge display.

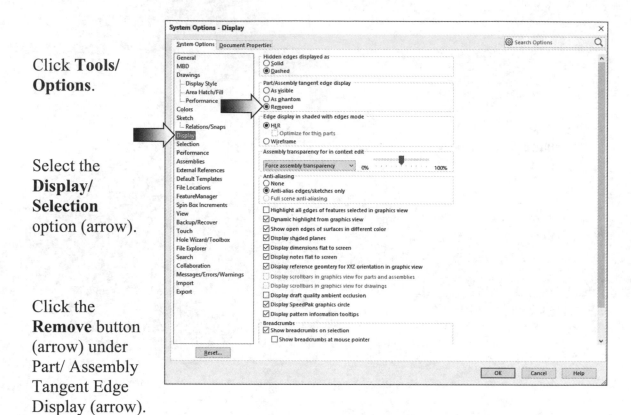

Click **Tools/ Options**.

Select the **Display/ Selection** option (arrow).

Click the **Remove** button (arrow) under Part/ Assembly Tangent Edge Display (arrow).

Click **OK**.

20. Saving your work:

Select **File / Save As**.

Keep the **same file name** and overwrite the previous document when prompted.

Click **Save**.

Exercise: <u>Surface Modifications</u>

1. Opening a part document:

Click **File / Open**.

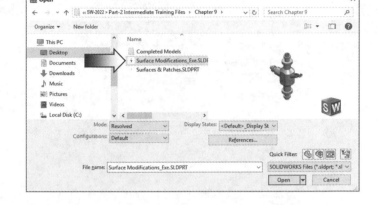

Browse to the Training Files folder and open the part document named: **Surface Modifications_Exe**.

The model was created in SOLIDWORKS but saved as a Parasolid; therefore, no feature history will be available for editing.

This exercise will demonstrate one of the techniques used to work with these types of models.

2. Converting to surface model:

By removing one of the faces of the solid model, it will be transformed into a surface model instantly.

Right-click the face as noted. Select the **Delete** option in the <u>Face section</u>.

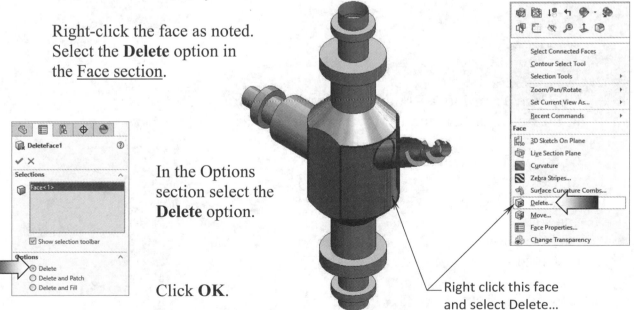

In the Options section select the **Delete** option.

Click **OK**.

Right click this face and select Delete...

The selected face is removed and the part changes into a surface model. This model has 2 surface bodies.

3. Rotating a surface body:

Select **Insert / Features / Move/Copy**.

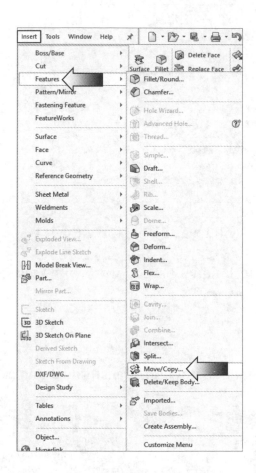

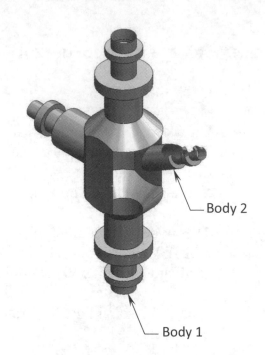

Body 2

Body 1

Press **Control + 1** to change to the Front view.

Click the **Translate / Rotate** button at the bottom of the tree to change to the Move / Rotate options.

Click in the Bodies to Move box to activate this section and select the **surface body** as indicated.

Expand the **Rotate** section and select the Origin as the center of rotation.

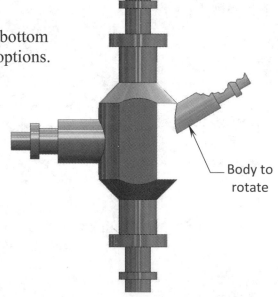

Body to rotate

More settings on the next page...

In the Z direction, enter **-30.00deg** for rotate angle (arrow).

The preview shows the selected surface body is being rotated downwards.

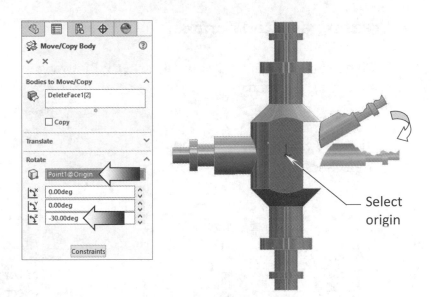

Select origin

Click **OK** to exit the Move/Rotate command.

4. Extending a surface body:

Click the **Extend Surface** command from the Surfaces toolbar.

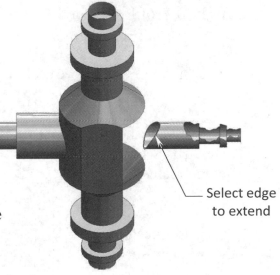

Select edge to extend

For Edges/Faces to Extend, select the **edge** as noted.

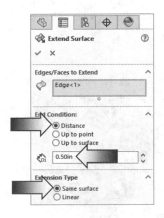

For End Condition, select the **Distance** option.

Enter **.500in** for distance.

For Extension Type, select the **Same Surface** option (arrow).

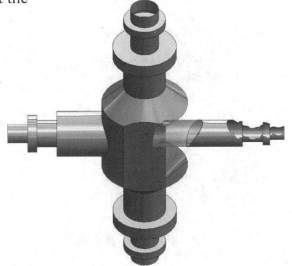

Click **OK**.

5. Creating a lofted surface:

The surface that was deleted earlier will now get recreated using its own existing geometry.

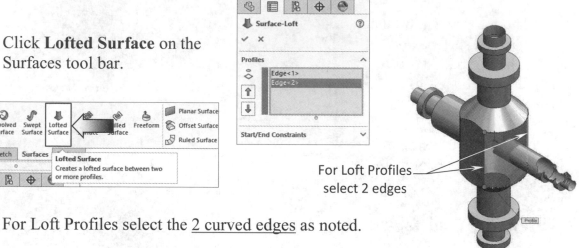

Click **Lofted Surface** on the Surfaces tool bar.

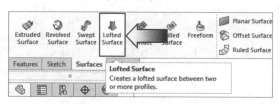

For Loft Profiles select the 2 curved edges as noted.

Click in the Guide Curve section and select the 2 vertical edges as noted.

The preview shows a new surface is being created.

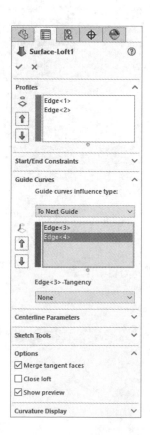

The grid lines on the preview surface is called the Mesh. To modify the mesh density or to remove them, right-click the preview surface, go to Mesh Preview and select **Clear All Meshed Faces**.

Keep other parameters at their default values.

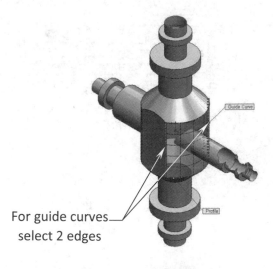

For Loft Profiles select 2 edges

For guide curves select 2 edges

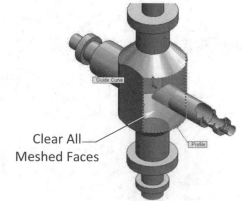

Clear All Meshed Faces

Click **OK**.

6. Repairing the broken surface body:

The broken area can be repaired with a revolved surface. This repair involves a few steps: Convert the broken edges into a new sketch, remove its relations, revolve the sketch as a new surface, delete the broken area, and then knit the revolved surface to the surface part.

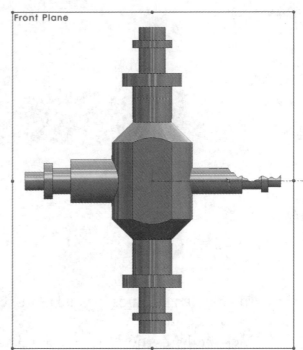

Select the <u>Front</u> plane and open a **new sketch**.

Sketch a <u>horizontal centerline</u> that will be used as the revolve line.

Hold the <u>Control</u> key and select the **8 edges** of the broken area as indicated in the image below.

Some of the converted lines are crossing the centerline; this will cause an overlapping error when revolving the sketch. Those lines need to be trimmed in the next step.

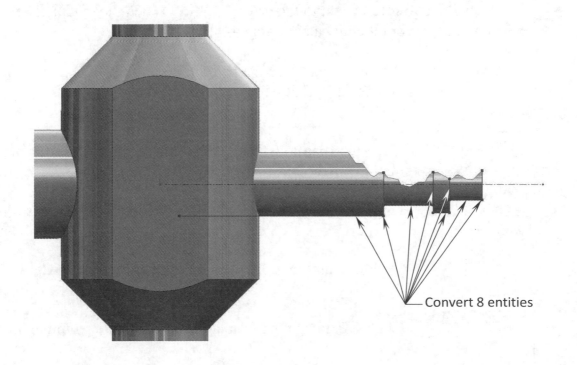

Convert 8 entities

Select the **Trim Entities** command from the Sketch toolbar.

Select the **Trim to Closest** option (arrow).

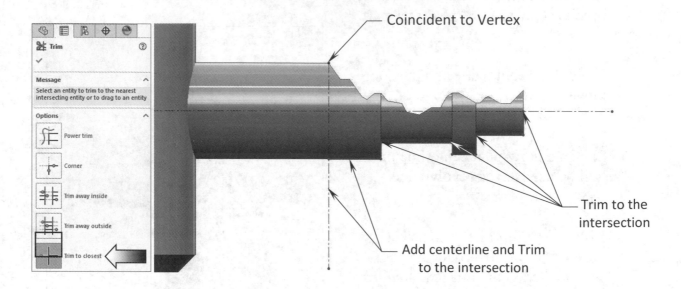

Coincident to Vertex

Trim to the intersection

Add centerline and Trim to the intersection

Add a <u>vertical centerline</u> and use it to trim one of the lines.

Click on the portions of the lines that need to be deleted. When completed, all of the lines must be connected continuously.

When an entity is converted from a model edge, SOLIDWORKS creates a relation called On-Edge to reference it to the original entity. When the model edge is changed, the converted entity also changed.

There are 8 on-edge relations that need to be removed, so that when the broken area is deleted, it will not cause the dangling errors.

Click the **Display/Delete Relations** command from the Sketch toolbar.

Under the relation section, select the option **All in this Sketch** (arrow).

Highlight all **On-Edge** relations and click **Delete All**.

Click **OK** to exit the Display/Delete Relation command.

Switch to the Surfaces toolbar and select the **Revolved Surface** command.

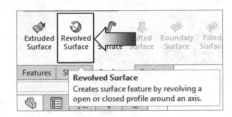

Select the **horizontal centerline** as the Axis of Revolution.

Use the default **360deg** revolve angle.

Click **OK**.

7. Creating a split line:

There are several ways to delete a surface, but we are going to try the split line approach instead. The Split Line divides a surface body into 2 bodies, and the broken area can be deleted without affecting the other half.

Select the <u>Front</u> plane and open a **new sketch**.

Sketch a **vertical line** starting at the vertex shown.

Switch to the **Feature** toolbar and select **Curves / Split Line**.

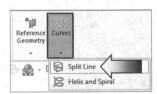

Use the default **Projection** Option.

Click the surface **body to split** as noted.

Click **OK** to complete and exit the split command.

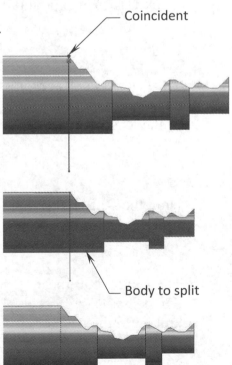

8. Deleting surfaces:

The broken surfaces can now be deleted.

Drag the **Roll-Back** line <u>up</u> one step to temporarily hide the Revolved Surface.

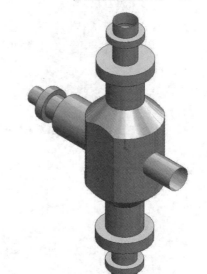

Delete 7 faces

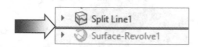

Right-click on one of the surfaces in the broken area and select the **Delete** option below the <u>Face</u> section.

Select the **7 faces** to delete as indicated.

Under the Options section click the **Delete** button.

Click **OK** to accept and exit the Delete Face command.

The selected surfaces are deleted leaving a sharp clean edge that matches the edge of the Revolved Surface.

Drag the **Roll-Back** line back <u>down</u> to the bottom.

The Revolved Surface reappears. It will get knitted to the surface model in the next couple of steps.

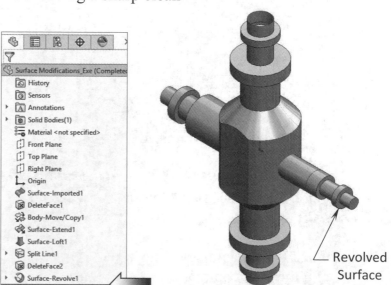

Revolved Surface

9. Creating a section view:

There are some areas that still need trimming. The section view will help us see and work with them a little easier.

Select the <u>Right</u> plane from the FeatureManager tree and click the **Section View** command on the View Heads Up toolbar.

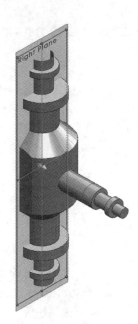

Rotate the view to see the left end of the Revolved Surface.

10. Trimming the surfaces:

Switch to the **Surfaces** tool tab.

Click the **Trim Surface** command.

Select **Mutual** under the Trim Type.

For Trimming Surface select the **cylindrical body** and the **lofted surface** as indicated.

Click the **Remove Selection** button and select the **two surfaces** as noted (the left end of the cylindrical body and the surface that intersects the lofted surface and the cylindrical body).

Click **OK** and inspect your surface model against the image shown here.

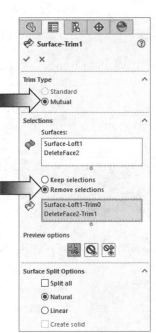

Delete 2 surfaces

Surfaces to Keep

Surfaces to remove

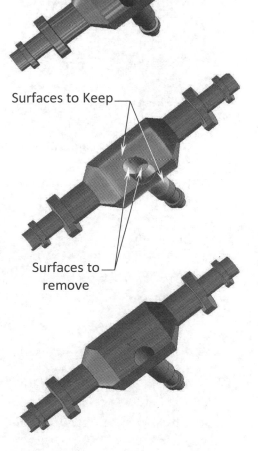

11. Creating the planar surfaces:

Before the surface model can be converted back into a solid model, all openings should be closed-off. One of the quickest ways to do this is to use the Planar Surface command.

Click the **Planar Surface** command from the Surfaces tool tab.

Select the **3 circular edges** as indicated for Bounding Entities.

A preview graphic of the new surfaces appears, indicating that they are being created.

Click **OK** to accept and exit the Planar Surface command.

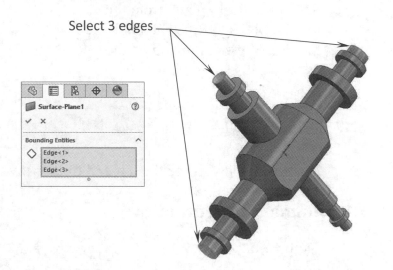

Select 3 edges

12. Knitting all surfaces:

After all surface bodies are knitted into a single body, it can be made into a solid model.

Click the **Knit Surface** command from the Surfaces tool tab.

Expand the Surface Bodies folder on the FeatureManager tree.

Select **all surfaces** inside of this folder; there should be a total of 6 surfaces.

Enable the **Try to Form Solid** checkbox and uncheck the Gap Control option.

Select all surfaces

Click **OK** to exit the Knit Surface command.

13. Verifying the solid model:

Create a section view to see if the surface model has been converted into a solid model.

Select the Front plane from the FeatureManager tree and click the **Section View** command.

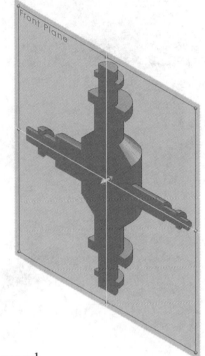

The preview shows the model is a solid part (solid blue color) at this point.

Click **Cancel** to exit the Section View command.

14. Saving your work:

Click **File / Save As**.

Enter **Surface Modification_Exe** for the name of the file.

Click **Save** and replace the existing file when prompted.

CHAPTER 10

Configure Features

Configure Features
Wooden Crate

The Configure Feature uses the Modify Configuration
dialog box to create and modify configurations for commonly configured
parameters in parts and assemblies.

To create or modify a configuration
do one of the following:
Right-click an item and select
Configure Dimension, Configure
Feature, Configure Component, or
Configure Material.

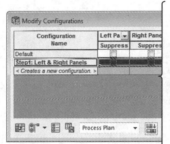

Only certain items can be configured in the Modify Configuration dialog box.
Right-click an item and start the Modify Configuration, or double-click and add
to Modify Configuration include:

* For parts: features, sketches, dimensions, and material.
* For assemblies: components, assembly features, mates, and
dimensions of assembly features and mates.

The Modify Configuration dialog box lists the Configurations of the model in one
column and configurable parameters in other columns.

New columns can be added by double-clicking items in the graphics area or
FeatureManager design tree. Click at the top of a column to select other parameters
of a feature to configure.

A column can be added for every configured parameter in the model at once by
clicking All-Parameters at the bottom of the dialog box.

Configure Features
Wooden Crate

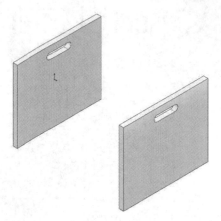

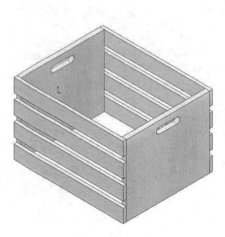

Dimensioning Standards: **ANSI**

Units: **INCHES** – 3 Decimals

Tools Needed:

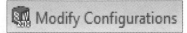

1. Opening a part document:

Select **File / Open**.

Locate the part document named:
Configure Features.sldprt and open it.

2. Configuring features:

The Configure Feature option provides a table whcrc you can create and modify configurations for commonly configured parameters in parts and assemblies.

In a part, you can configure dimensions of features and sketches, suppression states of features and sketches, material, and custom properties. Alternatively, you can create and modify configurations manually or with a design table.

Right-click the fcature named **Left Panel** and select **Configure Feature** (arrow).

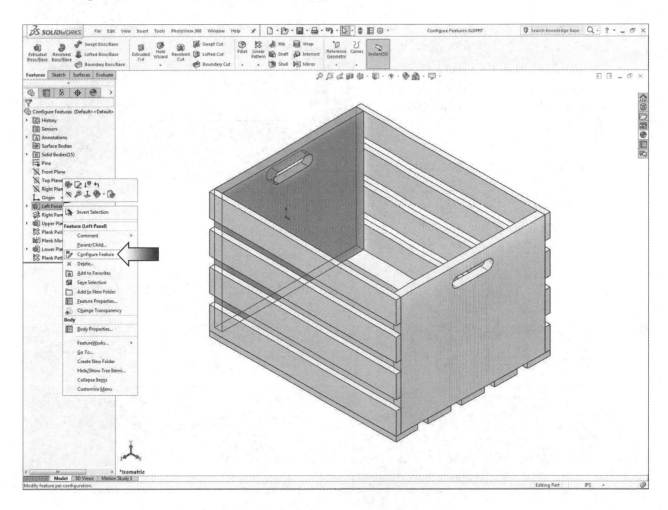

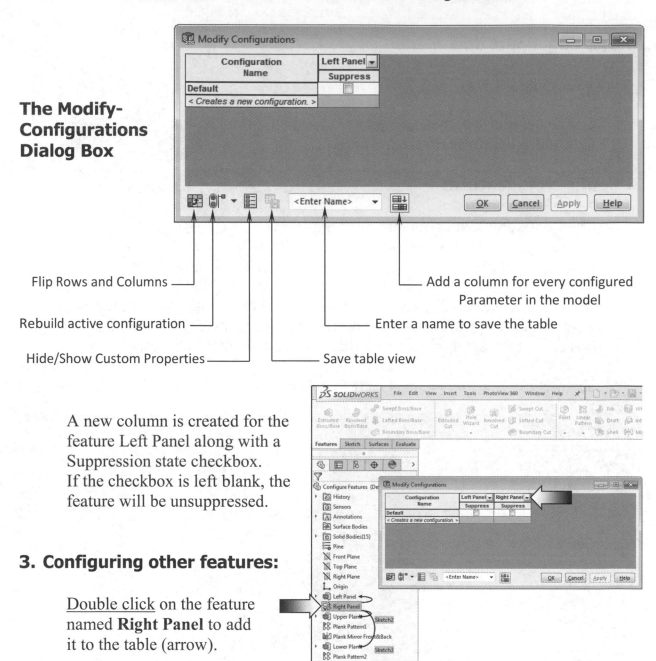

**The Modify-
Configurations
Dialog Box**

Flip Rows and Columns ————

Rebuild active configuration ————

Hide/Show Custom Properties ————

Save table view

Enter a name to save the table

Add a column for every configured
Parameter in the model

A new column is created for the
feature Left Panel along with a
Suppression state checkbox.
If the checkbox is left blank, the
feature will be unsuppressed.

3. Configuring other features:

Double click on the feature
named **Right Panel** to add
it to the table (arrow).

Repeat the same step to add all other features to the table (7 features all together).

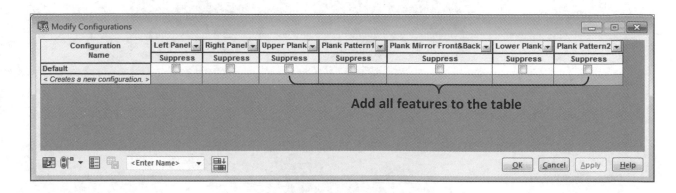

Add all features to the table

4. Saving the table:

Enter **Process Plan** for the name of the table and press **Save Table View.**

As good practice, always click the Save Table View after adding or changing the content of the table.

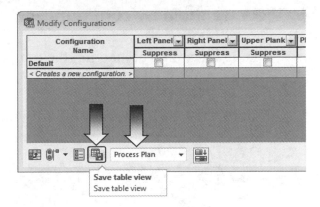

5. Adding a new configuration:

Click in the cell under the Default configuration and enter the name of the new configuration: **Step1: Left & Right Panels** (arrow).

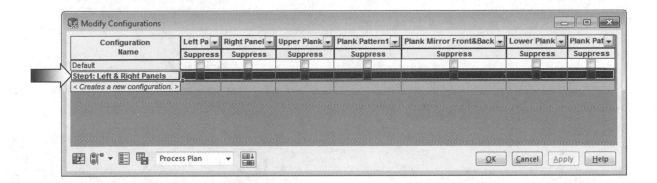

Select the Suppress checkboxes for the last 5 features as indicated below (arrow).

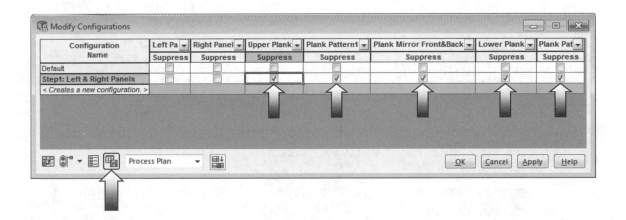

Click the **Save Table View** again to re-save the table and click **OK** to close it.

6. Viewing the new configuration:

Drag the split-tree-handle (the round dot) about halfway down the FeatureManager tree to split it into 2 sections (see note below).

Click the **Configuration tab** to switch to the lower section to the ConfigurationManager.

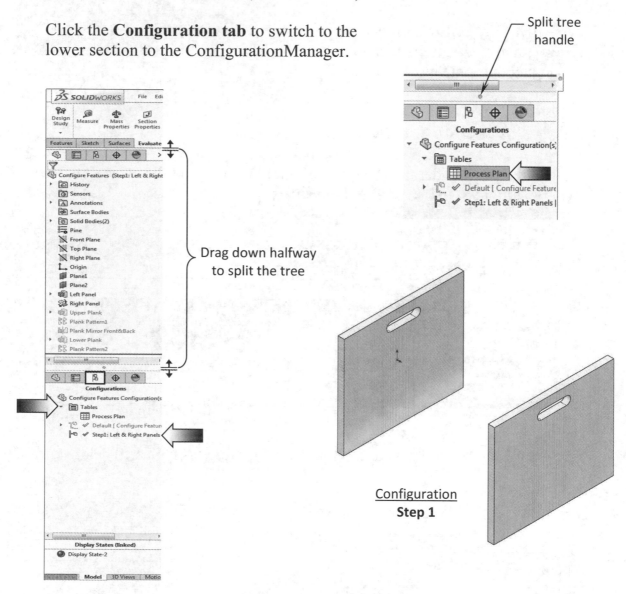

Configuration
Step 1

A folder is created automatically to store the table(s). Expand the folder to see the Process Plan table that you have just created.

The new configuration Step1 is displayed under the table. Double-click it to activate this configuration.

The new configuration displays only the features that did not have the check mark in their suppress boxes. The other features are suppressed in this configuration.

7. Adding more configurations:

Double-click the **Process Plan** table to re-activate it.

Enter the name of the new configuration: **Step 2: Upper Planks Front**.

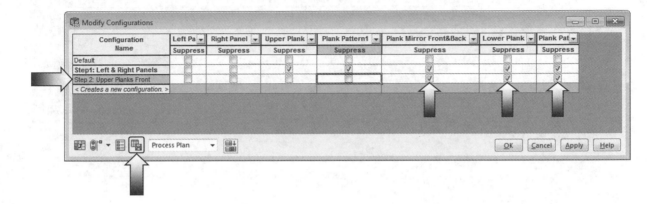

Leave the first 4 checkboxes blank and click **suppress** for the last 3 features.

Click **Save Table View** to update the table.

Click **OK** to close the table.

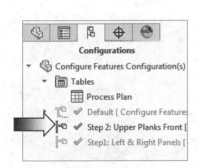

Locate the new configuration
Upper Planks Front and double-
click on its icon to activate it.

The 4 planks in the front of
the crate are displayed.

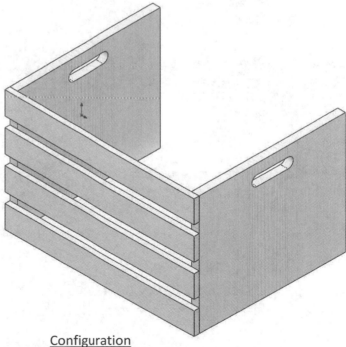

Configuration
Step 2: Upper Planks Front

Double click the **Process Plan** table to reactivate it.

Enter the name for the next configuration: **Step 3: Upper Planks Back**.

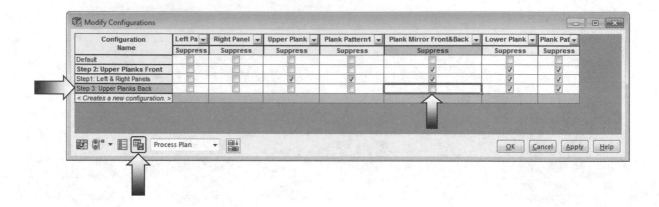

Leave the first 5 checkboxes blank and **suppress** the last 2 features.

Click **Save Table View** to update the table.

Click **OK** to close the table.

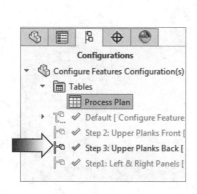

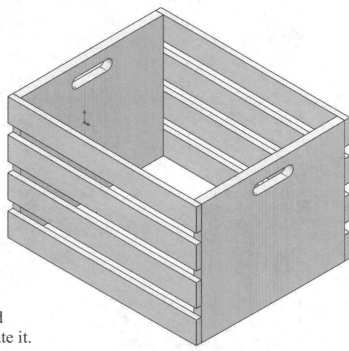

Locate the new configuration
Step 3: Upper Planks Back and
double-click on its icon to activate it.

The 4 planks in the back of the crate
are displayed.

Configuration
Step 3: Upper Planks Back

8. Renaming a configuration:

The Default configuration already has all the planks in it. It is quicker to rename it than to create another configuration.

To rename, either click+pause+click on the name of the configuration, or use the hot key **F2** to rename it.

Double-click on the **Default** configuration to activate it.

The active configuration will have a green check mark in front of its name; the inactive configuration will have the grey check mark instead.

Rename the **Default** configuration to **Step 4: Lower Planks** (arrow).

Rotate the model to see the planks from the bottom.

<u>NOTE:</u>

The order of the configurations is displayed either numerically or alphabetically.
The configurations should be renamed in such a way that they appear in the correct order when sorted.

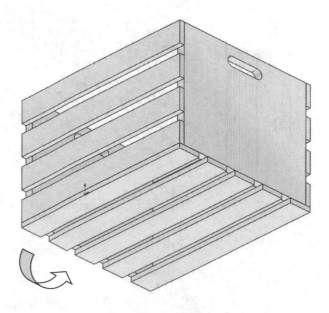

Configuration
Step 4: Lower Planks

Toggle between the configurations to see the changes in each one.

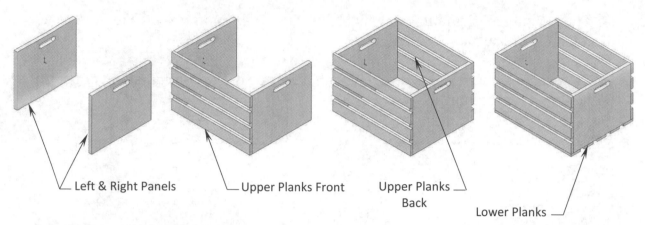

Left & Right Panels   Upper Planks Front   Upper Planks Back   Lower Planks

9. Inserting a design table:

Adding the Material Column in the Configure Table will allow you to change the material of any planks more quickly and easily.

Double-click the **Process Plan** table to activate it.

Double-click the **Material** option (arrow).

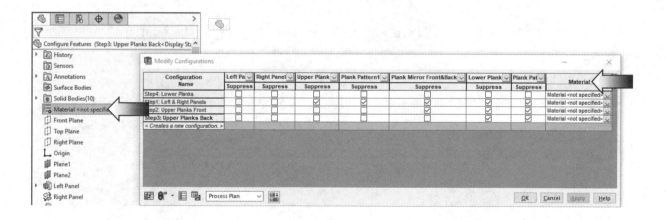

A new column is added to the table (arrow).

We will add different wood materials to the planks:

> The Lower Planks will have the Cedar wood material assigned to it.
> The Left & Right Panels will have the Pine material.
> The Upper Planks Front will have the Oak material.
> And the Upper Planks back will have the Maple material.

Click the drop-down arrow next to the 1st material and select **Browse More**...

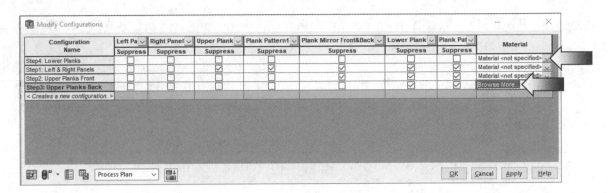

Scroll down to the Woods folder and select **Cedar** as the material for the Lower Planks.

Click **Apply** and **Close**.

Repeat the previous step and assign **Pine** to the **Left & Right Panels**.

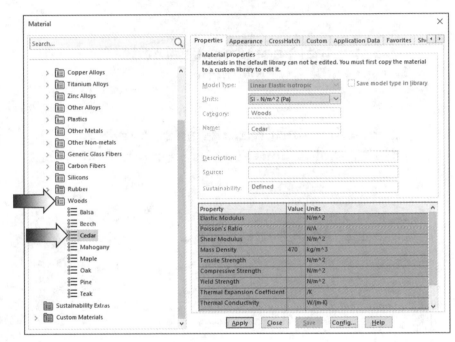

Assign **Oak** to the **Upper Planks Front**.

Assign **Maple** to the **Upper Planks Back**.

Click **Save Table View** and **OK** to exit the Modify Configuration dialog box.

10. Saving your work:

Click **File / Save As** and enter **Configure Features** for the name of the file. Overwrite the existing file if prompted.

Click **Save**.

Exercise: Design Tables & Tabulated Tables

A design table allows you to build multiple configurations of parts or assemblies by specifying parameters in an embedded Microsoft Excel worksheet.

The design table is saved in the model document and is not linked to the original Excel file. Changes you make in the model are not reflected in the original Excel file. However, you can link the model document to the Excel file if you wish.

This exercise will guide you through the creation of a Design Table and convert it to a Tabulated table used in a drawing.

1. Open the part document named:
   Design Table Example.sldprt

2. Right click Annotations and enable:
   Display Annotations and Show Feature Dimensions.

3. Select Insert, Tables, Design Table.
   Click the Auto-Create option and push OK.

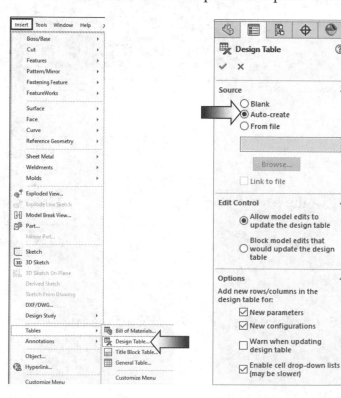

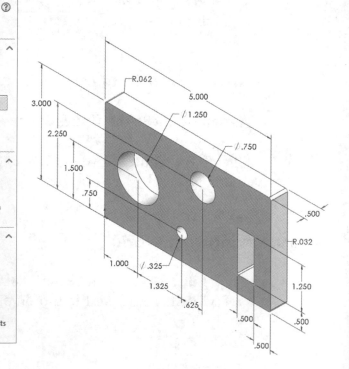

Select all dimensions in the dialog box (Shift + select the first and last) and click **OK**.

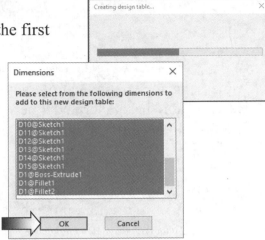

MS-Excel is launched and the selected dimensions are populated to the columns and rows automatically.

For the purpose of the exercise, we will only work with the overall dimensions (Length, Height, and Depth) and display them in the tabulation table.

4. Change the Configuration name from Default to Size1.

Add 3 other configurations and label them as Size2, Size3, and Size4.

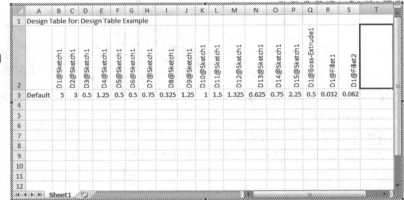

5. Enter the new dimension values:
for the next 3 sizes as indicated.

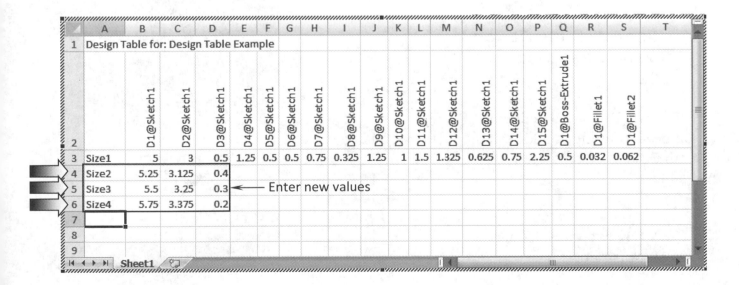

	D1@Sketch1	D2@Sketch1	D3@Sketch1	D4@Sketch1	D5@Sketch1	D6@Sketch1	D7@Sketch1	D8@Sketch1	D9@Sketch1	D10@Sketch1	D11@Sketch1	D12@Sketch1	D13@Sketch1	D14@Sketch1	D15@Sketch1	D1@Boss-Extrude1	D1@Fillet1	D1@Fillet2
Size1	5	3	0.5	1.25	0.5	0.5	0.75	0.325	1.25	1	1.5	1.325	0.625	0.75	2.25	0.5	0.032	0.062
Size2	5.25	3.125	0.4															
Size3	5.5	3.25	0.3															
Size4	5.75	3.375	0.2															

← Enter new values

Click **OK**.

A message appears indicating that the Design Table has generated 4 new configurations. Click **OK**.

The Default configuration is the same as the Size1 Configuration; it can be deleted to eliminate confusion.

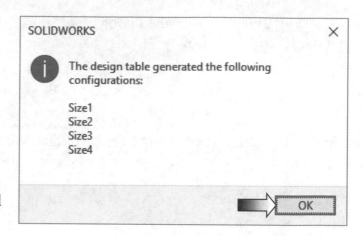

SOLIDWORKS

The design table generated the following configurations:

Size1
Size2
Size3
Size4

OK

6. Change to the ConfigurationManager:

Expand the **Tables** folder to see the Design Table.

The Magenta color dimensions indicate they have been linked to a Design Table. Changes can now be done bi-directionally.

We will customize the table by changing some of its parameters to make it look more like a Tabulated Table.

7. Right click the Design Table and select Edit Table.

The Add Rows and Columns appears offering to add other parameters to the table.

Click **OK** without selecting any new parameters.

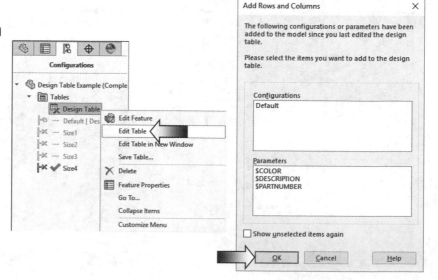

8. Right-click Row2 and select HIDE.

Right click **Row1** and select **Format Cells**.

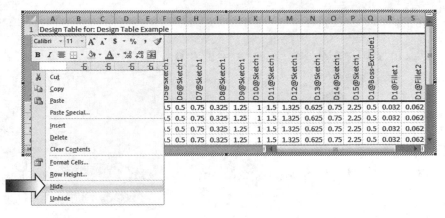

Click the **Alignment** tab in the Format Cell dialog box and enable the options **Wrap Text** and **Merge Cells** (arrows).

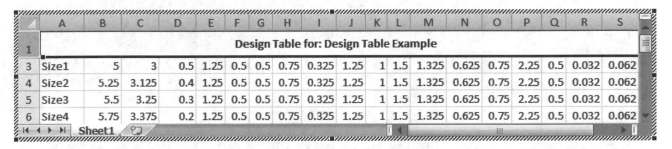

Click **OK**.

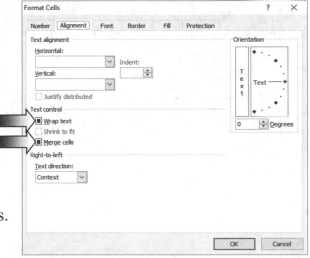

The Row1 now displays the title for the table as Design Table For: Design-Table Example. (The Title can be changed to the actual name of the product.)

Drag the handles inwards to hide the empty cells.

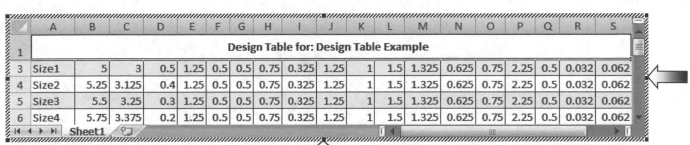

9. Transferring to a drawing:

Select **File, Make Drawing From Part**.

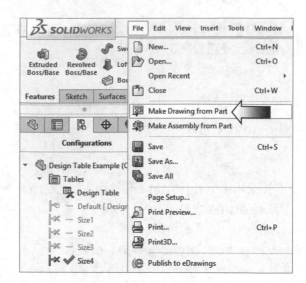

Click the Drawing template and push **OK**.

Select size **B (ANSI) Landscape**.
Click **OK**.

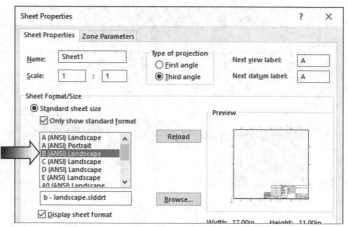

Expand the **View Palette** on the right.
Drag & drop the **Isometric** view to the
drawing approximately as shown.

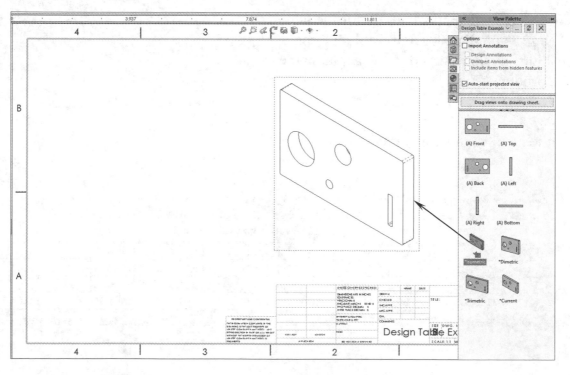

10. Inserting the Design Table.

Switch to the **Annotation** tab. Click the border of the Isometric view.

Select **Tables, Design Table** (arrow). Drag the table into the drawing.

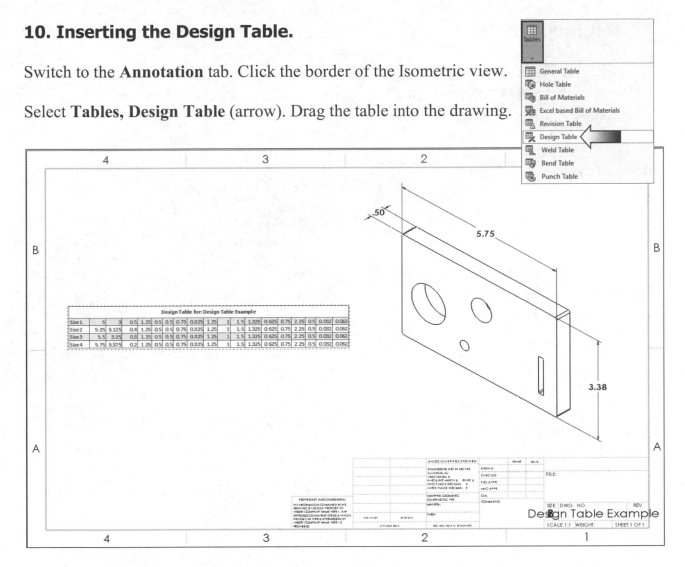

Design Table for: Design Table Example

Size 1	5	3	0.5	1.25	0.5	0.5	0.75	0.325	1.25	1	1.5	1.325	0.625	0.75	2.25	0.5	0.032	0.062
Size 2	5.25	3.125	0.4	1.25	0.5	0.5	0.75	0.325	1.25	1	1.5	1.325	0.625	0.75	2.25	0.5	0.032	0.062
Size 3	5.5	3.25	0.3	1.25	0.5	0.5	0.75	0.325	1.25	1	1.5	1.325	0.625	0.75	2.25	0.5	0.032	0.062
Size 4	5.75	3.375	0.2	1.25	0.5	0.5	0.75	0.325	1.25	1	1.5	1.325	0.625	0.75	2.25	0.5	0.032	0.062

11. Adding the overall dimensions:

Adds the Length, Height, and Depth reference dimensions to the Isometric view.

Change the Length dimension to **A** using the **Dimension Text** window (arrow).

Change the Height dimension to **B**, and the Depth dimension to **C**.

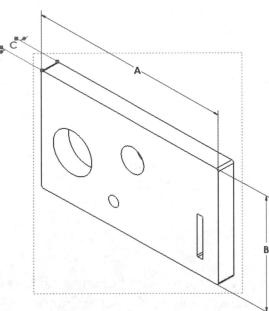

12. Returning to Excel:

Double-click inside the table
to launch MS-Excel for editing.

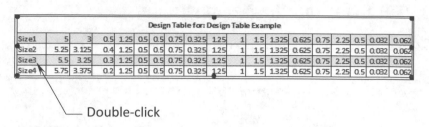

Double-click

Right click **Row3** and select
Insert.

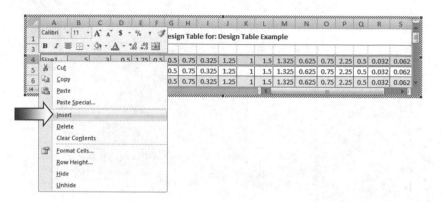

A new row is inserted above
Row4. It will be used to
enter the headers A, B,
and C.

Enter **A** in the cell **B3**.
Enter **B** in the cell **C3**.
Enter **C** in the cell **D3**.

Drag the horizontal and
vertical handles to hide the
other cells (arrow).

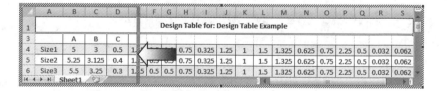

Change the **colors** of the Rows and Columns to make it easier to see the different values
in the table.

Add **borders** and change the text in the headers to **Bold**. This would
also improve the general appearance of the table.

13. Returning to the drawing:

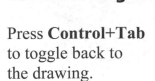

Press **Control+Tab** to toggle back to the drawing.

Click the dotted border of the Isometric view and re-insert the Design Table once again.

Place the new table and <u>delete</u> the old table.

Delete the old table

Save and close all documents.

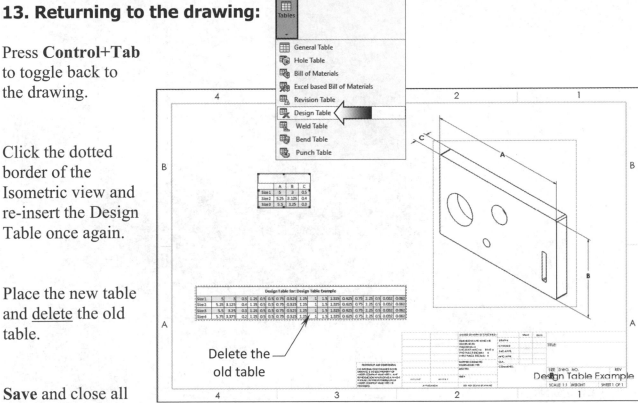

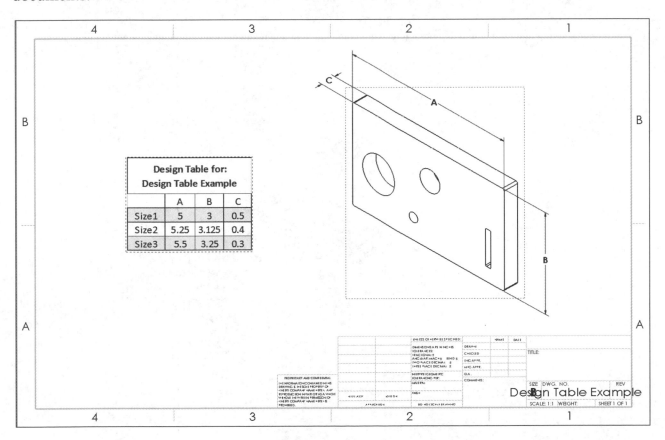

Design Table for:
Design Table Example

	A	B	C
Size1	5	3	0.5
Size2	5.25	3.125	0.4
Size3	5.5	3.25	0.3

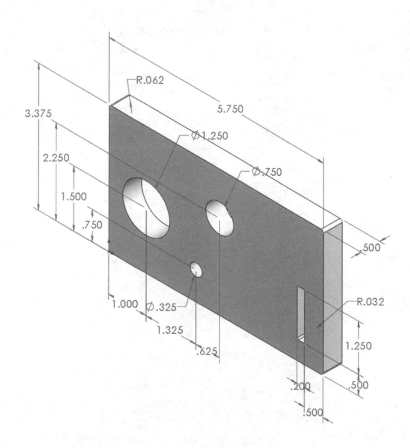

CHAPTER 11

Assembly Motions & Mates

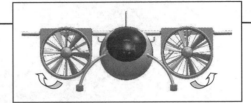

Assembly Motions & Mates
HeliDrone

Add mates to the components in an assembly to assemble them. Mates create geometric relationships between assembly components. When adding mates, the user defines the allowable directions of linear or rotational motion of the components. They can then be moved within their degrees of freedom, visualizing the assembly's behavior.

There are three types of mates available in SOLIDWORKS:

* **Standard mates:** include angle, coincident, concentric, distance, lock, parallel, perpendicular, and tangent mates.

* **Advanced mates:** include limit, linear/linear coupler, path, symmetry, and width mates.

* **Mechanical mates:** include cam-follower, gear, hinge, rack and pinion, screw, and universal joint mates.

A Mates folder is automatically included in the FeatureManager design tree to store the mates. Mates can be edited, suppressed, or deleted.

When a component is over mated there will be a plus sign (+) added to its name on the FeatureManager tree. If one or more degrees of freedom still exist in a component, a minus sign (-) appears next to its name, but a fully mated component does not have any sign next to its name.

This lesson discusses the use of some of the options in the three mate types: Standard, Advanced, and Mechanical Mates.

Assembly Motions & Mates
HeliDrone

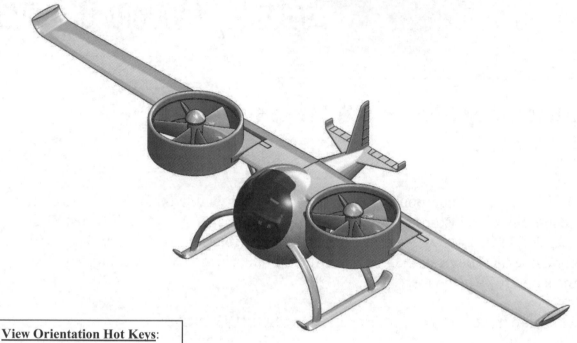

View Orientation Hot Keys:

Ctrl + 1 = Front View
Ctrl + 2 = Back View
Ctrl + 3 = Left View
Ctrl + 4 = Right View
Ctrl + 5 = Top View
Ctrl + 6 = Bottom View
Ctrl + 7 = Isometric View
Ctrl + 8 = Normal To Selection

Dimensioning Standards: **ANSI**

Units: **INCHES** – 3 Decimals

Tools Needed:

 Mate

 Width Mate

 Concentric Mate

 Coincident mate

 Limit Angle Mate

 Parallel Mate

 Gear Mate

 Pattern Driven Component Pattern

 Radial Explode

Assembly Motions and Mates

1. Opening an existing assembly document:

Browse to the Training Files location and open the assembly document named: **HeliDrone Assembly.sldasm**

This assembly contains one component and two sub-assemblies.

The sub-assemblies are set to **Rigid** by default; they will need to be changed to **Flexible** so that the blades inside the fan housings can rotate, when needed.

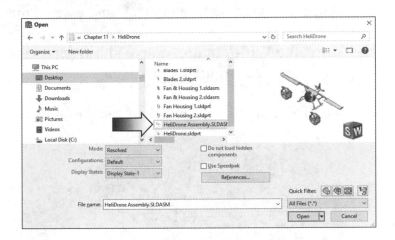

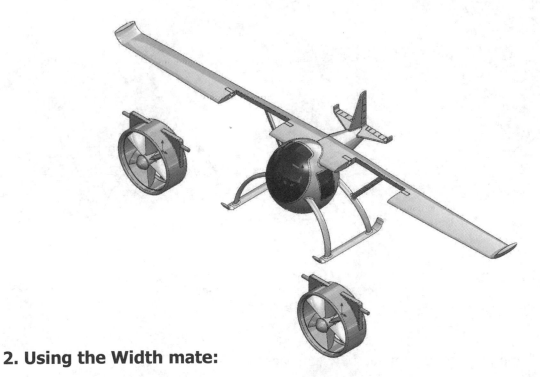

2. Using the Width mate:

Use the following options to define the limits of the mate: (Note: A width mate must be selected prior to choosing one of the options below.)

    * **Centered:**   Centers a tab within the width of a groove.
    * **Free:**   Lets the components move freely within the limits of the selected faces or planes with respect to the components.

* **Dimension:**   Sets a distance or angle dimension from one selection set to the closest opposing selection set of faces or planes.

* **Percent:**   Sets the distance or angle based on a percentage value dimension from one set of the selection set to the center of the other selection set.

Select the **Mate** command 🗓 and expand the **Advanced Mate** section.

Click the **Width** button. For Width-Selection, select the <u>2 planar faces</u> on the left and right sides of the Fan Housing as noted.

For Tab-Selection, select the <u>2 planar faces</u> on the left and right side of the wing.

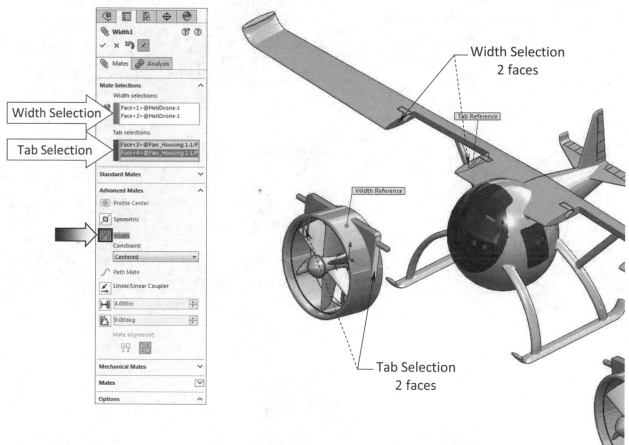

The 4 selected faces are centered between each other. Click **OK**.

3. Adding a Concentric mate:

Concentric mates can be added between a circular edge, a cone or a cylinder, a line, a point, or a sphere.

Switch back to the **Standard Mates** section (arrow).

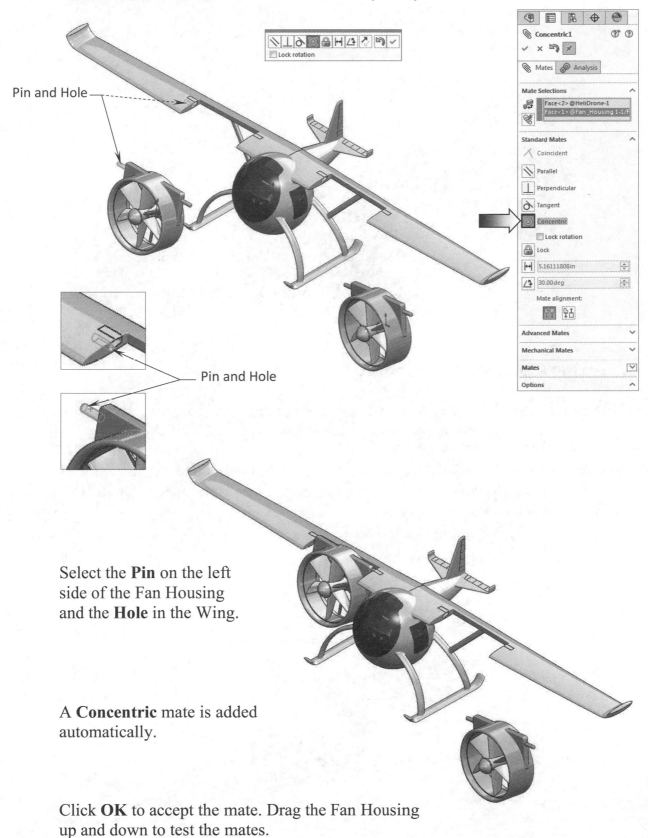

Pin and Hole

Pin and Hole

Select the **Pin** on the left side of the Fan Housing and the **Hole** in the Wing.

A **Concentric** mate is added automatically.

Click **OK** to accept the mate. Drag the Fan Housing up and down to test the mates.

4. Adding another Concentric mate:

Repeat the last step and mate the second Fan Housing to the opposite wing.

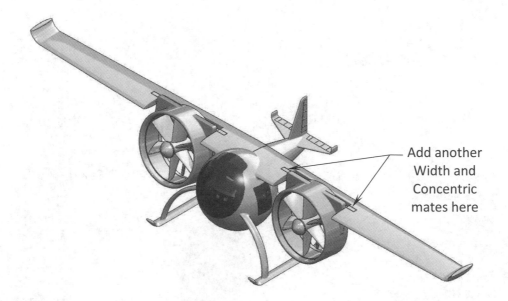

Add another
Width and
Concentric
mates here

5. Testing the assembly motions:

Drag the 2 Fan Housings up and down. They are moved independently from each other. The Fan Blades are also rotated independently at this point.

More mates need to be added so that the fan blades will move and rotate together. We will take a look at the other mate options such as Limit mates, Parallel mates, and Gear mates in the next few steps.

6. Adding a Limit-Angle mate:

Limit Angle mate allows components to move within a range of specified distances or angles.

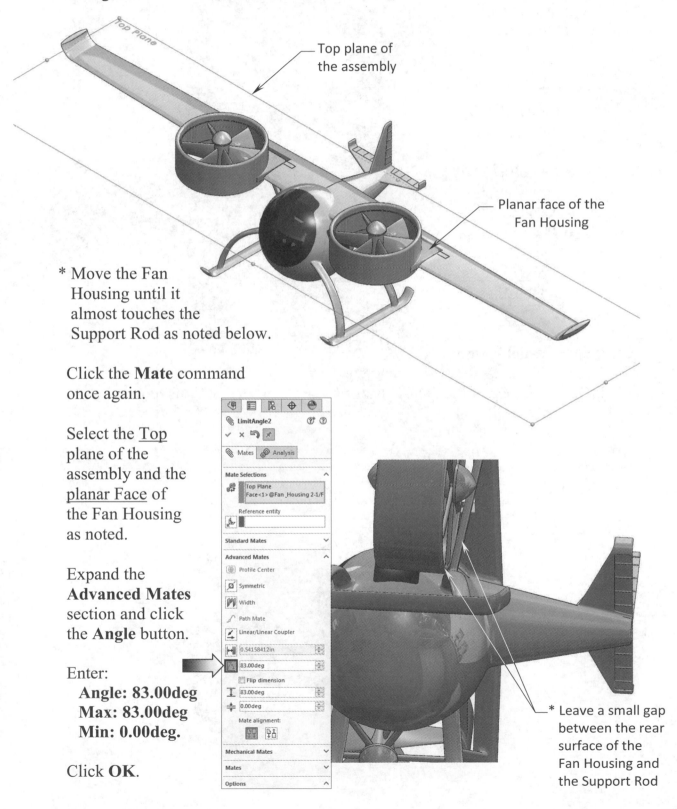

Top plane of the assembly

Planar face of the Fan Housing

* Move the Fan Housing until it almost touches the Support Rod as noted below.

Click the **Mate** command once again.

Select the Top plane of the assembly and the planar Face of the Fan Housing as noted.

Expand the **Advanced Mates** section and click the **Angle** button.

Enter:
   Angle: 83.00deg
   Max: 83.00deg
   Min: 0.00deg.

Click **OK**.

* Leave a small gap between the rear surface of the Fan Housing and the Support Rod

Drag the Fan Housing up and down to test its limit motions.

When moving upward it should stop at zero degree, and when moving downward it should stop right before it hits the Support Rod.

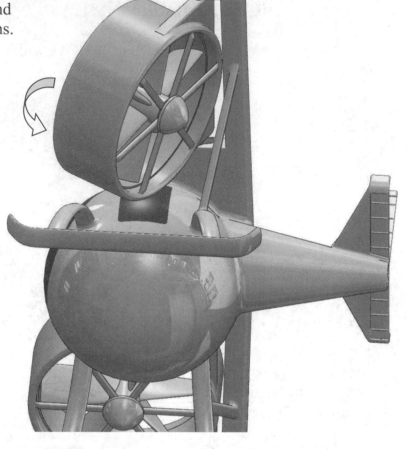

7. Adding a Parallel mate:

Switch back to the **Standard Mates** section.

Select the 2 planar faces of the 2 Fan Housings as indicated.

Click the **Parallel** button.

Click **OK**. Drag one of the Fan Housings up and down to check its motions.

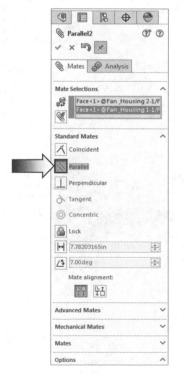

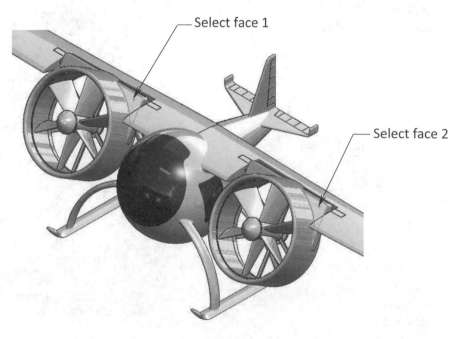

Select face 1

Select face 2

8. Adding a Gear mate:

A Gear mate forces two components to rotate relative to one another about selected axes.

Switch to the **Mechanical Mates** section (arrow).

Click the **Gear** button and select the circular edges of the 2 Blades as noted below.

The **Ratio** of the 2 selected edges should be the same (0.825in X 0.825in).

Click the **Reverse** checkbox if necessary.

Click **OK** to exit the Mate command.

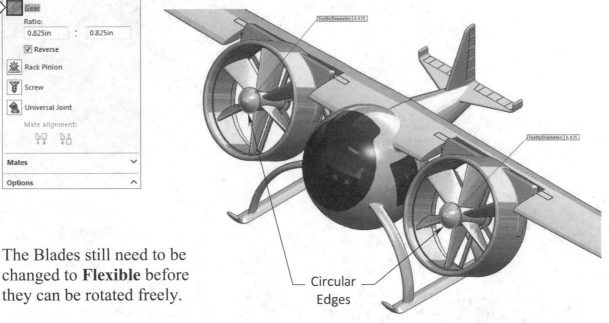

Circular Edges

The Blades still need to be changed to **Flexible** before they can be rotated freely.

Click on the **Fan & Housing** icon and select the option **Make Sub Assembly Flexible** (arrow).

Make both Fan & Housing flexible.

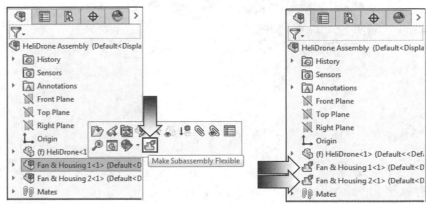

9. Testing the assembly motions:

Drag one of the Fan Housings up and down; both of the Fan Housings should move together this time.

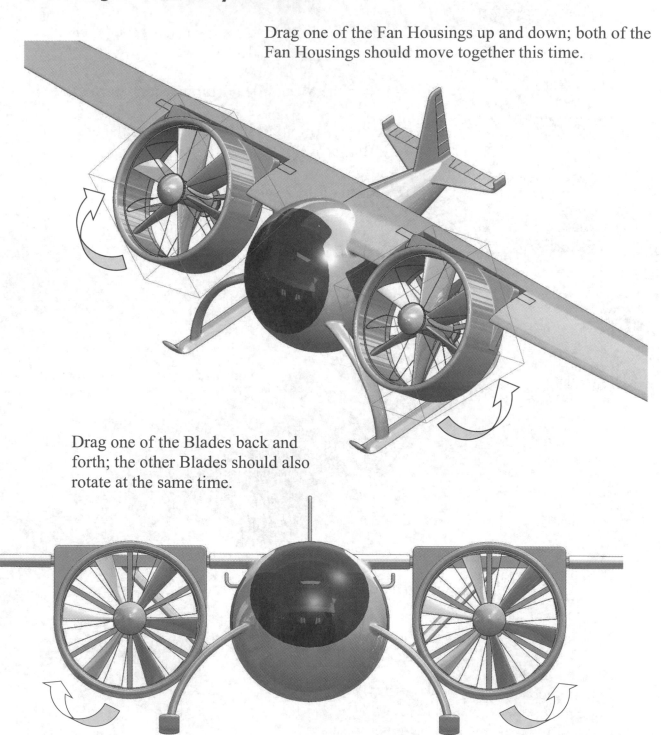

Drag one of the Blades back and forth; the other Blades should also rotate at the same time.

10. Saving your work:

Press **Save** (or Control + S) and click **Yes** to overwrite the existing assembly document with your completed one.

Radial Explode

Radial Explode allows you to explode components aligned cylindrically about an axis in one step.

You can explode components aligned radially or cylindrically about an axis. The radial explode steps are saved in the assembly. In the animation the steps play back, showing the components rotating around their center axis. When you open an eDrawings file and click explode, the exploded view appears.

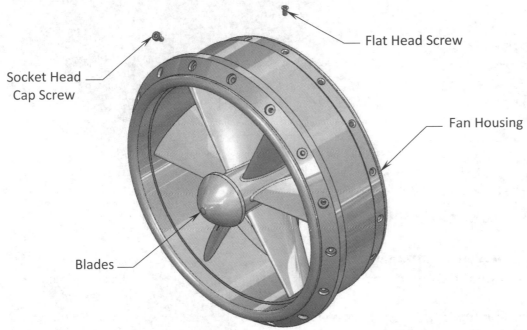

Flat Head Screw

Socket Head
Cap Screw

Fan Housing

Blades

1. Opening an existing assembly:

Browse to the Training Files and open an assembly document named:
Fan Assembly.sldasm.

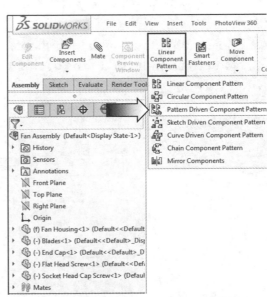

2. Creating the first Pattern Driven:

Select **Pattern Driven Component Pattern** from the Linear Component drop down (arrow).

Expand the FeatureManager tree and select the **Flat Head Screw** to use as the Component-To-Pattern.

For Driving Feature or Component, select **one of the holes** in the existing pattern as noted.

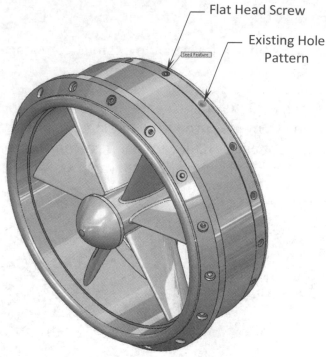

Flat Head Screw

Existing Hole Pattern

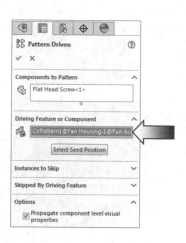

Click **OK**.

3. Creating the second Pattern Driven:

Select the **Driven Component Pattern** option once again.

Select the **Socket Head Cap Screw** for Component to Pattern.

Select **one of the holes** in the existing pattern as indicated for Driving Feature.

Socket Head Cap Screw

Existing Hole Pattern

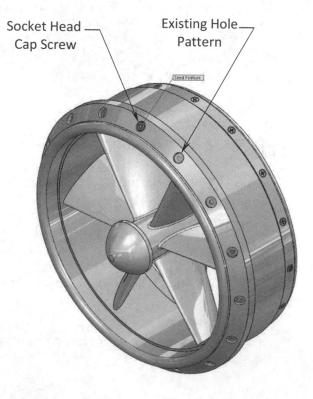

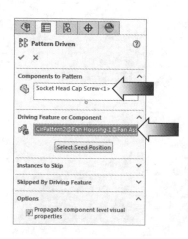

Click **OK**.

4. Creating the first radial exploded view:

From the Assembly toolbar, click **Exploded View**.

Select the **Radial Step** option (arrow).

Select the **Flat Head Screw** and all **15 Instances**.

Drag the **Direction-Arrow** outward to approximately **.700in**.

The Flat Head Screws are exploded radially.

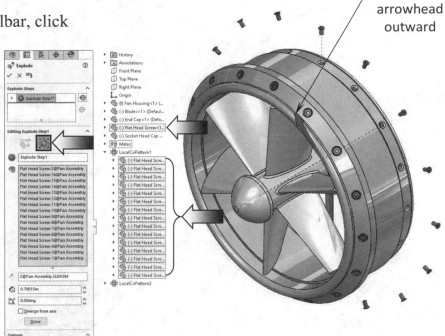

Drag the arrowhead outward

Do not click OK just yet; we will need to explode the Socket Head Cap Screws as the next explode step.

5. Creating the second radial exploded view:

The Radial Step option should be selected already.

Select the **Socket Head Cap Screw** and its **15 Instances**.

Drag the **Direction-Arrow** outward to approximately **1.250in**.

The Socket Head Cap Screws are exploded radially.

Click **OK** to exit the exploded view mode.

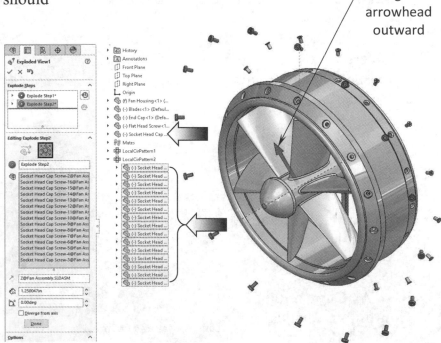

Drag the arrowhead outward

6. Verifying the exploded view:

Change to the **Front** orientation (Control + 1).

Your exploded view should look similar to the image shown on the right.

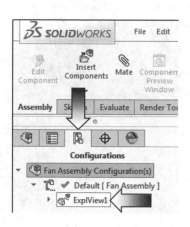

Switch to the **Configuration-Manager** tree (arrow) and expand the **Default** configuration to see or to edit the Exploded View (arrow).

Double-click on the **ExplView1** to Collapse or to Explode the assembly.

7. Saving your work:

Click **File / Save As**.

Enter **Radial Explode** for the name of the document.

Select the option **Save As Copy and Continue**, and click **Save**.

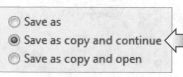

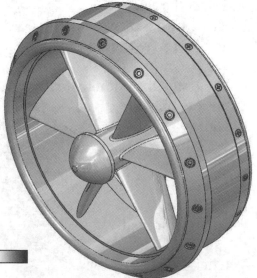

Exercise: Assembly Motions

1. Opening an assembly document:

Click **File / Open**.

Locate the Training Files folder
and open the assembly
document named:
Twisted Wires.sldasm

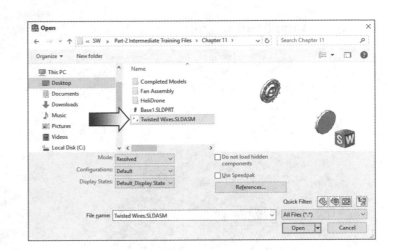

This assembly document has 2
instances of the **Base1** and they
have already been positioned
ahead of time to help focus on
creating the wires.

2. Creating a new part:

From the **Assembly** tab, click the **Insert Component** drop down arrow and select:
New Part (arrow).

The mouse pointer changes to a Green Checkmark  asking
for a plane or a face
to reference the new
part; select the <u>face</u>
of the pin as
indicated.

Select
face

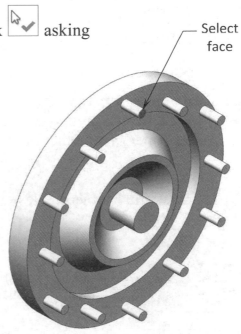

A new component is
created and a default
name is displayed
on the Feature tree.
A new sketch is also
opened on the selected
face automatically.

Exit the 2D Sketch. We need to be in 3D Sketch mode to sketch a 3D Spline.

Select **3D Sketch** under the Sketch drop-down arrow.

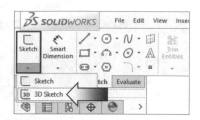

Select the **Spline** command 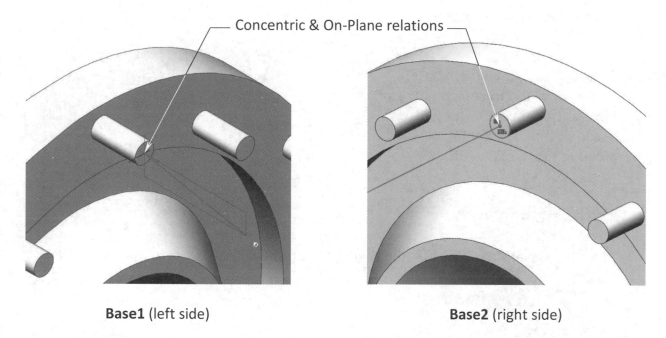.

Hover the cursor over the circular edge of the pin to "wake up" its center point.

Concentric & On-Plane relations

Base1 (left side)   **Base2** (right side)

Start sketching the **first point** of the spline on the <u>center</u> of the Pin.

Rotate the assembly to the opposite side and
place the second point of the spline also
on the center of the other pin.

Change to the Isometric view
(Control+7).

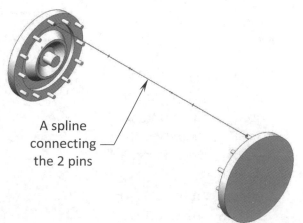

A spline
connecting
the 2 pins

The 2 ends of the spline must be
constrained to the <u>centers</u> of the 2 pins.

Enable the **Temporary Axis** under the
View Heads-Up toolbar (arrow).

Add a **Tangent** relation
between the Spline and
the Temporary Axis of
the Pin as noted below.

Add another **Tangent** relation
between the Spline and the
Temporary Axis of the
second Pin as indicated.

The Tangent relations will
keep the spline normal
to the 2 pins when the
bases are rotated.

Exit the 3D Sketch.

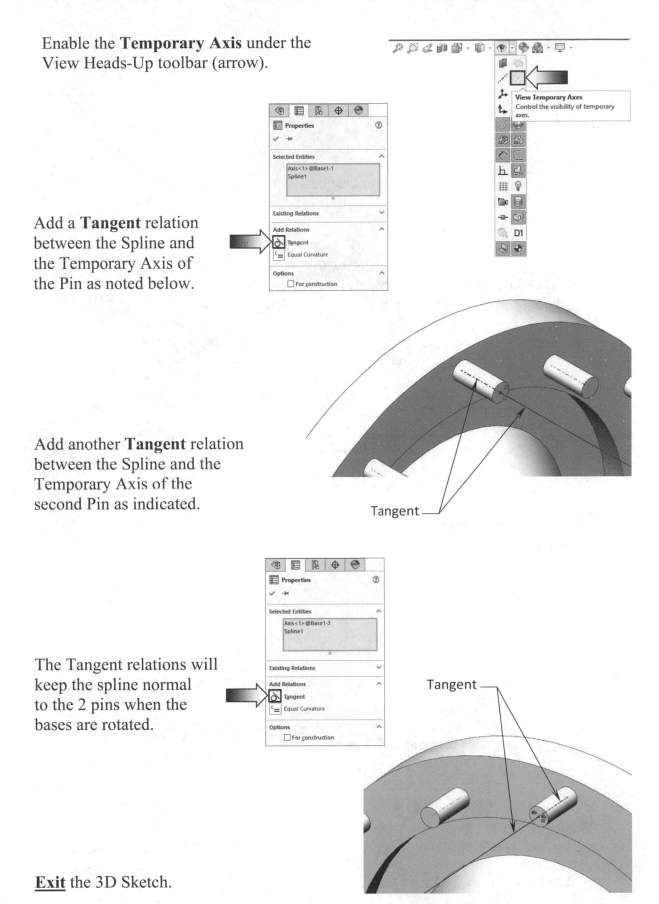

Turn-off the Temporary Axis.

3. Creating a Swept feature:

Switch to the **Features** tool tab and select the **Swept Boss Base** command (arrow).

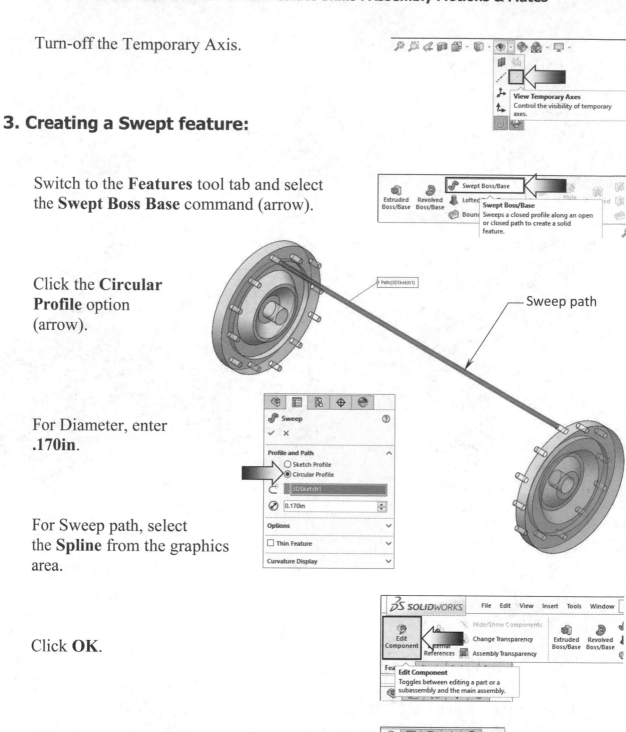

Path(3DSketch1)

Sweep path

Click the **Circular Profile** option (arrow).

For Diameter, enter **.170in**.

For Sweep path, select the **Spline** from the graphics area.

Click **OK**.

Click-Off the **Edit Component** command (arrow).

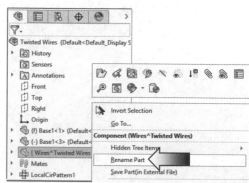

Right-click the name of the new part and renamed it to **Twisted Wires** (arrow).

4. Testing the Wire:

The Base on the left side is fixed. Drag the Base component on the right counter-clockwise, approximately one revolution.

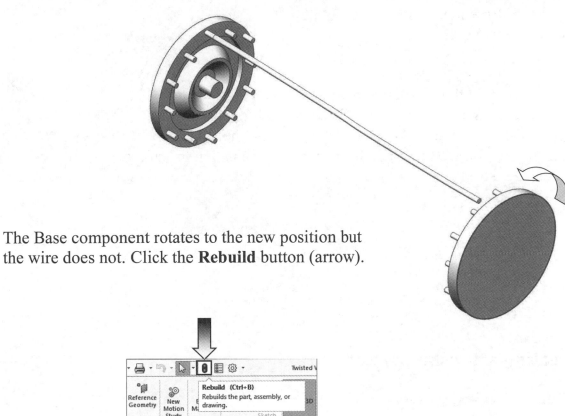

The Base component rotates to the new position but the wire does not. Click the **Rebuild** button (arrow).

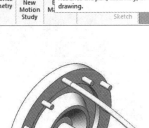

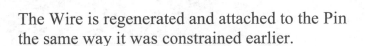

The Wire is regenerated and attached to the Pin the same way it was constrained earlier.

This was possible because we used a spline as the sweep path instead of a line.

5. Creating the second Wire:

Either repeat step 1 thru
step 3 to create the second
wire, or edit the component
Twisted Wire and create
a second sweep feature.

Be sure to add the On-Plane relation between the
endpoint of the spline and the planar face of the Pin,
and a Tangent relation between the spline and the temporary
axis of the pin. Do this to both sides to fully constrain the spline
to the 2 bases.

6. Creating a circular pattern:

Change to the **Assembly** tool tab.

Select edge for
Pattern Direction

Create a **Circular Component
Pattern** of the 2 wires.

Use one of the circular
edges of the Base for pattern
Direction.

Create **6 instances** with **Equal** spacing.

Click **OK**.

Drag the 1st wire about 1 revolution and click Rebuild to test it out.

7. Assigning materials:

Expand the FeatureManager tree.

Expand the **Base1** and the **Twisted Wires** components.

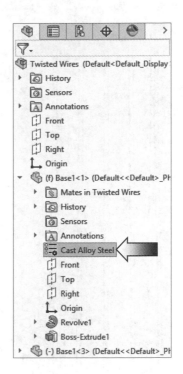

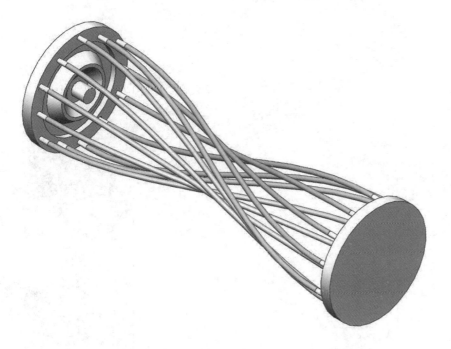

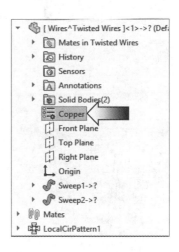

Right-click the material feature for the **Base1** component and select **Cast Alloy Steel** (arrow).

Right-click the material feature for the **Twisted Wires** component and select **Copper** (arrow).

Enable the RealView Graphics if applicable to display a more realistic and dynamic representation without the need to render the assembly.

8. Saving your work:

Save your work as **Twisted Wires (completed).sldasm**.

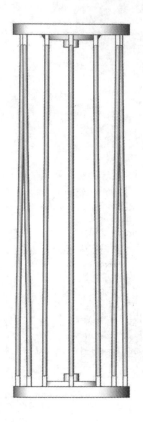

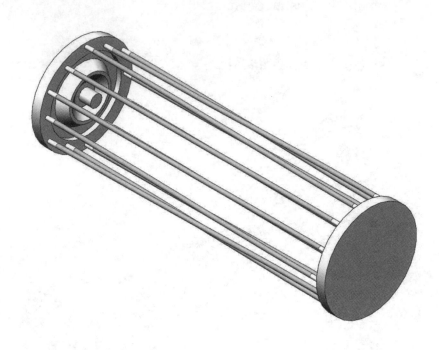

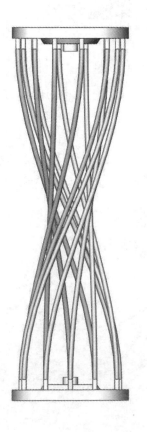

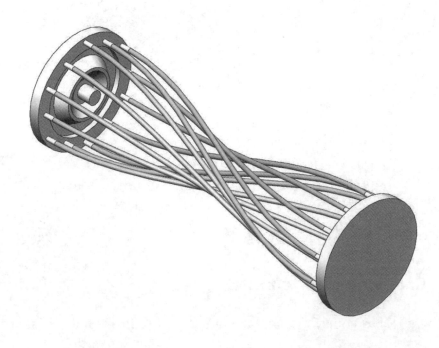

CHAPTER 12

Using Smart-Mates

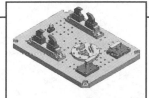

Using Smart-Mates
Fixture Assembly

Beside the standard mates such as Concentric, Coincident, Tangent, Parallel, etc., which must be done one by one, SOLIDWORKS has an alternative option that is much more robust called Smart-Mates.

There are many advantages for using Smart-Mates. It can create several mates at the same time, depending on the type of entity selected. For example: If you drag a circular edge of a hole and drop it on a circular edge of another hole, SOLIDWORKS will automatically create two mates, a concentric mate between the two cylindrical faces of the holes and a coincident mate between the two planar faces.

Using Smart-Mates, you do not have to select the Mate command every time. Simply hold the Alt key and drag an entity of a component to its destination, a smart-mate symbol will appear to confirm the types of mates it is going to add. At the same time, a mate toolbar will also pop up offering some additional options such as Flip Mate Alignment, change to a different mate type, or simply undo the selection.

When using the Alt + Drag option, the Tab key is used to flip the mate alignments. This is done by releasing the Alt key and pressing the Tab key, while the mouse button is still pressed. This option works well even if your assembly is set to lightweight.

In addition to the Alt + Drag, if you hold the Control key and Drag a component, SOLIDWORKS will create an instance of the selected component and apply the smart mates to it at the same time. Using this option you will have to click the Flip Mate Alignment button on the pop-up toolbar to reverse the direction of the mate, the Tab key does not work.

This lesson will teach us how to use the Smart-Mate options.

Using Smart-Mates
Fixture Assembly

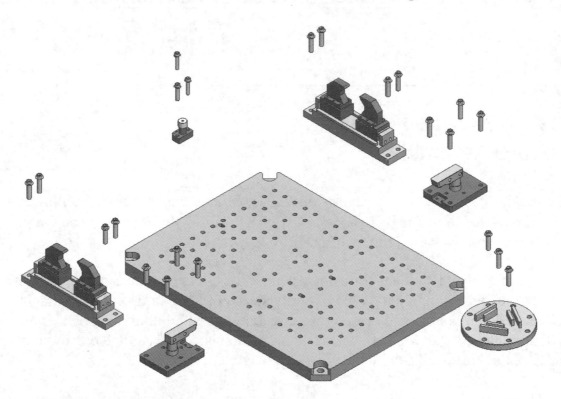

View Orientation Hot Keys:

Ctrl + 1 = Front View
Ctrl + 2 = Back View
Ctrl + 3 = Left View
Ctrl + 4 = Right View
Ctrl + 5 = Top View
Ctrl + 6 = Bottom View
Ctrl + 7 = Isometric View
Ctrl + 8 = Normal To Selection

Dimensioning Standards: **ANSI**

Units: **INCHES** – 3 Decimals

Tools Needed:

Concentric & Coincident	Concentric two cylindrical faces	Coincident two edges
Coincident two planar faces	Coincident two vertices	Coincident two origins

1. Opening an existing assembly document:

Select **File / Open**.

Browse to the Training Files folder, locate and open the document named:
Fixture Assembly.

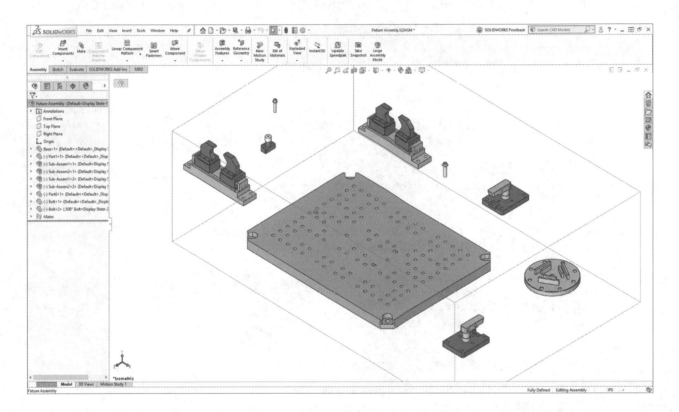

2. Enabling the Selection options:

Select **Tools / Options / Selection**
and enable the two checkboxes:

* Allow selection in wireframe...
* Allow selection in HLR and...

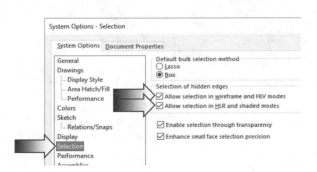

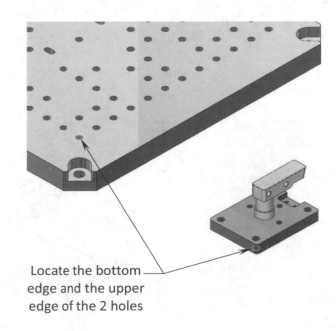

Locate the bottom
edge and the upper
edge of the 2 holes

3. Exploring the Smart-Mate options:

Some types of Smart-Mates can be created in an assembly by dragging and dropping one entity to another.

Depending on the entity that you drag, you will get a different mate result. Use either a linear or circular edge of the model, a planar or cylindrical face, or a vertex to drag with Smart-Mates.

 Concentric & Coincident 2 circular edges

 Concentric 2 cylindrical faces

 Coincident 2 linear edges

 Coincident 2 planar faces

 Coincident 2 vertices

 Coincident 2 origins or coordinate systems

> Start Here

<u>Hold</u> the **ALT** key, drag the <u>hidden edge</u> of the hole to the <u>upper edge</u> of the mating hole but <u>do not</u> release the mouse. Hover the cursor over the edge of the hole until the Smart-Mate symbol appears, then release the mouse button.

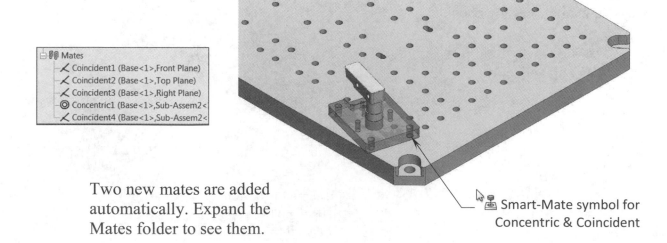

```
□·〇〇 Mates
   ∠ Coincident1 (Base<1>,Front Plane)
   ∠ Coincident2 (Base<1>,Top Plane)
   ∠ Coincident3 (Base<1>,Right Plane)
   ◎ Concentric1 (Base<1>,Sub-Assem2<
   ∠ Coincident4 (Base<1>,Sub-Assem2<
```

Smart-Mate symbol for Concentric & Coincident

Two new mates are added automatically. Expand the Mates folder to see them.

4. Using Smart-Mate Concentric:

A concentric mate is created when two cylindrical surfaces are dragged and dropped onto one another.

Hold the **ALT** key, drag the cylindrical face of the 1st hole, and drop on the cylindrical face of the 2nd hole. Note: the mouse cursor must hover over an entity and must not be moving, in order for the Smart-Mate symbol to appear.

Release the mouse cursor when the concentric symbol pops-up, and click **OK** on the pop-up toolbar to accept the concentric mate.

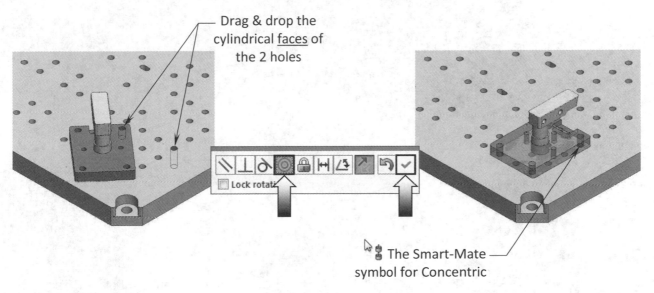

Drag & drop the cylindrical faces of the 2 holes

The Smart-Mate symbol for Concentric

The sub assembly becomes fully defined. There is no minus sign (-) before the name Sub-Assem2.

The Base component is considered the parent part. It was previously mated to the 3 planes of the assembly, and it is fully defined. There is no minus sign in the front of its name.

The rest of the components and sub-assemblies are under defined; they still have 6 degrees of freedom to move or rotate about any direction.

We are going to explore the Smart-Mates some more and assemble the entire assembly using this dynamic mate option.

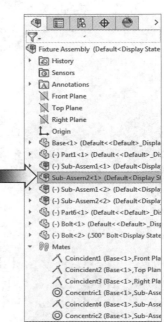

5. Creating a Smart-Mate Concentric & Coincident:

Zoom closer between the two components that will get mated next.

Hold the **ALT** key, and drag and drop the bottom edge of hole1 onto the top edge of hole2 as indicated.

Keep the mouse cursor steady until the Smart Mates symbol appears, and then release the mouse button.

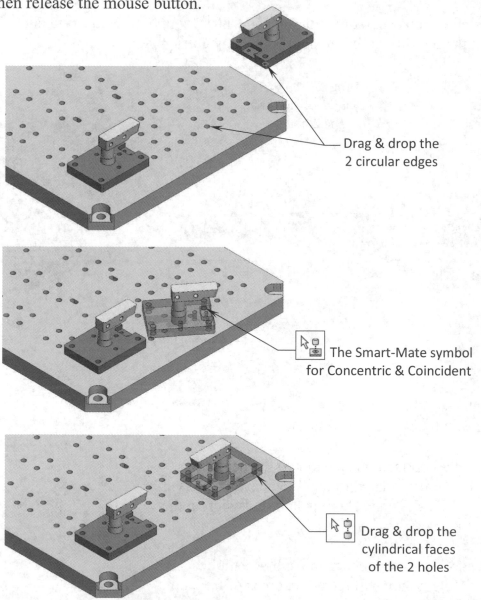

Drag & drop the
2 circular edges

The Smart-Mate symbol
for Concentric & Coincident

Drag & drop the
cylindrical faces
of the 2 holes

Next, drag and drop the cylindrical face of hole1 to the cylindrical face of hole2, wait for the Smart-Mate Concentric symbol to appear, then click the checkmark to accept it.

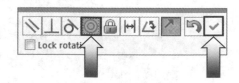

Lock rotati

6. Repeating the previous Mate:

Move the cylindrical part above the base as shown.

We are going to use the same mate techniques to assemble this part.

(The Explode Lines are added for clarity only; you do not have to add them).

<u>Hold</u> the **ALT** key and drag & drop the <u>bottom edge</u> of the hole in the cylindrical part onto the <u>upper edge</u> of the mating hole in the base.

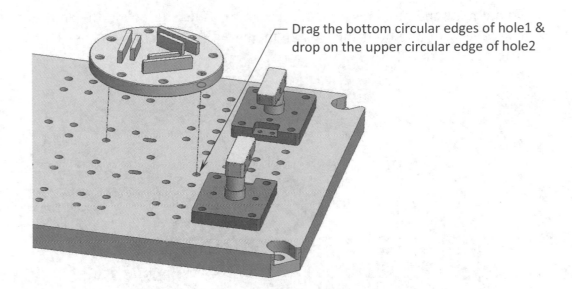

Drag the bottom circular edges of hole1 & drop on the upper circular edge of hole2

If done correctly, there should be two mates added to the selected entities already. Check the Mates group and look for the new Concentric and Coincident mates.

<u>Repeat</u> the same step and mate one other hole of the same components, to fully define their positions.

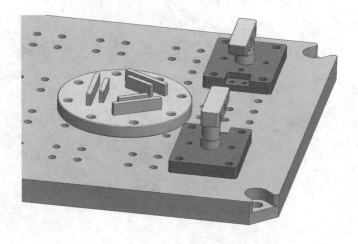

7. Mating other components:

Since the holes in the components were patterned and have the exact same diameters, two mates per component would be enough to fully define them.

(The Explode Lines are added for clarity again; you do not need to add them).

Use the same techniques and add the Smart-Mates to the other components, use the explode lines for reference locations, and only create two mates for each component.

Do not mate the bolts just yet. We are going to take a look at the configurations that come with the bolts and switch them to different sizes and use them in different locations.

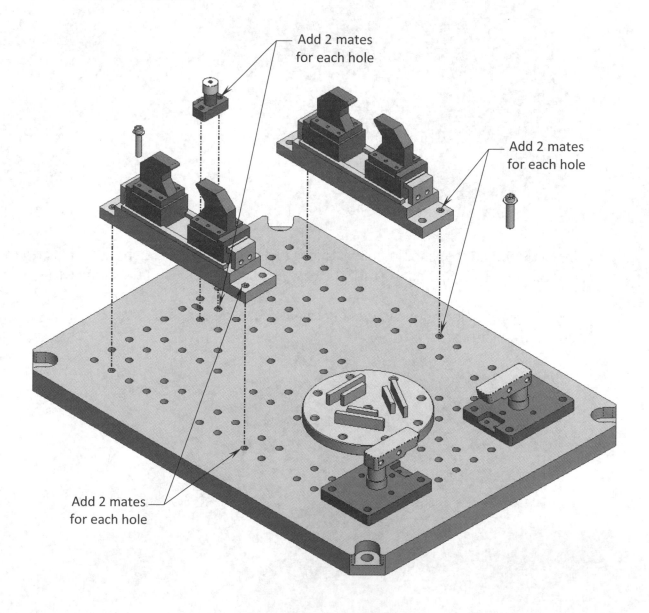

Add 2 mates for each hole

Add 2 mates for each hole

Add 2 mates for each hole

8. Checking the status of the components:

After adding two mates to each component, the FeatureManager updates and shows only the minus signs (-) before the bolts.

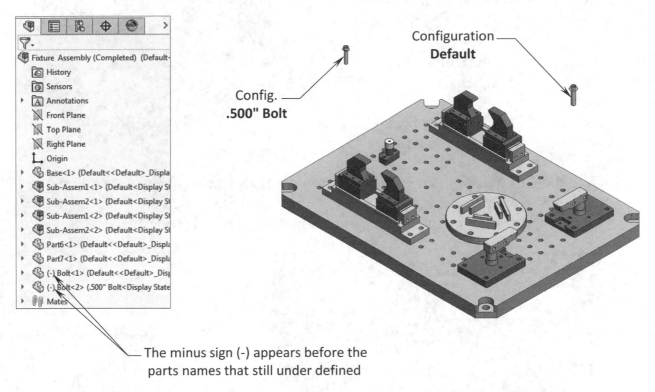

Configuration —
Default

Config. —
.500" Bolt

— The minus sign (-) appears before the parts names that still under defined

9. Switching Configuration:

The Bolt has 2 configurations, the **Default** and the **.500"**.

We have been using the Default size for most of the components, but the next part will need to be switched to the .500".

Click the part **Bolt** and select the **Component-Properties** button (arrow).

Select the **.500"** configuration (arrow).

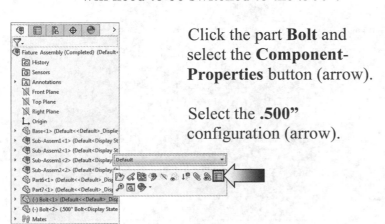

10. Creating an instance of the .500" bolt:

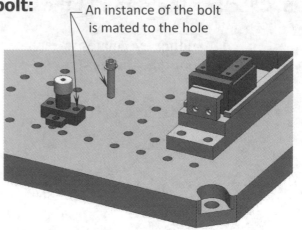

An instance of the bolt is mated to the hole

When holding the Alt key and dragging a component, the Smart Mate option is activated to assist with the mating of the 2 components.

But if you hold the Control key and drag a component, SOLIDWORKS will make a copy of the component and smart mates it to the other component at the same time.

Hold the **Control** key, drag and drop (the shaft body) an instance of the bolt to the mating hole as pictured below. Release the mouse button when the smart-mate symbol appears. Click the Flip Alignment button if needed (arrow).

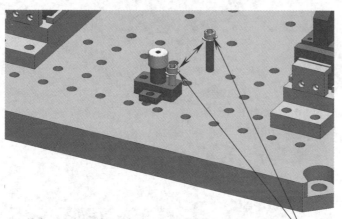

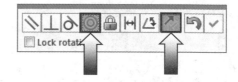
☐ Lock rotati

At this point, it might be difficult to select the 2 faces to apply the next mate. Let us use the standard mates instead.

Hold the Control key drag & drop a copy of the bolt to the hole

Click the **Mate** command from the Assembly toolbar.

Select the bottom face of the flange and the top face of the **Part6**.

A **coincident** mate is added. Click **OK** to exit the mate.

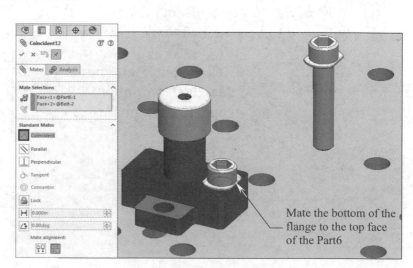

Mate the bottom of the flange to the top face of the Part6

11. Creating another instance of the bolt:

Move the Default-Bolt closer to the Sub-Assembly1 on the right as noted.

Hold the **Control** key, drag and drop an instance of the (default) bolt to the mating hole as noted. Look for the smart-mate symbol before releasing the mouse button.

Click the Flip Mate Alignment button if needed (arrow).

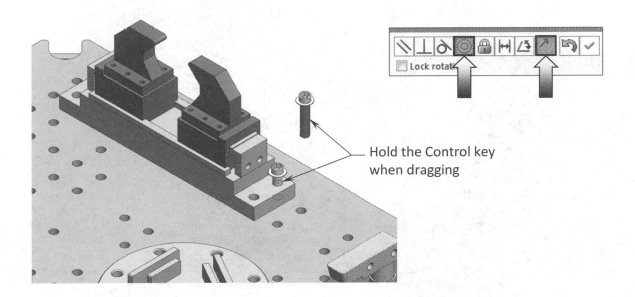

Hold the Control key when dragging

Click **OK** to close the mate toolbar.

It would be easier to use the standard mates again for the next mate.

Click the **Mate** command ⬛ from the assembly toolbar.

Select the <u>bottom face</u> of the flange and the <u>top face</u> of the **Part2** as noted.

A **coincident** mate is added automatically.

Click **OK**.

The bolt still has one degree of freedom left.

Select 2 faces

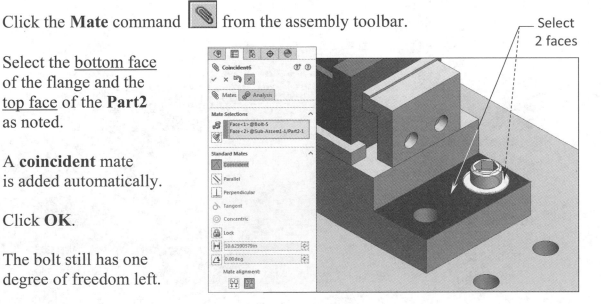

12. Adding more bolts:

Rotate around to the other side of the sub-assembly1.

<u>Hold</u> the <u>CONTROL</u> key and drag the default-bolt to make a copy and place it on the right of the Part2.

Hold the Control key again and drag the copy of the default bolt to the mating hole as indicated. Look for the smart-mate symbol before releasing the mouse button. Click the Flip Mate Alignment if needed (arrow).

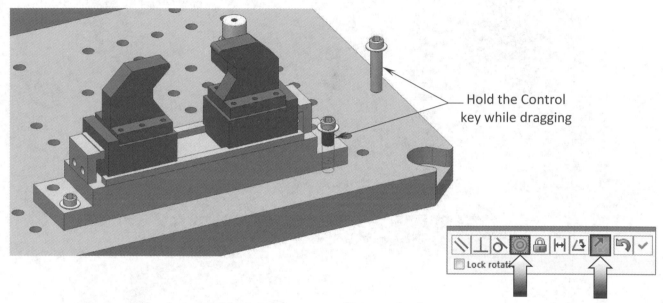

Hold the Control key while dragging

Once again, it would be easier to use the standard mates for the next step, since we cannot really see the two mating faces very clearly.

Click **Mate** .

Select the <u>bottom face</u> of the flange and the <u>top face</u> of the **Part2**.

Another **coincident** mate is added.

Click **OK** to exit the mate mode.

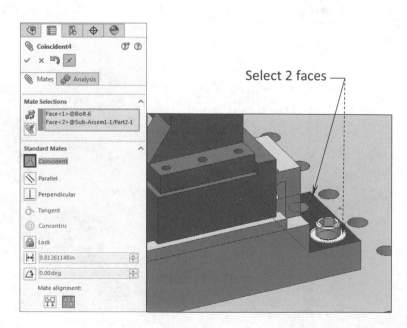

Select 2 faces

13. Repeating:

For practice purposes, we are going to add a few more instances of the bolts and mate them to the other components.

Keep in mind that fasteners do not have to be fully defined. Each bolt only needs two mates, and even though they still rotate, the assembly will not be affected at all.

By using one less mate for each component you can reduce the file size of this assembly by about 10%. It will also help reduce the time that it takes to rebuild the assembly every time a change is applied. In other words, the less mates (and less components), the faster the computer performance.

14. Saving your work:

Save your work as **Fixture Assembly.sldasm** and overwrite the existing file when prompted.

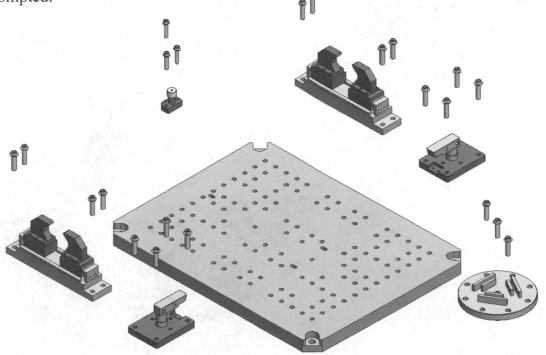

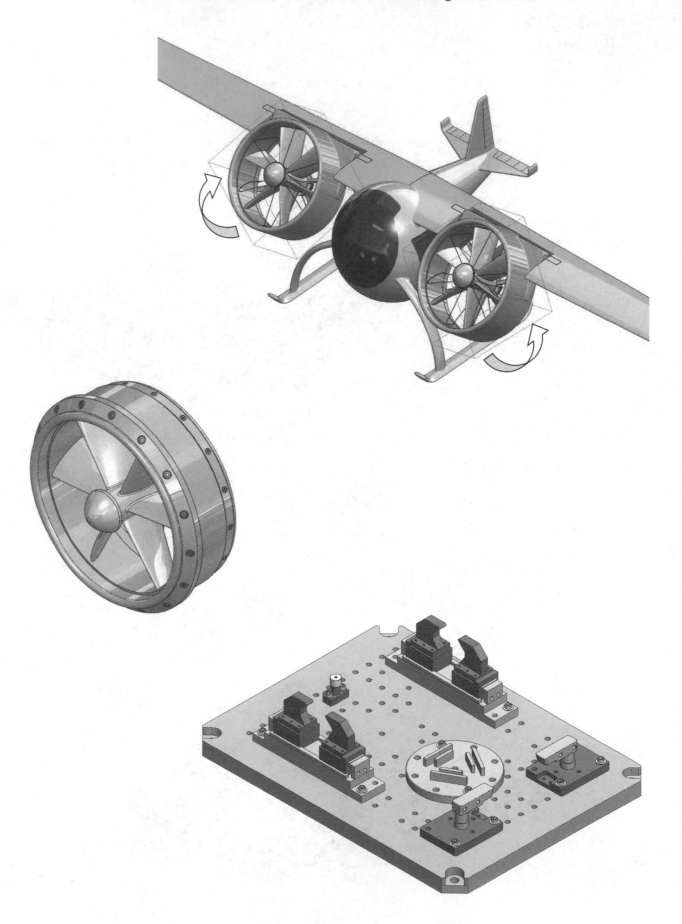

Using Mate Controller
Assembly Mates

Mate Controller

This exercise discusses how Mate Controller is used to show and to save the positions of assembly components.

Mate Controller lets you manipulate specific mates that control the degrees of freedom for a design. You can save and recall saved positions and mate values. You can also create animations based on the saved positions.

In Mate Controller, you can show and save the positions of assembly components at various mate values and degrees of freedom without using configurations for each position. You can create simple animations between those positions and save the animations to .avi files (Audio Video Interleaving).

Mate Controller is integrated with Motion Studies, so you can use Animation to create animations based on the positions you define in Mate Controller.

Supported mate types include:

* Angle
* Distance
* LimitAngle
* LimitDistance
* Slot (Distance Along Slot, Percent Along Slot)
* Width (Dimension, Percent)

This chapter discussed the three basic steps when using the Mate Controller:
* Collect All Supported mates and rearrange them
* Create new mate positions
* Calculate and create animation from the mate positions

Mate Controller
Assembly Mates

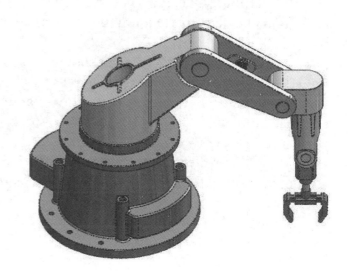

View Orientation Hot Keys:

Ctrl + 1 = Front View
Ctrl + 2 = Back View
Ctrl + 3 = Left View
Ctrl + 4 = Right View
Ctrl + 5 = Top View
Ctrl + 6 = Bottom View
Ctrl + 7 = Isometric View
Ctrl + 8 = Normal To
          Selection

Dimensioning Standards: **ANSI**

Units: **INCHES** – 3 Decimals

Tools Needed:

 Assembly Document   Mate Controller

1. Opening an assembly document:

Click **File / Open**.

Browse to the Training Files location and open the assembly document named:
Mate Controller Assembly.sldasm.

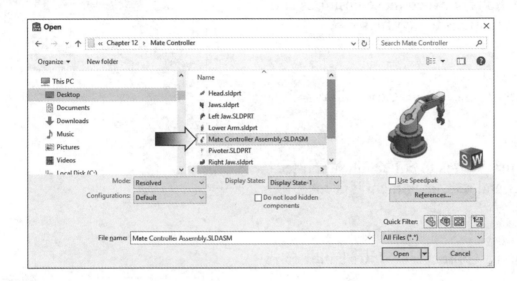

2. Examining the existing mates:

Locate the Mates Group at the bottom of the
FeatureManager tree and expand it (arrow).

The assembly has several mates that were
used to position the components but
only the supported mates can be
used to create and save the new
positions in Mate Controller.

The supported mates are:

    * Angle   * Distance

    * LimitAngle   * LimitDistance

    * Slot   * Width

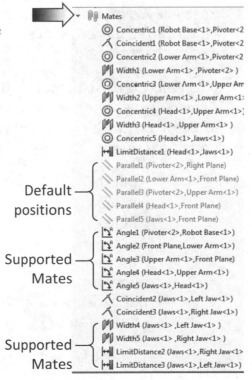

3. Using the Mate Controller:

Instead of using configurations to save each position, you can show and save the positions of assembly components at various mate values and degrees of freedom with Mate Controller. You can create simple animations between those positions and save the animations to .avi files.

Select **Insert / Mate Controller** .

The first step is to collect all supported mates and rearrange them if needed.

Click the **Collect All Mates** button (arrow).

Eight mates are found. The mate types are displayed in the Mates dialog along with their dimension values.

Select the **LimitDistance1** mate and move it down the list using the up and down arrows to create the order shown on the lower right image. The order should show:

> Angle1
> Angle2
> Angle3
> Angle4
> Angle5
> LimitDistance1
> LimitDistance2
> LimitDistance3

This same mate order will be presented for each new position created.

4. Adding new positions:

Position 1 is the default starting position (**225°**) and will have no motion.

Click the **Add Position** button to create position number 2.

Use the default name "**Position 2**" and click **OK**.

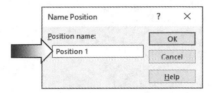

Keep the mate value of **Angle2** at **90°** to create a 2-second delay. Also leave all other dimensions at their default values.

Click the **Add Position** button 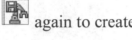 again to create position number 3.

Use the default name "**Position 3**" and click **OK**.

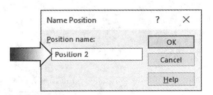

Change the mate value of **Angle2** to **70°** and **Angle3** to **45°**.

Create a total of **9 new positions** using the angles and distances provided in the dialog boxes below.

Position 1	Position 2	Position 3

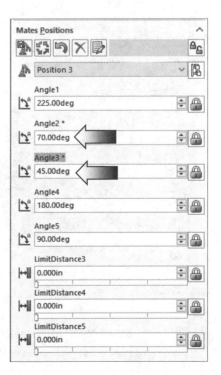

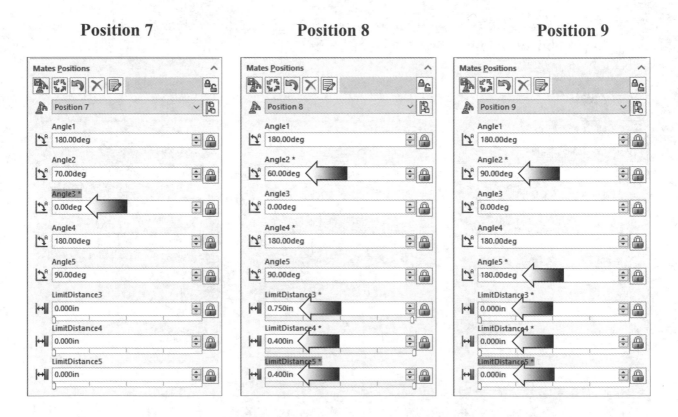

Position 4

Mates Positions	∧
Position 4	
Angle1	225.00deg
Angle2	70.00deg
Angle3	45.00deg
Angle4	180.00deg
Angle5	90.00deg
LimitDistance3 *	0.750in
LimitDistance4 *	0.400in
LimitDistance5 *	0.400in

Position 5

Mates Positions	∧
Position 5	
Angle1	225.00deg
Angle2	70.00deg
Angle3	45.00deg
Angle4	180.00deg
Angle5	90.00deg
LimitDistance3	0.750in
LimitDistance4 *	0.000in
LimitDistance5 *	0.000in

Position 6

Mates Positions	∧
Position 6	
Angle1 *	180.00deg
Angle2	70.00deg
Angle3	45.00deg
Angle4	180.00deg
Angle5	90.00deg
LimitDistance3 *	0.000in
LimitDistance4	0.000in
LimitDistance5	0.000in

Position 7

Mates Positions	∧
Position 7	
Angle1	180.00deg
Angle2	70.00deg
Angle3 *	0.00deg
Angle4	180.00deg
Angle5	90.00deg
LimitDistance3	0.000in
LimitDistance4	0.000in
LimitDistance5	0.000in

Position 8

Mates Positions	∧
Position 8	
Angle1	180.00deg
Angle2 *	60.00deg
Angle3	0.00deg
Angle4 *	180.00deg
Angle5	90.00deg
LimitDistance3 *	0.750in
LimitDistance4 *	0.400in
LimitDistance5 *	0.400in

Position 9

Mates Positions	∧
Position 9	
Angle1	180.00deg
Angle2 *	90.00deg
Angle3	0.00deg
Angle4	180.00deg
Angle5 *	180.00deg
LimitDistance3 *	0.000in
LimitDistance4 *	0.000in
LimitDistance5 *	0.000in

Make any modifications as needed to match the dimension values for each step as indicated by the arrows.

The total duration for the current animation is
18 seconds (2 second x 9 positions = 18 seconds).
These times can be changed to 3 or 4 seconds to
simulate slow play back, or 1 second per position
to simulate a faster playback.

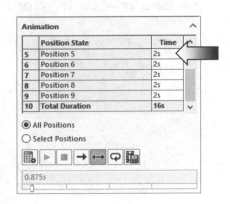

Select the **All Positions** option to include all of
the positions in the animation but choose **Select
Positions** to select only the positions you select to include in the animation.

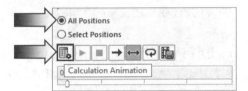

Select the **All Position** radio button.

Click **Calculate Animation** (arrow) to calculate the current animation. If you alter
the positions or animation parameters, you must recalculate the animation before
replaying it.

The animation plays from beginning to end once. Make any adjustments needed
such as changing the angles, distances, and durations.

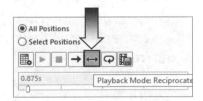

For Play Back Mode, select the **Reciprocate** (arrow)
to play back continuously beginning to end, and end
to beginning, until you click Stop.

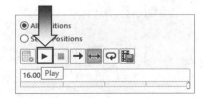

Click the **Play** button (arrow) to view the animation.
Play it back and forth a few times then click the **Stop**
button to stop the animation.

12-21

5. Saving the animation:

Click the <u>Save Animation</u> button (arrow) in the Motion-Manager area.

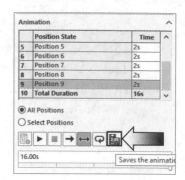

For **Save As Type**, select
AVI. (To save as a series of still images, select .bmp or .tga formats.)

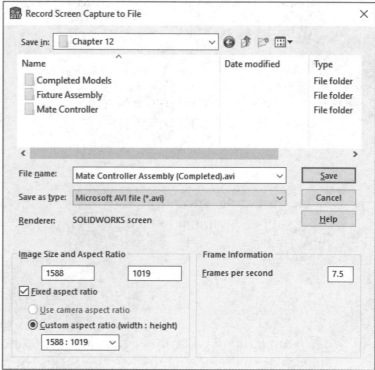

* **SOLIDWORKS Screen:**
Saves the on-screen display as a video or series of images.

* **PhotoView:** (Only available with PhotoView 360 added in) Saves and enhances the graphics quality of the recording to include features such as shadowing, true reflections, and anti-aliasing.

* **Width and Height:** These settings originally correspond to the size of the SOLIDWORKS screen.

* **Fixed aspect ratio:** Retains the original proportions of the image when you change Width or Height.

* **Use camera aspect ratio:** If the view is through a camera, the aspect ratio of the camera is used.

* **Custom aspect ratio (Width & Height):** Sets the aspect ratio to one of a set of preset values, to match the active window, or to match the background image.

* **Frame Information:** Sets the number of Frames per second.

Click **Save**.

The Video Compression dialog box appears, select **Full Frames** (Uncompressed).

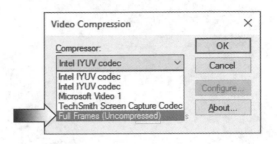

Record the animation a couple times around and then press **Stop** when the animation reaches its start position (**0 second**).

Start Position (0 second)

Click **OK** to exit the mate Controller.

6. Saving your work:

Click **File / Save As**.

Enter **Mate Controller Assembly.sldasm** for the file name and press **Save**.

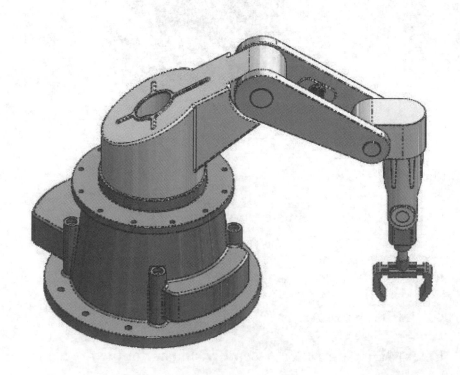

Using Copy with Mates

Copy with mates allows you to create additional instances of a component in an assembly; you do not have to create the mates manually for each new instance. Use this PropertyManager to include mates from the original instance.

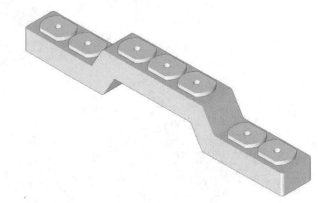

Only Standard mates can be used. Advanced, mechanical, or lock mates are not yet supported.

1. Opening an assembly document:

Click **File, Open**.

Locate the Training Files folder and open an assembly document named: **Copy with Mates.sldasm**

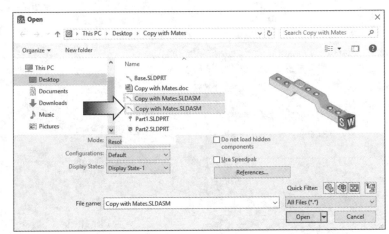

This assembly document has 3 components in it and they have been fully mated.

The components can be patterned or copied and mated manually, but this exercise will guide you through the use of Copy with Mates instead.

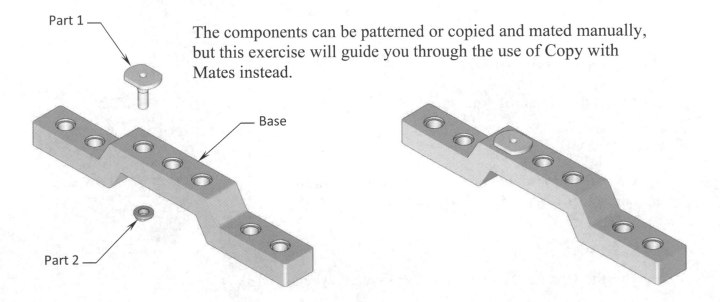

Part 1

Base

Part 2

There are **2 Concentric** mates, **2 Coincident** mates, and **2 Parallel** mates have already been created to constrain these components.

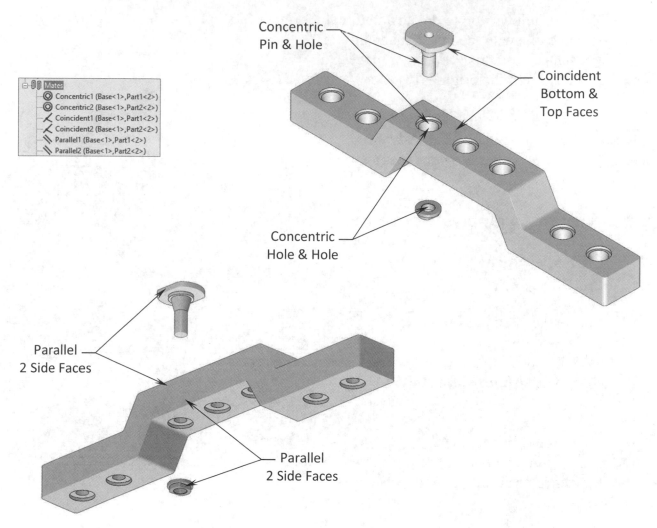

Concentric
Pin & Hole

Coincident
Bottom &
Top Faces

Concentric
Hole & Hole

Parallel
2 Side Faces

Parallel
2 Side Faces

2. Using Copy with Mates:

Select the **Assembly** tool tab.

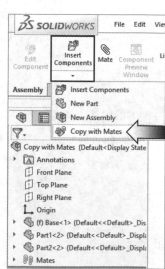

Click the drop-down arrow under **Insert Component** and select **Copy with Mates** (arrow).

Use the **Up** and **Down** arrow keys to rotate when needed.

For Selected Components, select the **Part1** and
Part2 either from the FeatureManager tree or
directly from the graphics area.

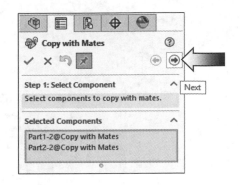

Click the **Next** button (arrow).

For **Concentric1**
select the upper
conical hole as
noted.

The **Concentric1**
mate is between the
Pin and the conical Hole.

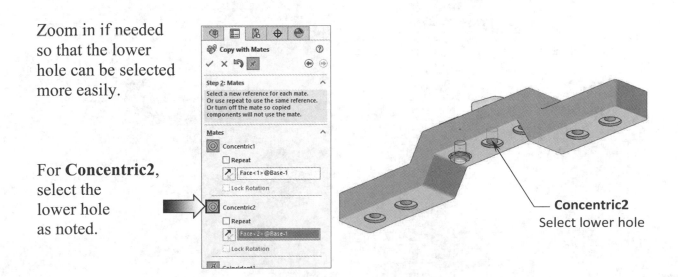

Concentric1
Select upper hole

Concentric2
Select lower hole

Press the **Up-Arrow** key to rotate the assembly.

Zoom in if needed
so that the lower
hole can be selected
more easily.

For **Concentric2**,
select the
lower hole
as noted.

The preview graphics of the Part1 and Part2 appear. Do not click OK just yet.

For **Coincident1**, click the **Repeat** checkbox.

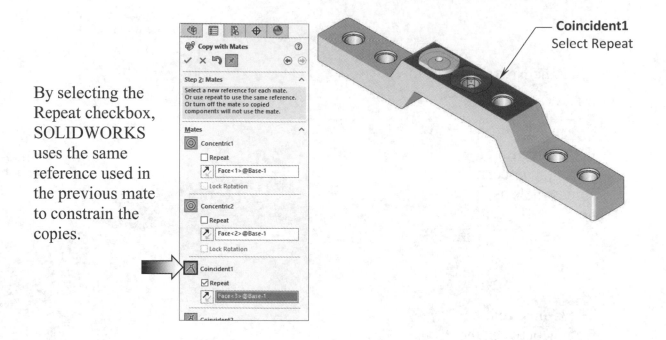

By selecting the Repeat checkbox, SOLIDWORKS uses the same reference used in the previous mate to constrain the copies.

For **Coincident2**, click the **Repeat** checkbox.

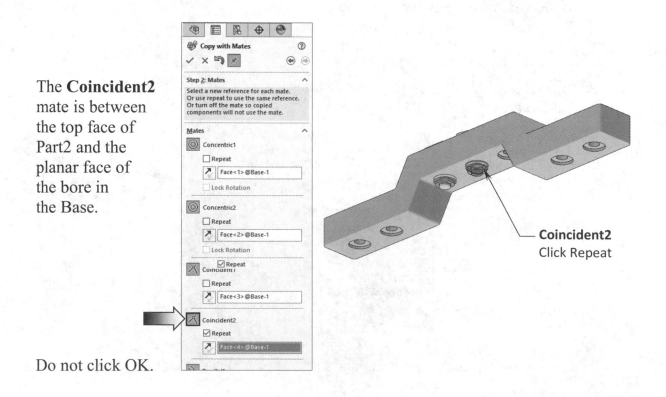

The **Coincident2** mate is between the top face of Part2 and the planar face of the bore in the Base.

Do not click OK.

For **Parallel1**, click the **Repeat** checkbox.

The **Parallel1** mate is between the flat side of Part1 and the flat side of the Base.

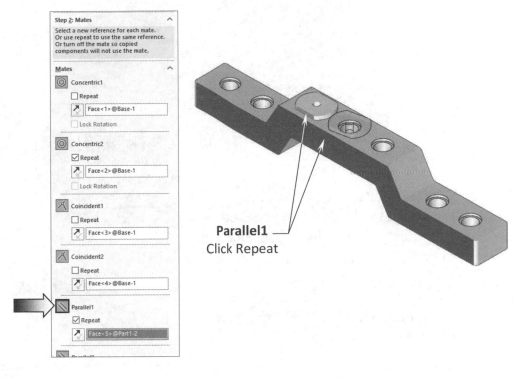

Parallel1
Click Repeat

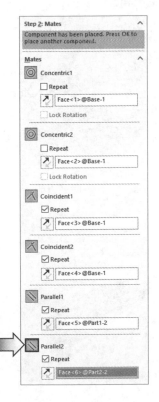

For **Parallel2**, click the **Repeat** checkbox.

The **Parallel2** mate is between the flat side of the Part2 and the flat side of the Base.

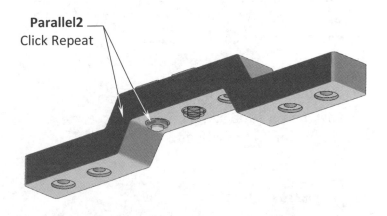

Parallel2
Click Repeat

A "green message" appears indicating all components are placed.

Click **OK** (the Green Check). All the mates are solved at this point.

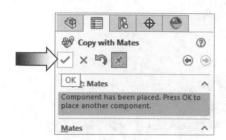

3. Placing other instances:

After all mates are solved, only the mates that did not get marked as Repeat will need to be reselected, in order to place new components.

For **Concentric1**, select the upper conical Hole.

For **Concentric2**, rotate the assembly and select the lower Hole as noted.

Click **OK**. New components are copied and all mates are solved.

Repeat the same steps to place more instances of the same components on the left and right sides.

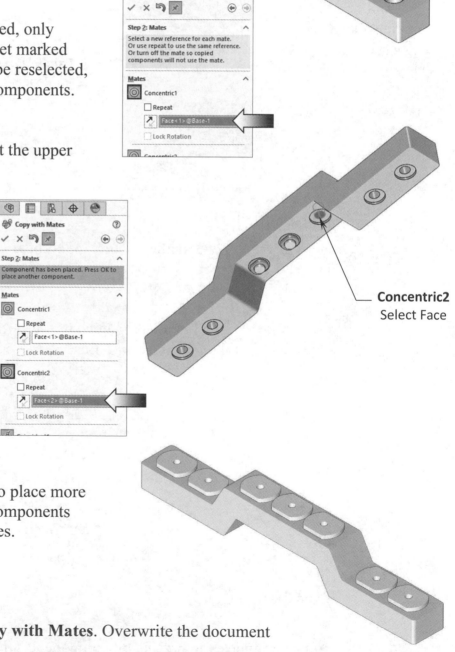

4. Saving your work:

Save your work as **Copy with Mates**. Overwrite the document when prompted.

CHAPTER 13

Introduction to Top Down Assembly

Intro to Top Down Assembly
Car Remote Control Housing

Top Down Assembly drives multiple part designs using a single "parent" part and creates in-context features. Top Down approach allows the SOLIDWORKS user to quickly put together the parametric models without much effort and time.

For example, you can build a shaft in the context of an assembly. Then if a new plate with a mating hole to the shaft is required, the mating hole can be built by referencing the edge (diameter) of a shaft. Thus, it establishes the parent and child relationship of the shaft and the mating hole. And if the shaft diameter changes, it will automatically change the corresponding mating hole diameter as well.

There are 3 methods in Top Down Assembly mode:

  * Individual features can be designed top down by referencing other parts in the assembly, as in the case of the shaft and hole described above.

  * Complete parts can be built with top down methods by creating new components within the context of the assembly. The component you build is actually mated to another existing component in the assembly.

  * An entire assembly can be designed from the top down as well, by first building a layout sketch that defines component locations, key dimensions, etc. Then build 3D parts using one of the methods above, so the 3D parts follow the sketch for their size and location.

Whenever you create a part or feature using top down technique, some external references are created to the geometry you referenced. These references can be locked, broken, or deleted if needed.

Introduction to Top Down Assembly
Car Remote Control Housing

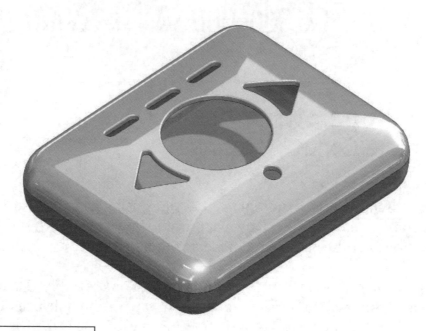

View Orientation Hot Keys:

Ctrl + 1 = Front View
Ctrl + 2 = Back View
Ctrl + 3 = Left View
Ctrl + 4 = Right View
Ctrl + 5 = Top View
Ctrl + 6 = Bottom View
Ctrl + 7 = Isometric View
Ctrl + 8 = Normal To Selection

Dimensioning Standards: **ANSI**

Units: **INCHES** – 3 Decimals

Tools Needed:

 Make Assembly   New Part   Convert Entities

 Extruded Boss/Base   Shell   Fillet

 Offset Entities   Explode View   Section View

1. Opening a part document:

Browse to the training folder and
open a part document named:
Square Remote Top.sldprt.

Individual features can be designed
by referencing other parts in the assembly.
These external references are created when one document is dependent on
another document for its solution. If the referenced document changes, the
dependent document changes also.

2. Making an assembly from part:

The construction of the new part will be created in the context of an assembly.
The existing geometry of the Remote Top will be used to define the shape and
size of the new Remote Bottom.

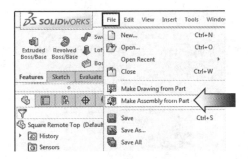

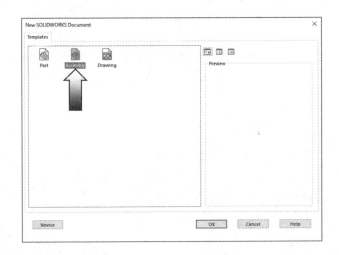

Select **File / Make Assembly from
Part** (arrow).

Select the **Assembly Template** (arrow) if prompted.

3. Placing the first component:

If the Origin is not visible in the new assembly document, click **View, Hide / Show, Origin**.

Hover the cursor over the origin, look for the Inference Origin symbol (the double arrows symbol) and place the first part there*.

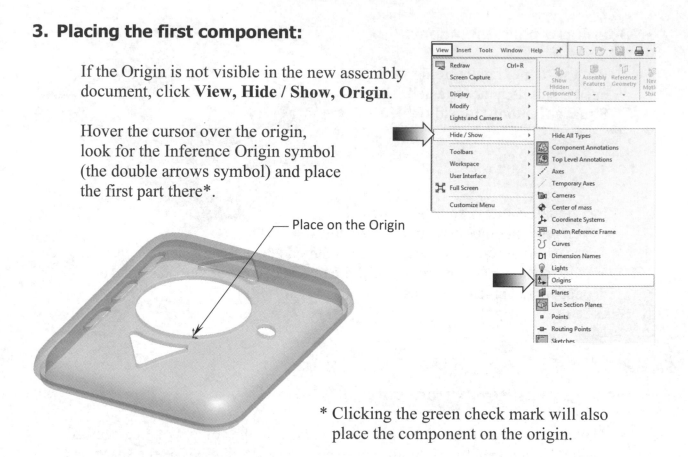

Place on the Origin

* Clicking the green check mark will also place the component on the origin.

4. Creating a new part:

Switch to the **Assembly** tool tab and select **New Part** under the **Insert-Component** drop down.

Select the **Top** plane to reference the new part (arrow). A new **InPlace** mate is created under the Mates group (arrow).

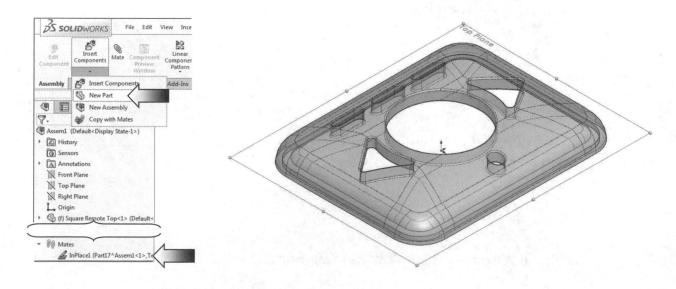

The **Edit Component** command is activated, and a new sketch is also opened on the <u>Top</u> plane.

Change to the Bottom orientation (Control+6).

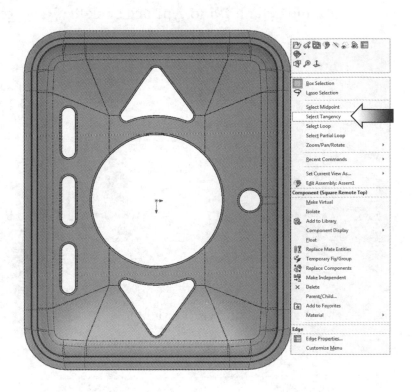

5. Converting the entities:

One of the quick ways to create new sketch entities is to use the Convert Entities tool.

<u>Right-click</u> one of the outer edges of the Remote Top and pick **Select Tangency**.

All the edges that are tangent to each other are selected.

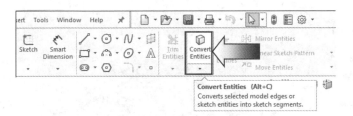

Click the **Convert Entities** command 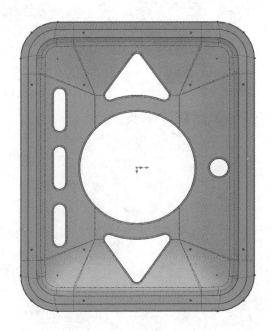.

The selected edges are converted to sketch entities. An **On-Edge** relation is added to each entity, and the sketch becomes fully defined automatically.

6. Extruding the sketch:

Change to the **Features** tool tab and click **Extruded Boss/Base** .

Use the default **Blind** type and click the **Reverse** button (arrow).

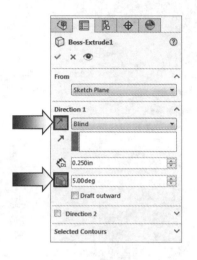

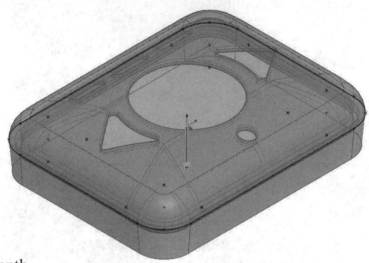

Enter **.250in** for extrude depth.

Enable the **Draft On/Off** button and enter **5.00deg** for draft angle.

Click **OK**.

7. Adding the .1875in fillet:

Click the **Fillet** command and enter **.1875in** for radius.

Use the default **Constant Size Radius** option.

Select the **bottom edge** as noted.

Click **OK**.

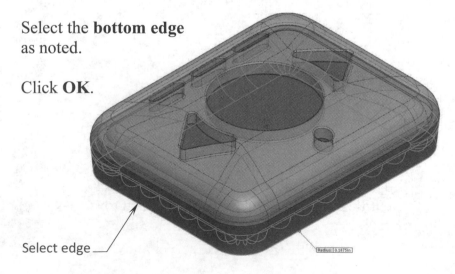

Select edge

8. Shelling the part:

The Shell command hollows out a part, but leaves open the selected faces, and creates thin-walled features on the remaining faces.

Select the **Shell** command from the **Features** tool tab.

For Wall Thickness, enter **.040in**.

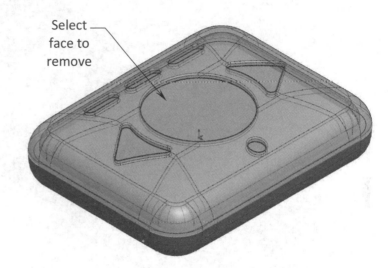

Select face to remove

For Faces to Remove, select the **top face** of the Remote Bottom.

Click **OK**.

9. Hiding a component:

Select the component **Square Remote Top** from the Feature tree and click **Hide Component** (arrow).

Hide

10. Adding features to the part:

Select the <u>top face</u> of the Remote Bottom and open a **new sketch**.

Sketch face —

Right-click one of the <u>inner edges</u> and pick **Select Tangency** (arrow).

Right-click one of the
inside edges and pick
Select Tangency

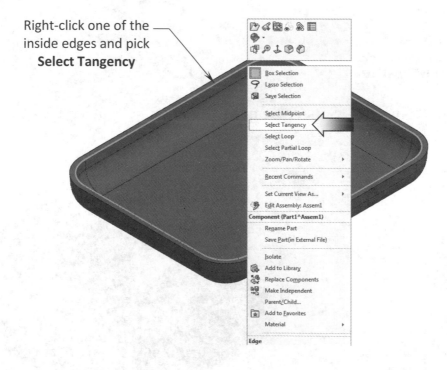

All inside edges are selected. They will be converted to new sketch entities
and used to create a boss feature on the inside of the part.

Click the **Convert Entities** command (arrow).

The selected edges are converted to new sketch entities.

Converted edges

The new entities are fully linked to the model where they were converted from.
If an edge in the model is changed, the converted entity will also change.

11. Creating offset entities:

We will use the offset entities option to define the thickness of the feature.

Right-click one of the converted entities and pick **Select Chain**.
All connected entities are selected.

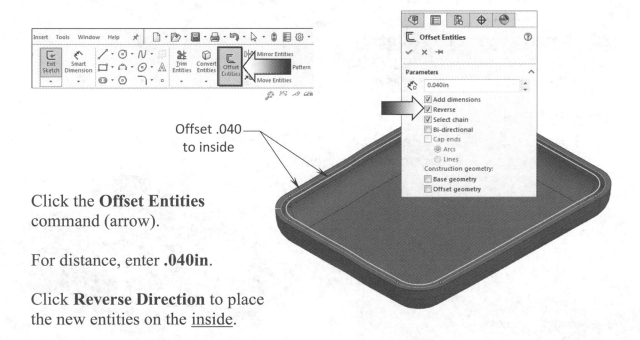

Offset .040 to inside

Click the **Offset Entities** command (arrow).

For distance, enter **.040in**.

Click **Reverse Direction** to place the new entities on the <u>inside</u>.

Click **OK**.

12. Extruding the sketch:

Switch to the **Features** tool tab. Select the **Extruded Boss/Base** command .

For <u>Extrude From</u>, select the **Offset** option and enter **.060in.** for distance (arrow).

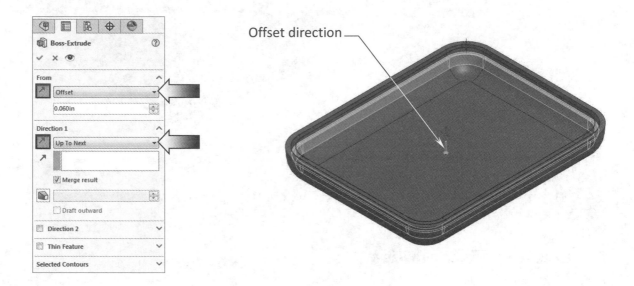

Offset direction

For Direction 1, click the **Reverse** button <u>first</u> then select **Up-to-next** from the list.

Click **OK**.

13. Adding the .125" fillet:

Select the **Fillet** command  and enter **.125in.** for radius.

Right-click one of the edges on the inside of the component and pick **Select Tangency**.

The review graphics shows the fillet is being propagated to all connecting edges.

Click **OK**.

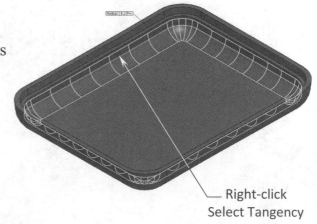

Right-click
Select Tangency

14. Showing a component:

Select the component **Square Remote Top** from the Feature tree and click **Show Component** (arrow).

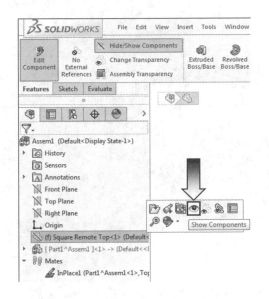
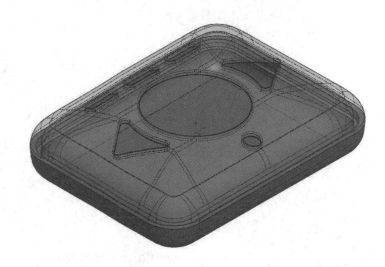

15. Creating a section view:

Click the **Section View**  command from the **View Heads-Up** toolbar (arrow).

For Section 1, select the **Top** plane as the cutting plane (arrow).

Inspect the area where the lip meets the groove (circle).

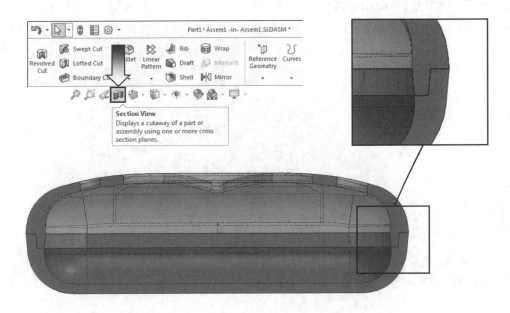

Click-off the Section View commands .

16. Applying dimension changes:

Click off the **Edit Component** button from the **Assembly** tool tab.

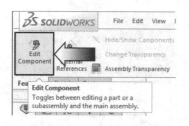

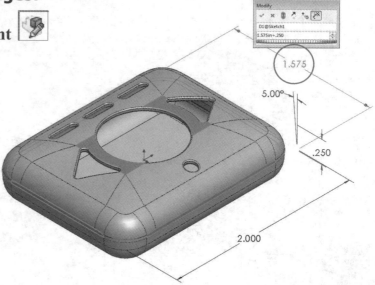

Expand the component **Square Remote Top** from the Feature tree and double-click the feature named **Boss-Extrude1** to reveal its dimensions.

Double-click the dimension **1.575** (circle) and **add .250in** to it.

Push the **Rebuild** button (the stop light) to update the model.

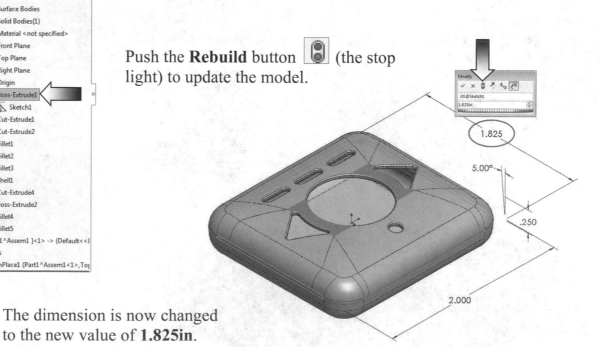

The dimension is now changed to the new value of **1.825in**.

Double-click the dimension **2.000** and <u>**add**</u> **.250in** to it.

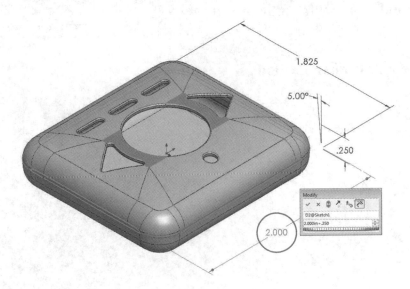

Push the **Rebuild** button 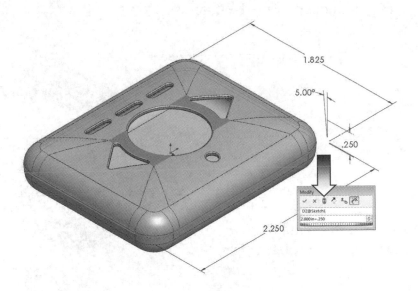 to regenerate the change.

The dimension is now changed to a new value of **2.250in**.

Notice both components were updated. This is one of the benefits of designing new components in the context of an assembly. Any features that have external references will be updated at the same time when changes are done to them.

17. Creating an exploded view:

Click **Exploded View** from the **Assembly** tool tab.

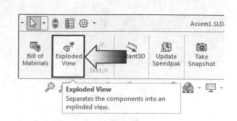

Use the default **Linear Explode Step** option.

Select the component **Square Remote Top** either in the graphics area or from the Feature tree.

Drag the arrowhead upward

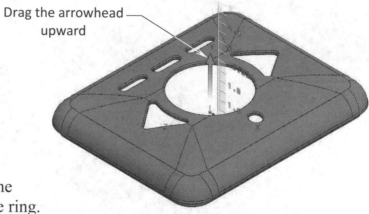

The **Triad** appears showing the 3 direction arrows and a rotate ring.

Hover the mouse cursor over the **vertical arrowhead** and drag it upward.

Click **OK**.

18. Playing the animation:

SOLIDWORKS creates an animation for each exploded view and stores it on the ConfigurationManager tree.

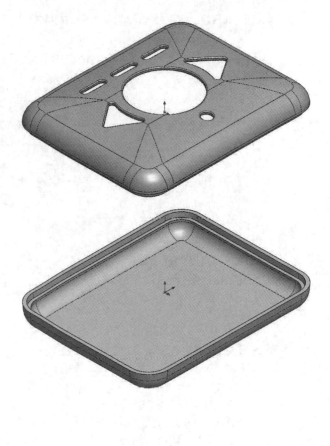

Switch to the **ConfigurationManager** tree (arrow).

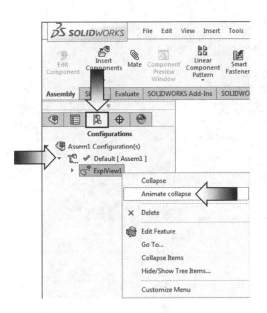

Expand the **Default** Configuration (click the ▼ symbol).

Right-click on **ExpView1** and select **Animate Collapse**.

Locate the **Animation Controller** toolbar and select the 2 options:

     * **Reciprocate**

     * **½ speed**

Click the **Play** button to play back the animation simultaneously.

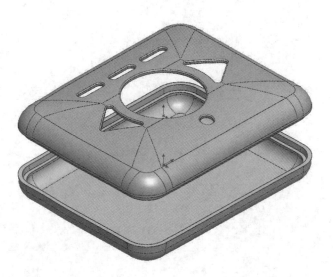

19. Changing the display mode:

The appearance of the models can be improved using several methods, but in this lesson we will take a look at the option called **Tangent Edges Remove**.

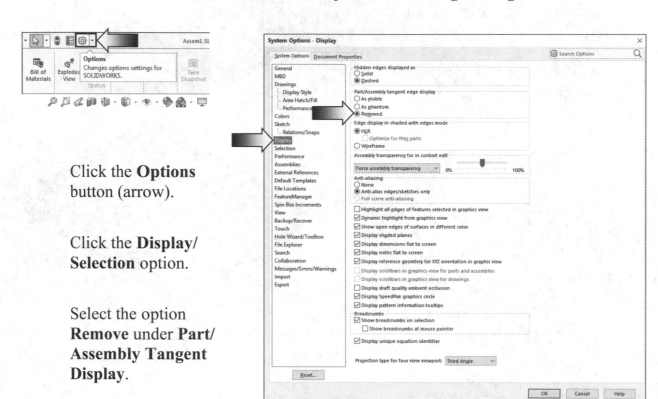

Click the **Options** button (arrow).

Click the **Display/ Selection** option.

Select the option **Remove** under **Part/ Assembly Tangent Display**.

Click **OK**.

<u>Shaded With Edges Visible</u>   <u>Shaded With Edges Removed</u>   <u>RealView Graphics Enabled</u>

20. Saving your work:

Select **File / Save As**.

Enter **Top Down Assembly.sldasm** for the name of the file and click **Save**.

Top Down Assembly
Creating a Flat Spring Assembly

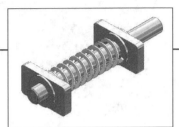

Creating a Flat Spring Assembly

One of the methods for creating a spring is to use the sweep option. The sweep command uses one sketch profile and one path to define the 3D feature.

The profile gets moved along the path, and depending on the shape of the feature, one or more guide curves are used to help control the twisting and how the profile is moved along the path.

In this lesson, the Flat Spring is created as a separated part, then inserted into an assembly document and mated to other components.

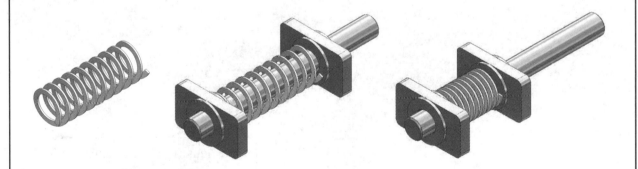

The Core Pin and the Compression Block on the left side are fixed. On the right, an instance of the Compression Block is still moveable along the longitudinal direction of the Core Pin.

The two ends of the Flat Spring will be mated to the two inner faces of the Compression Blocks, and the length of the Spring will be altered a couple times. When it is expanded, the Compression Block will move outwards, and when the Spring is compressed, the block will move inwards.

Top Down Assembly
Creating a Flat Spring Assembly

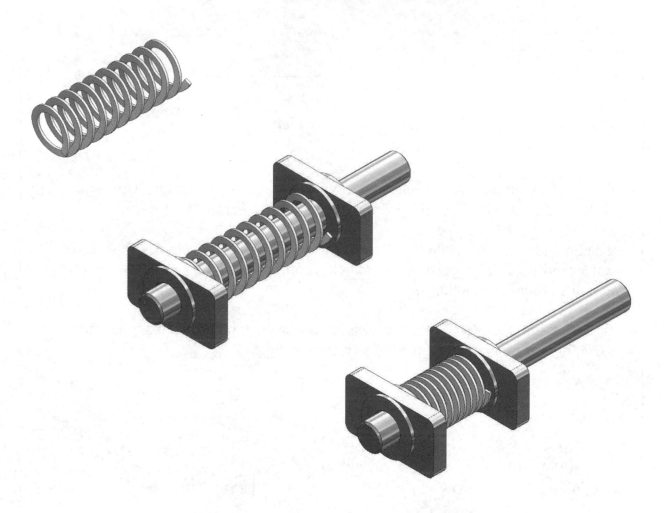

| Dimensioning Standards: **ANSI** |
| Units: **INCHES** – 3 Decimals |

Tools Needed:

Circle	Center Rectangle	Helix Spiral
Swept Boss Base	Axis	Mate

1. Starting a new Part template:

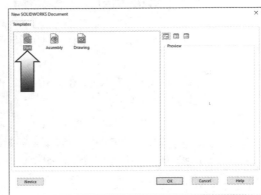

The Flat Spring is going to be created as a new part, and then inserted into an assembly document later on.

In the assembly, the Flat Spring gets mated to other components, and its overall length will get changed to simulate the expand / collapse motions. The components that were mated to the Flat Spring will also move in / out with each change.

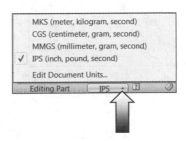

Click **File / New / Part**. Select the option **IPS** (Inch, Pound, Second) from the right corner of the screen.

2. Creating the first Sketch:

A helical coil, or a helix, is created from a circle, then the pitch, the number of revolutions, and the starting angle are entered to convert the circle into a helix. We will start by sketching a circle first.

Select the <u>Top</u> plane and open a **new sketch**.

<u>NOTE:</u> *To automatically rotate the view normal to the screen every time a sketch is activated, go to **Tools / Options / Sketch** and enable the checkbox:*
***Auto-Rotate View Normal to Sketch Plane on Sketch Creation** (arrow).*

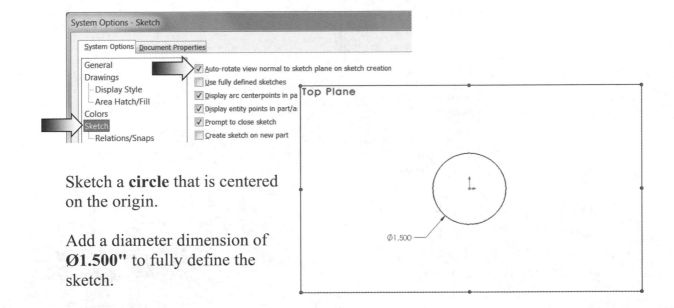

Sketch a **circle** that is centered on the origin.

Add a diameter dimension of **Ø1.500"** to fully define the sketch.

3. Creating the sweep path (the helix):

Switch to the **Features** toolbar.

Click the **Curves** button and select **Helix / Spiral**.

Under Defined by, use the default **Pitch & Revolution** option.

Under Parameters, use the default **Constant Pitch** option.

Enter **.500in** for Pitch.

Enter **10** for revolutions.

Enter **0** (zero) for Start Angle.

Click **OK**.

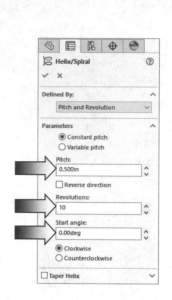
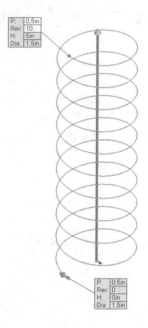

4. Creating the sweep profile:

The profile can be any shape. For this particular design only, the profile is a rectangle, drawn on a plane that is normal to the helix.

Select the <u>Right</u> plane from the FeatureManager and open a **new sketch**.

Sketch a **Center Rectangle** approximately as shown, away from the helix.

Add the width and height dimensions (.080" x .215").

The profile is still under defined at this point. The center of the profile needs to be pierced to the helix.

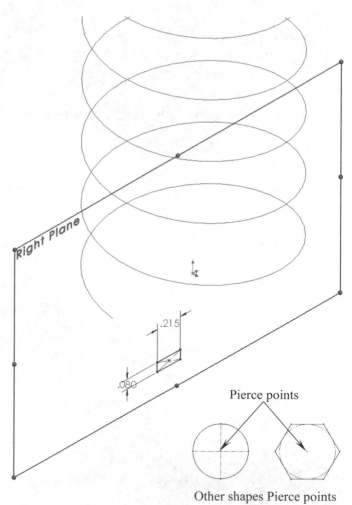

Pierce points

Other shapes Pierce points

5. Adding the Pierce relation:

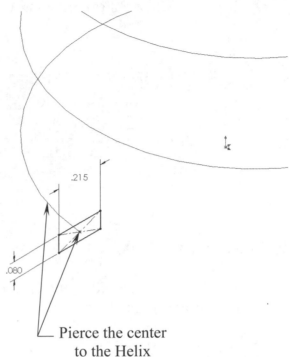

The helix is considered a 3D curve, and the rectangle is a 2D sketch. This is one of the instances where a Pierce relation is used to snap a 2D point onto a 3D curve.

Hold the **Control** key, select the **Center Point** of the rectangle and anywhere along the first revolution of **the helix** (but not at its endpoint).

On the Properties tree, there should be two relations: Coincident and Pierce. Click the **Pierce** relation.

Pierce the center to the Helix

The center of the rectangle should snap to the end of the helix and become fully defined (black color).

<u>Exit</u> the sketch (or press the hot key Control + Q). The sweep command is only available when the sketch mode is off.

6. Creating a swept feature:

The sweep feature is created by moving the profile along the path. If the profile is created as a closed skctch, a solid feature is made from it, but if the profile is created as an open sketch, a surface feature can be made from it.

Switch to the Features toolbar and click the **Swept Boss/Base** command.

For Profile, select the **rectangle** either from the feature tree or from the graphics area.

For Path, click the **helix** also from the feature tree or from the graphics area.

The preview appears indicating that the sweep feature is being created; click **OK**.

Rotate to different orientations to verify the result of the swept feature.

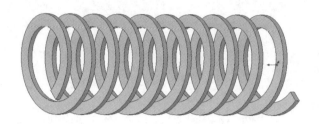

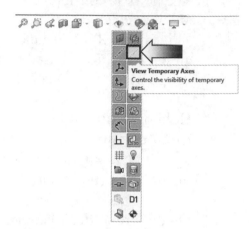

7. Creating an Axis:

An axis is often used as the center of a circular pattern, or simply used to mate the centers of the two parts.

A cylindrical part, a hole, or a fillet will automatically get a temporary axis created. Simply toggle this option on or off under the View pull down menu (arrow).

For practice purposes, we will create an axis in the center of the helical coil and use it as a mating entity later on.

From the **Features** toolbar, select the **Reference Geometry** button, and then click the **Axis** command.

There are five options available to create an axis; select the second option, the **Two Planes** (arrow).

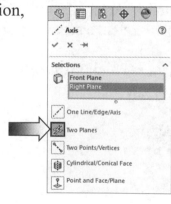

Expand the FeatureManager tree, select the **Front** plane and the **Right** plane from the feature tree.

A preview of an axis appears; Click **OK**.

<u>Save</u> the part as **Flat Spring**.

Next, we are going to insert the part into an assembly document for some additional assembly work.

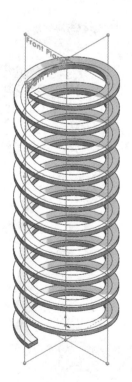

8. Opening an existing assembly:

Click **File / Open**.

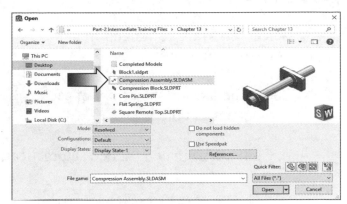

Browse to the <u>Training Files</u> and open the assembly named **Compression Assembly**.

From the **Assembly** tool tab, click the **Insert / Components** button.

Select the part **Flat Spring** from the Part/Assembly-To-Insert window.

Place the flat spring on the left side of the assembly.

"Right-drag" the flat spring and rotate it approximately as shown.

9. Showing the axis:

Select the **Visibility** pull-down menu and enable the **Axis** and **Temporary Axis** options (arrows).

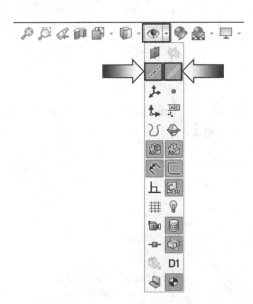

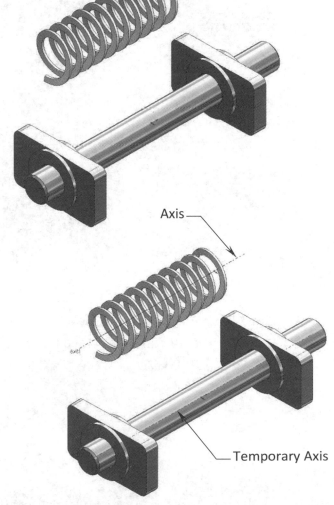

Both Axis and Temporary Axis appear in the centers of the components.

10. Adding mates:

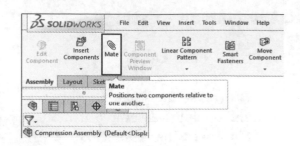

Select the **Mate** command from the assembly toolbar.

Select the **axes** in the center of each component as noted.

A **Coincident** mate is added automatically between the two axes.

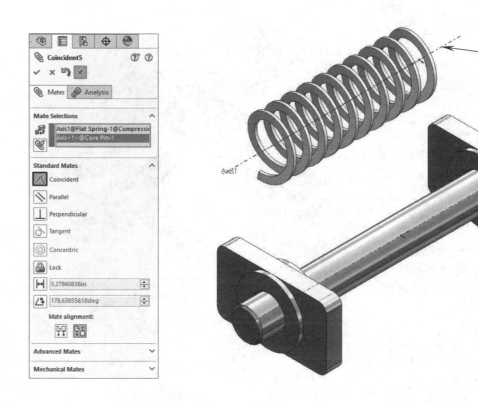

Select the 2 Axes

The Coincident mate removes 4 out of 6 degrees of freedom, and at this point, the Flat Spring can only move and rotate back and forth. Drag the flat spring to see how it moves.

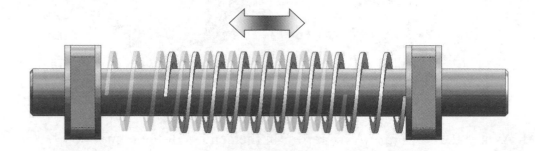

The goal is to mate the two ends of the Flat Spring to the two inner faces of the two Compression Blocks, so that when the spring is expanded or compressed, one of the Compression Blocks will move back and forth with the change.

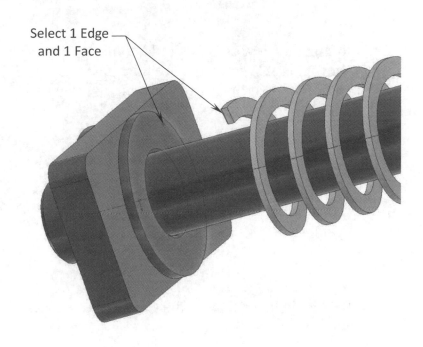

Select 1 Edge and 1 Face

The Mate command should still be active, if not, select it again.

Rotate the view so it is similar to the image shown here.

Select the end face of the Compression Block and the outer edge of the Flat Spring as illustrated.

Another **Coincident** mate is added, and the edge of Flat Spring moves forward and touches the surface or the Compression Block.

Click **OK** to accept the mate, but do not close the dialog just yet; the other end of the Flat Spring also needs to be constrained the same way.

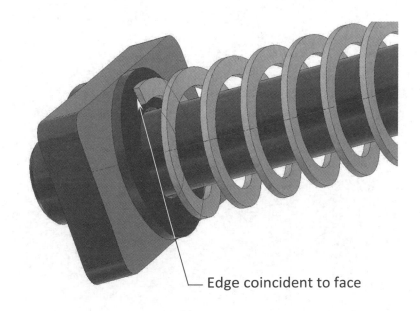

Edge coincident to face

Rotate the view so it is similar to the image shown here. We need to see the right end of the Flat Spring to constrain it.

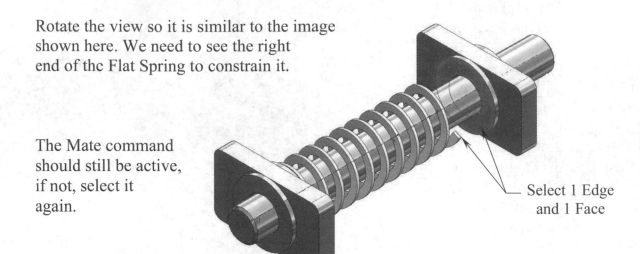

The Mate command should still be active, if not, select it again.

Select 1 Edge and 1 Face

Select the <u>inner face</u> of the Compression Block2 and the <u>outer edge</u> of the Flat Spring.

Another **Coincident** mate is added, and the inner surface of the Compression Block2 moves forward and touches the edge of the Flat Spring.

Click **OK** to accept the mate.

At this point, the two inner surfaces of the two Compression Blocks are mated to the two ends of the Flat Spring; they should move in and out when the Flat Spring is expanded or compressed. This is done in the next couple of steps.

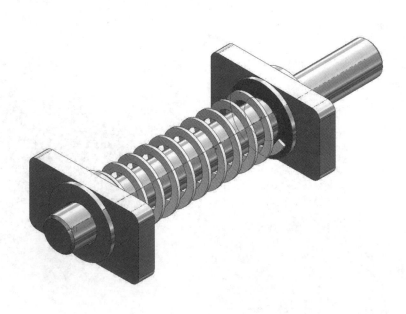

11. Applying the dimension changes:

Double-click on the Flat Spring to see its dimensions.

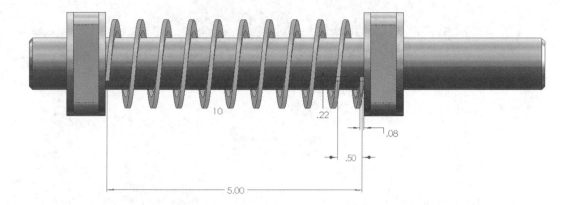

Double-click the pitch dimension **.50"** and change it to **.375"**.

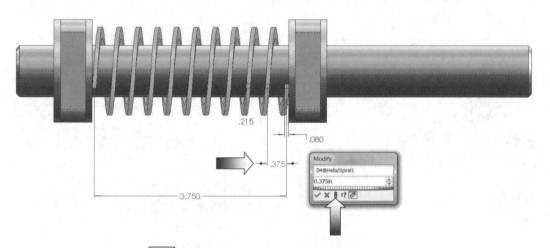

Press the **Rebuild** button either on the dialog box or on the General toolbar.

The length of the Flat Spring is reduced and this dimension change causes the Compression Block on the right to move with it.

Correct any rebuild errors if necessary before moving forward.

Double-click the pitch dimension **.375"** and change it to **.250"**.

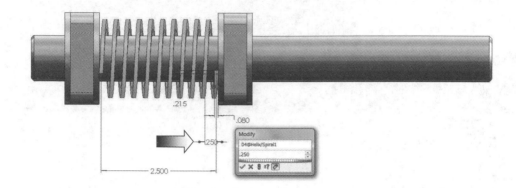

Press the **Rebuild** button either on the dialog box or press Control+Q.

Once again, the dimension change causes the length of the Flat Spring to be reduced and the Compression Block is also moved with the change.

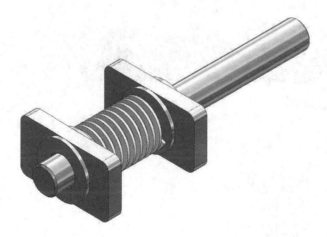

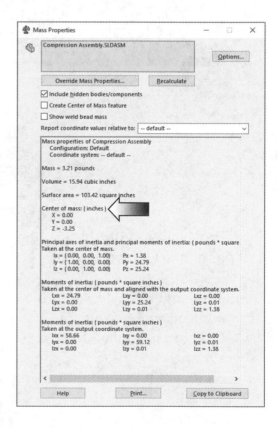

12. Calculating the Center of Mass:

Click **Tools / Mass Properties**.

Enter the **Center of Mass here**:

X= _____   Y= _____   Z= _____

Save and **Close** the Assembly.

___

<u>NOTE:</u> *Refer to the completed part saved in the Training Files folder for reference or to compare your results against it.*

Exercise: Spring Assembly

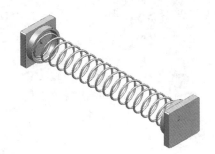

1. Creating a new assembly:

This exercise will guide you through one of the easier methods of creating new components in the context of an assembly, called Top Down Assembly.

Click **File / New Assembly**.

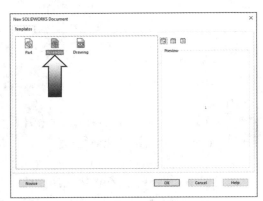

Select the default **Assembly Template** and click OK.

Click **Cancel** to close the Begin Assembly mode.

Select the **IPS** option at the lower right corner. Also change the Drafting Standard to ANSI, if not yet selected.

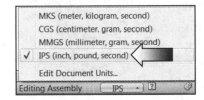

2. Saving the Assembly document:

Click **File / Save As**.

Enter **Spring Assembly_Exe** for the name of the file.

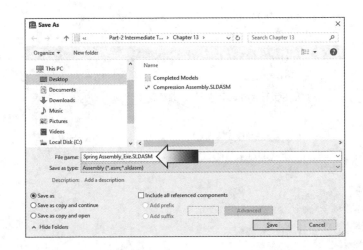

Click **Save**.

This will be the name of the top level assembly.

3. Creating a new component:

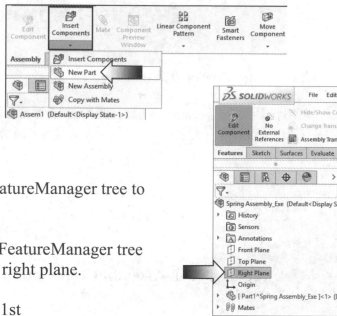

Switch to the **Assembly** tool tab.

Select the **New Part** option under the **Insert-Component** command (arrow).

Select the Right plane from the FeatureManager tree to reference this new part.

A new component is added to the FeatureManager tree and a new sketch is opened on the right plane.

The name **Part1** is created for the 1st component by default.

Sketch a **Center Rectangle** as shown.

Add the horizontal and vertical dimensions to fully define this sketch.

<u>Note:</u> *An equal relation can also be used to replace one of the two dimensions.*

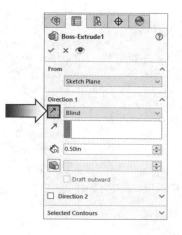

Switch to the **Features** tab.

Click **Extruded Boss-Base**.

Use the default **Blind** type.

Click the Reverse direction button.

Enter **.500"** for depth.

Click **OK**.

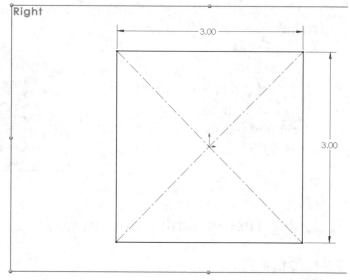

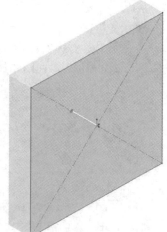

4. Adding a boss feature:

This boss feature will be used as a spring retainer to hold the spring in place.

Select the **face** as noted and open a <u>new sketch</u>.

Sketch **2 circles** centered on the origin.

Add the 2 diameter dimensions to fully define the sketch.

Switch to the **Features** tab.

Click **Extruded Boss-Base**

Use the default **Blind** type.

Enter **.500"** for depth.

Extrude direction is outward.

Click **OK**.

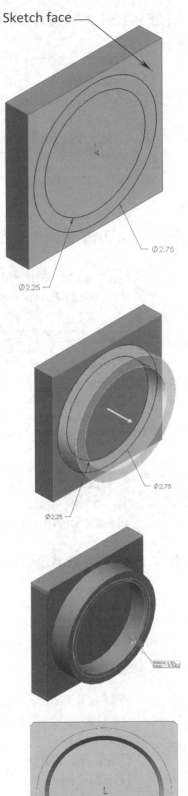

5. Adding the chamfers:

Click the **Chamfer** command under the Fillet drop-down.

Select the <u>4 edges</u> of the rectangular block and the <u>inside edge</u> of the circular boss, **5 edges** total.

Enter **.08in** for depth.

Use the default **45°** angle.

Click **OK**.

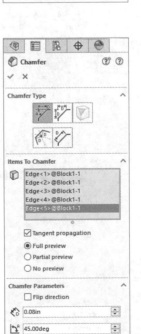

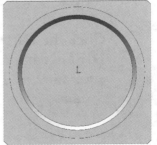

6. Exiting the Edit Component mode:

The first component is completed at this point. Before new components can be created, the Edit Component must be turned off.

Click the **Edit Component** command again to turn it off. The first component is no longer active; its color changes back to the default grey color.

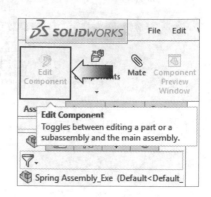

7. Creating a new component:

Select the **New Part** command from the Insert **Component** drop-down.

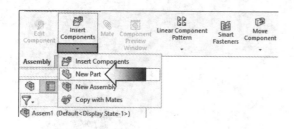

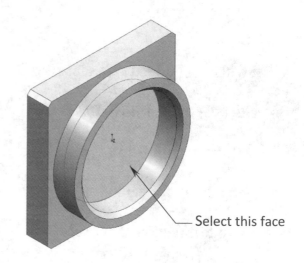

Select this face

Select the <u>face</u> as indicated to reference this new part. A new sketch is created automatically, but we are not going to sketch anything just yet; click **Exit** sketch (or Control+Q).

8. Creating a new plane:

Select the **Plane** command from the <u>Reference Geometry</u> drop down list.

Select the **face** as noted for the <u>First Reference</u>.

Click the **Offset-Distance** button and enter **.0625in** for distance.

Click **OK**.

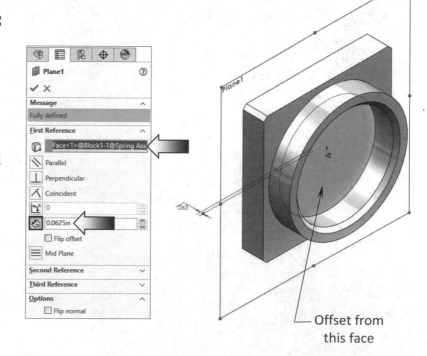

Offset from this face

Open a <u>new sketch</u> on the new **Plane1**. We are going to take advantage of the Top Down method to create the spring.

Select the <u>circular edge</u> as noted and click the **Offset Entities** command.

Enter **.125in** for the offset distance.

Click the **Reverse** checkbox to place the new circle on the <u>inside</u>.

Click **OK** to close the Offset command.

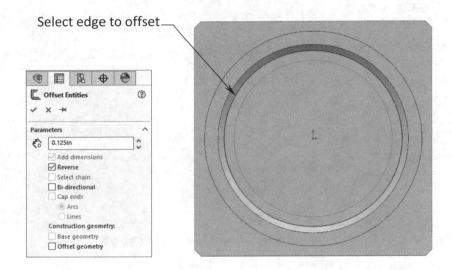

Select edge to offset

Switch to the Features tool tab and click the **Helix** command under the <u>Curves</u> drop-down list (arrow).

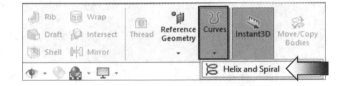

Enter the following parameters:

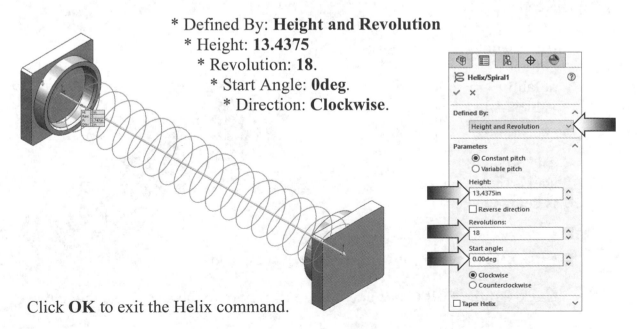

                * Defined By: **Height and Revolution**
                   * Height: **13.4375**
                      * Revolution: **18**.
                        * Start Angle: **0deg**.
                          * Direction: **Clockwise**.

Click **OK** to exit the Helix command.

9. Creating a swept feature:

An option called Circular Profile creates a solid rod or hollow tube along a sketch line, edge, or curve directly on a model, without having to sketch the profiles (this option is only available in SOLIDWORKS 2016 or newer).

Switch to the **Features** tool tab and click the **Swept Boss-Base** command.

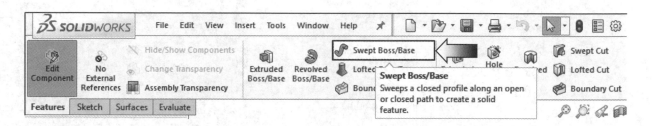

Click the **Circular Profile** option (arrow).

Enter **.125in** for Diameter.

For sweep path, select the **Helix**.

Keep all other parameters at their default values.

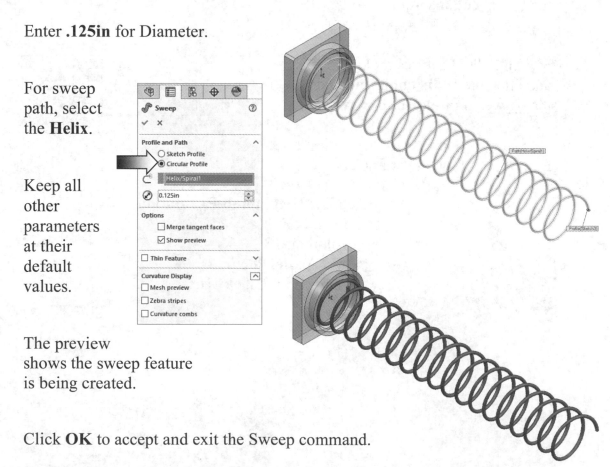

The preview shows the sweep feature is being created.

Click **OK** to accept and exit the Sweep command.

Click-Off the **Edit Component** button to return to the Edit Assembly mode.

10. Making a copy of a component:

One of the quicker methods to make a copy of a component(s) is to hold the Control key and drag it.

<u>Hold</u> down the **Control** key and <u>drag</u> the first component to the right to make a copy from it.

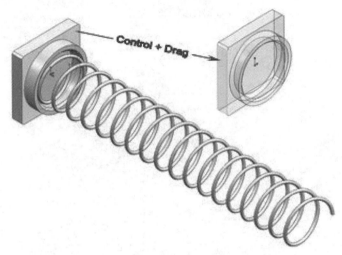

Place the copy on the right side of the spring approximately as shown.

In the assembly environment, the left mouse button is used to drag and move a component, and the right button is used to drag and rotate it.

Right + drag the copied component to reorient it. To correctly position the component we will have to constrain it with some mates.

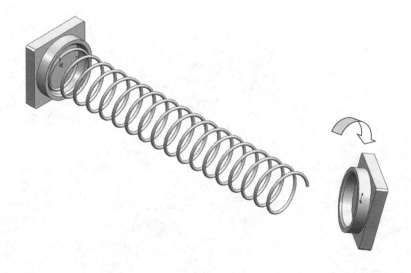

11. Adding mates between the components:

Click the **Mate** command from the Assembly tool tab.

Select the **2 cylindrical faces** of the two components as indicated.

A **Concentric** mate is created automatically.

Click **OK** but do not exit the mate mode.

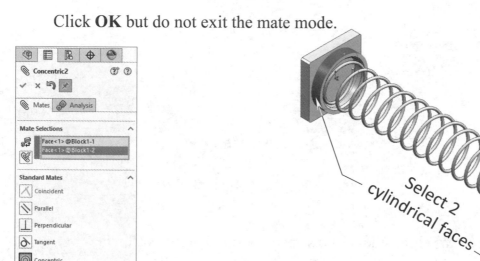

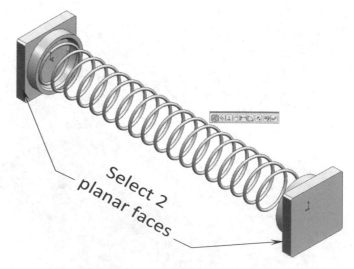

Select 2 cylindrical faces

Next, select the 2 planar faces on the front of the two components. A **Coincident** mate is added, and the copied component is moved to the correct orientation.

Click **OK** twice to exit the mate command.

Select 2 planar faces

Before adding the last mate we will need to move the copied component outward, and show the sketch of the spring's diameter.

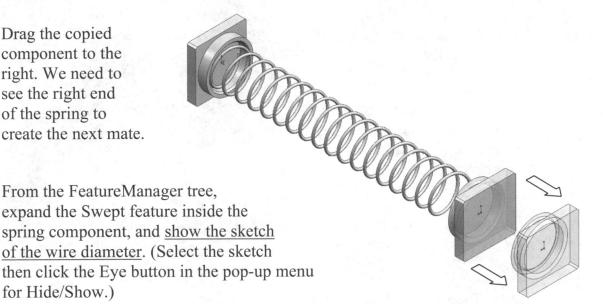

Drag the copied component to the right. We need to see the right end of the spring to create the next mate.

From the FeatureManager tree, expand the Swept feature inside the spring component, and show the sketch of the wire diameter. (Select the sketch then click the Eye button in the pop-up menu for Hide/Show.)

Click the **Mate** command again.

Select the Circular Edge of the Spring and the planar face on the inside of the copied component as noted.

A **Tangent** mate is added between the two selected entities.

Click **OK** twice to accept and exit the Mate command.

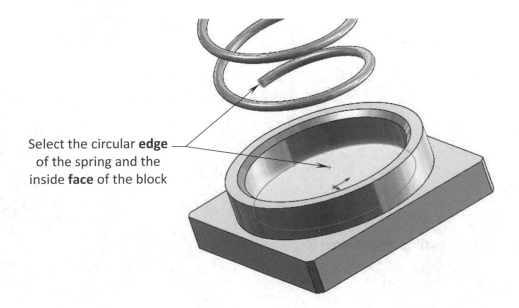

Select the circular **edge** of the spring and the inside **face** of the block

12. Changing dimensions:

Since the 3 components are mated to one another, changing the location of one component will affect the others.

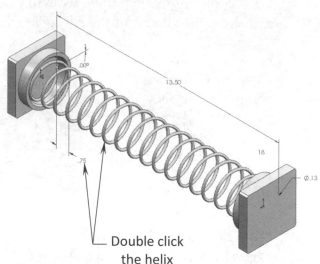

Double click
the helix

We will change the length and pitch dimensions of the spring to simulate its expand or compress modes.

Double-click on the body of the spring to display all of its dimensions.

Locate the **13.4375"** length dimension and change it to **6.750"** and change the pitch **.75"** to **.375"**.

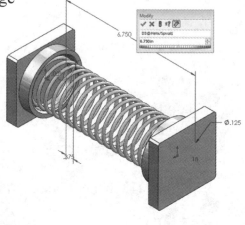

Press the **Rebuild** button (the stop light) to execute the change.

The spring is compressed and the component on the right moved inwards, following the dimension change.

Double-click the **6.750"** length dimension and change it to **4.500"** and also change the pitch from **.375"** to **.250"**.

Click **Rebuild** to regenerate the change.

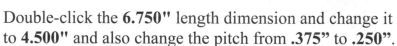

The spring is compressed even more, and the component on the right moved along with the dimension change.

13. Saving your work:

Save and close all documents.

CHAPTER 14

Using the Lip and Groove Options

Using the Lip & Groove Options

The Fastening Features toolbar provides tools for creating fastening features used in mold and sheet metal products. They help streamline the creation of common features for plastic and sheet metal parts.

The available tools are Lip / Groove, Mounting Boss, Snap Hook, Snap Hook Groove, and Vent.

This chapter discusses the use of the Lip / Groove and the Mounting Boss features.

* The Lip / Groove option creates fastening features using two solid bodies where the lip is added to one body and the groove is created by subtracting material from the other body.

   For the groove, material is removed from the inside edges, but the lip gets material added to the outside edges.

   Clearance or gap and draft angle can be added to the lip / groove to help close and open the two solid bodies more easily.

* The Mounting Boss, on the other hand, creates a tapered boss on each side of the two solid bodies, with or without the support ribs.

   A 3D sketch point is created to position the mounting boss. Relations and dimensions are used to precisely position the mounting boss.

   The options for the mounting boss are Hardware-Boss and Pin-Boss. The hardware-boss can accept a screw to fasten the two solid bodies while the pin-boss option uses a pin and aligns it to a hole.

Using the Lip & Groove Options
and the mounting Bosses

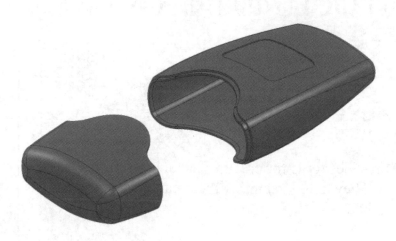

Dimensioning Standards: **ANSI**

Units: **INCHES** – 3 Decimals

Tools Needed:

 Edit Component

 Mounting Boss

 Lip / Groove

 Section View

Top Down Assembly
Using the Lip & Groove Options

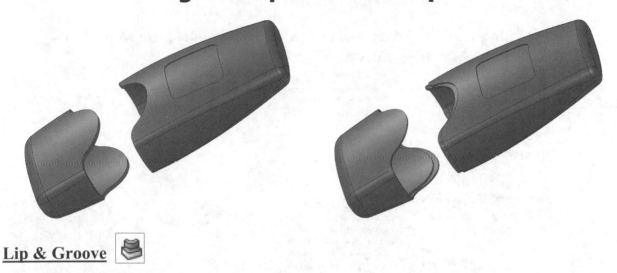

Lip & Groove

You can create lip and groove fastening features to align, mate, and hold together two plastic parts.

* The Lip option adds material to the part
* The Groove option removes material from the part

This session discusses the in-depth details of creating a set of Lip & Groove for the two plastic parts pictured above.

Fastening features streamline the creation of common features for plastic and sheet metal parts. You can create: **Lips and Grooves**. Align, mate, and hold together two plastic parts. Lip and groove features support multibodies and assemblies.	
Mounting bosses. Create a variety of mounting bosses. Set the number of fins and choose a hole or a pin.	
Snap hooks and **snap hook grooves**. Customize the snap hook and snap hook groove. You must first create a snap hook before you can create a snap hook groove.	
Vents. Create a variety of vents using a sketch you create. Set the number of ribs and spars. Flow area is calculated automatically.	

1. Opening the existing assembly:

Select **File / Open**.

Browse to thc Training Folder and open an assembly document named:
Lip & Groove Assembly.sldasm.

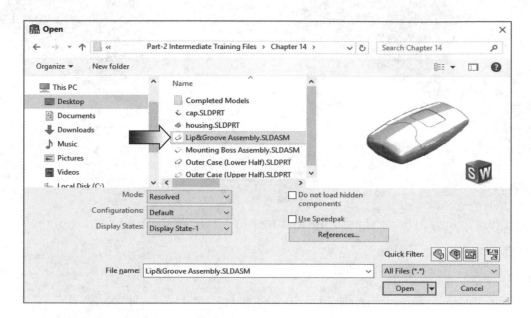

2. Editing the Housing:

Select the component **Housing** and click **Edit Component** .

From the **Insert** menu, select **Fastening Feature / Lip /Groove** .

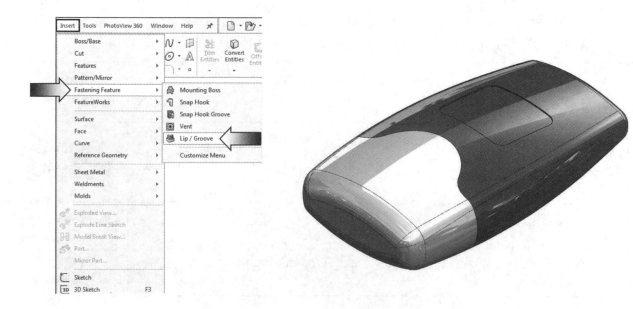

Select the following from the FeatureManager tree:

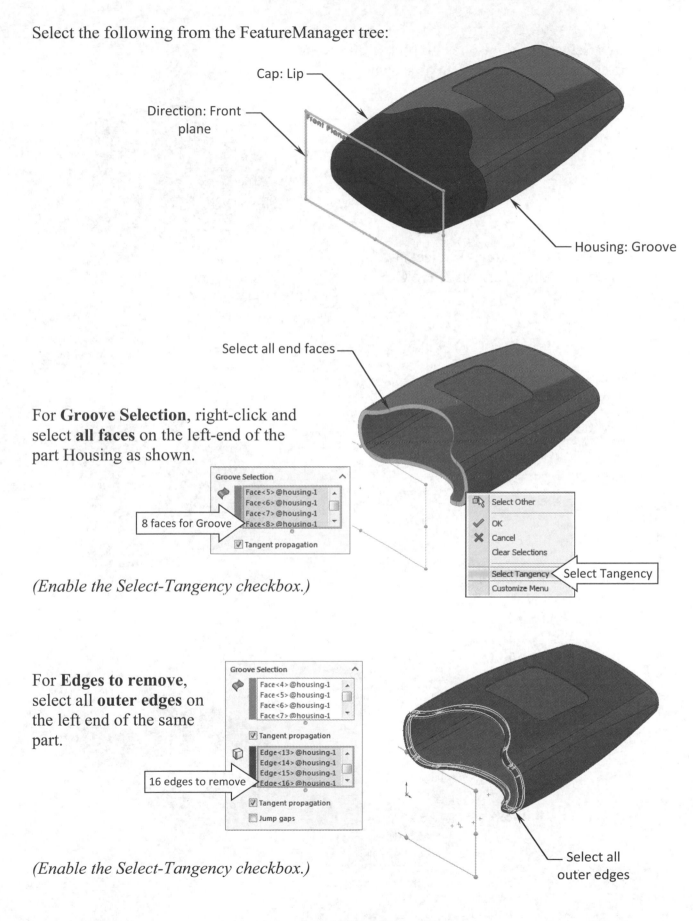

Cap: Lip

Direction: Front plane

Housing: Groove

Select all end faces

For **Groove Selection**, right-click and select **all faces** on the left-end of the part Housing as shown.

Groove Selection

Face<5> @housing-1
Face<6> @housing-1
Face<7> @housing-1
Face<8> @housing-1

8 faces for Groove

☑ Tangent propagation

Select Other

✓ OK
✗ Cancel
Clear Selections

Select Tangency   Select Tangency

Customize Menu

(Enable the Select-Tangency checkbox.)

For **Edges to remove**, select all **outer edges** on the left end of the same part.

Groove Selection

Face<4> @housing-1
Face<5> @housing-1
Face<6> @housing-1
Face<7> @housing-1

☑ Tangent propagation

Edge<13> @housing-1
Edge<14> @housing-1
Edge<15> @housing-1
Edge<16> @housing-1

16 edges to remove

☑ Tangent propagation
☐ Jump gaps

Select all outer edges

(Enable the Select-Tangency checkbox.)

For **Lips Selection**, right-click and select **all faces** on the right-end of the part Cap as shown.

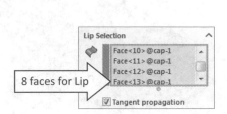

8 faces for Lip

For **Edges to Add Material**, select all **outer edges** on the right-end of the Cap.

Select all end faces

Select Tangency

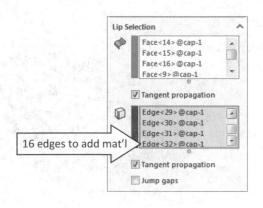

16 edges to add mat'l

Under **Parameters**, set the following:

* Groove Width: **.031in**.

* Spacing: **0.005in**.

* Groove Draft: **3.00deg**.

* Upper Gap: **0.005in**.

* Lip Height: **.062in**.

* Lip Width: **.031in**.

* Lower Gap: **0.005in**.

Click **OK**.

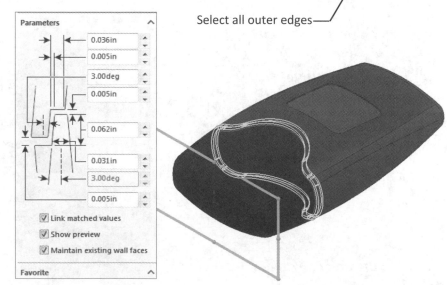

Select all outer edges

Click-off the **Edit Component** button .

3. Creating a section view:

Create a Section View using the
Right plane as the cutting plane.

Zoom in on the Lips & Grooves details
and verify the gaps between them.

Click-off the section view when finished.

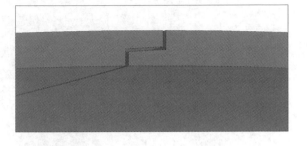

(Section View)

4. Saving your work:

Select **File / Save As**.

Use the same file name and overwrite the original document.

Click **Save**.

Top Down Assembly
Using the Mounting Boss Tool

Mounting Boss

You can create common mounting features for plastic parts like Mounting Bosses, Snap Hooks, Snap Hook Grooves, Vents, and Lips/Grooves.

The mounting bosses come with a variety of options to help streamline the creation of common Fastening features.

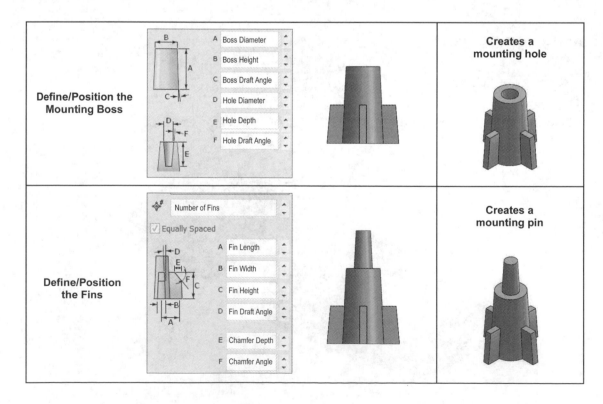

Using the Mounting Boss Tool

1. Opening the existing assembly:

Select **File / Open**.

Browse to the Training Folder and open an assembly document named: **Mounting Boss Assembly**.sldasm.

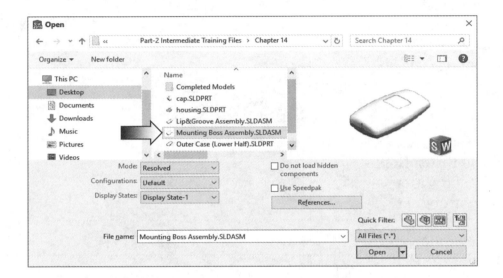

2. Editing Part: Select the part **Outer Case (Lower Half)** and click **Edit Component**.

From the **Insert** menu, select **Fastening Feature / Mounting Boss** .

3. Setting the Parameters:

Select Face

The options to set the parameters for the mounting boss appear on the Feature tree.

For **Position Face** to attach the mounting-boss, select the **face** as indicated.

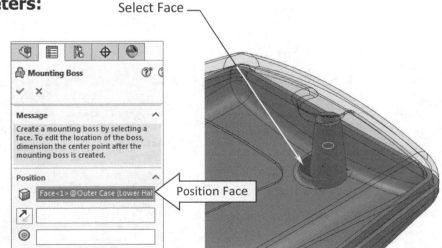

Position Face

For **Position Edge** to align the mounting boss, select the **circular edge** as noted.

(Use the Top plane here for direction only if required)

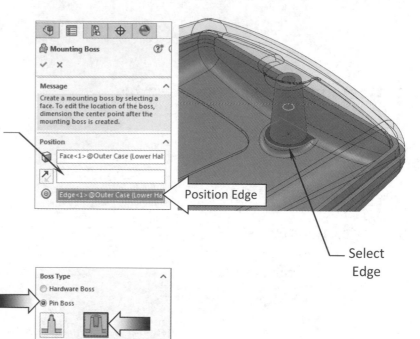

Position Edge

Select Edge

For the **Boss Type** section, select the **Pin Boss** and the **Hole** option (arrow).

For the Boss section, click the **Enter Boss Height** and **Enter Diameter** options.

Enter the following dimensions:

* **A = .200in**.

* **B = .165in**.

* **C = 5.00deg**.

* **D = .093in**.

* **E = .175in**.

* **F = 2.00deg**

* **Clearance = .005in**.

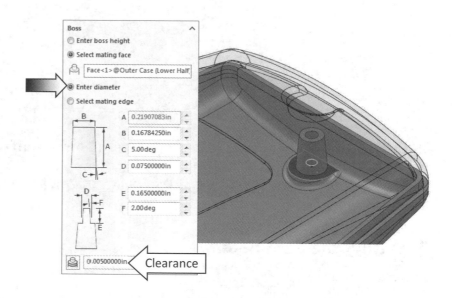

Clearance

Click in the **Fins** section, change the number of fins to **0** (zero).

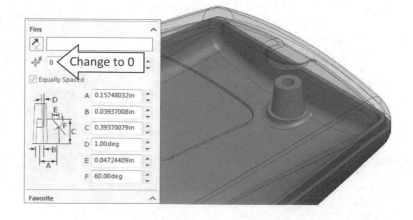

Change to 0

Click **OK**.

4. Adding two more mounting bosses:

Repeat the previous steps and create two other mounting bosses on the opposite side.

Exit the **Edit Component** 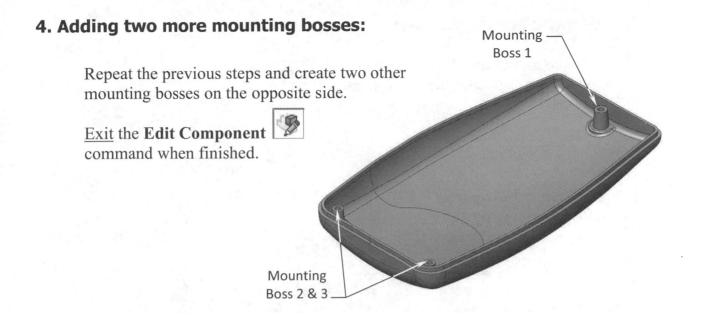 command when finished.

Mounting Boss 1

Mounting Boss 2 & 3

5. Toggling between the Explode and the Collapse views:

Right-click the name of the assembly and select **Explode***.

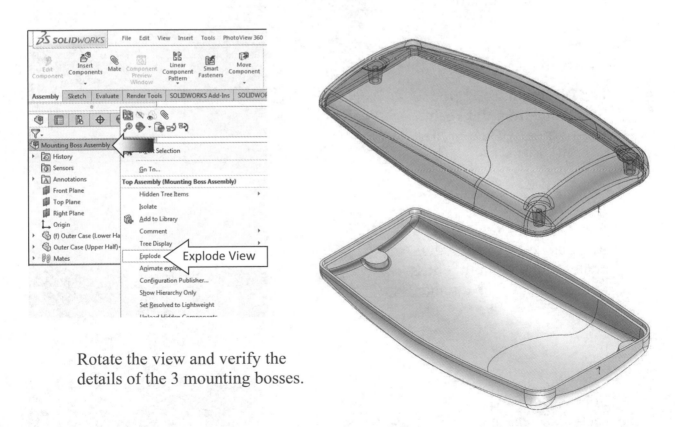

Rotate the view and verify the details of the 3 mounting bosses.

* Remember to **<u>Collapse</u>** the assembly prior to editing the next component.

6. Creating the Mating Bosses:

Select the part **Outer-Case (Upper Half)** and click the **Edit Component** command.

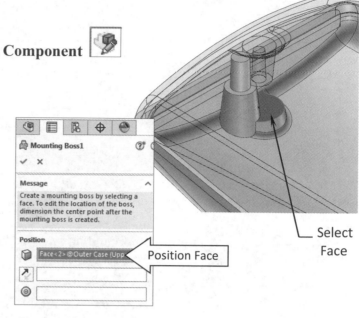

Select **Insert/Fastening Feature/ Mounting Boss**.

For **Position Face**, select the **Face** as indicated. (To select a hidden face, right-click in the area where you wish to select and click **Select Other**, then pick the face from the list.)

For **Direction**, select the **Top** plane from the Feature tree (use only if required).

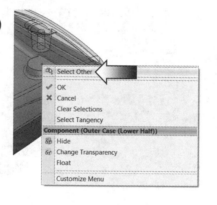

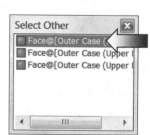

For **Position Edge**, select the **Edge** as noted to align the center of the mounting boss to the center of the circular edge.

For the **Boss Type** section, select the **Pin Boss** option.

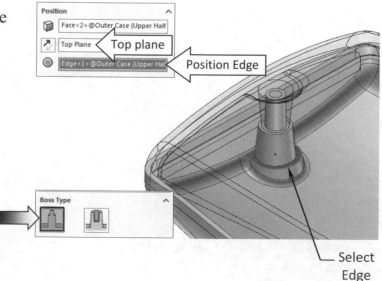

Select the options
Select Mating Face
and **Enter Diameter**.

Select the **upper surface** of the
mounting boss of the Lower Half
as noted.

Enter the following dimensions:

* A = **Defined by mating face**.

* B = **.1678 in.**

* C = **5.00deg.**

* D = **.075in.**

* E = **.165in.**

* F = **2.00deg.**

Click **OK**.

Click off the **Edit Component** button .

7. Creating a Section View:

Create a section view using the <u>Right</u> plane as the cutting plane.

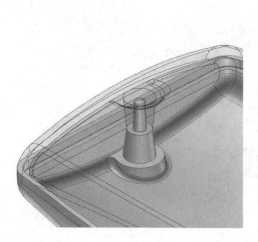

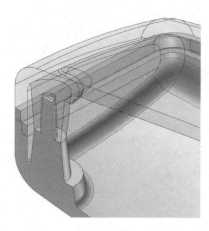

<u>Section View</u>

8. Repeating:

Create two additional mounting bosses; use the same settings as the first one.

9. Saving your work as:

Select **File / Save As**.

Use the <u>same file name</u> and overwrite the original document when prompted.

Click **Save**.

CHAPTER 15

Assembly Drawings & BOM

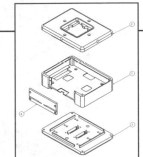

Assembly Drawings & BOM
Aluminum Enclosure

An assembly drawing normally consists of drawing views, balloons, and a bill of materials.

In a simple assembly that has only a few components, creating one exploded view to show all the components, and one isometric view to show how they are assembled should be sufficient. But in a more complex assembly that has many components, other drawing views such as Section views and Detail views are also added to enhance the clarity and details of the components.

If an assembly Section view is created, the crosshatch pattern for each component should be altered so that they will have different angles to help differentiate the components. In addition, the pattern that represents the material for each component should also be applied to help eliminate any confusion (see example).

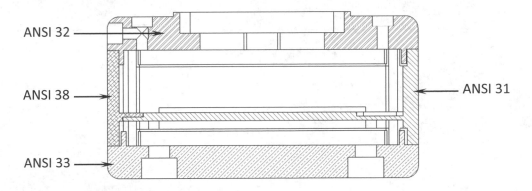

A Bill of Materials is added to show the item numbers, quantities, part numbers, and custom properties in assembly drawings.

Balloons will then be inserted into the drawing. The item numbers and quantities in the balloons correspond to the numbers in the Bill of Materials.

Assembly Drawings & BOM
Aluminum Enclosure

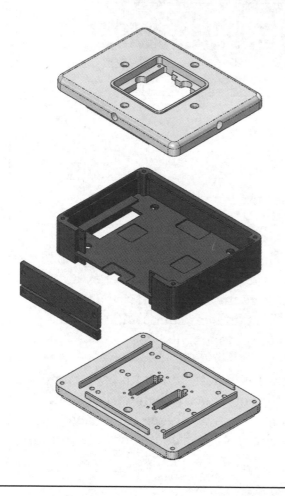

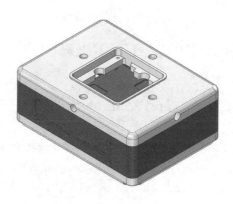

Dimensioning Standards: **ANSI**

Units: **INCHES** – 3 Decimals

Tools Needed:

New Drawing   Bill of Materials

Balloons   Magnetic Line    View Palette

Link to Property   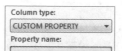

Column type:

CUSTOM PROPERTY

Property name:

1. Starting a new drawing document:

Select **File / New / Drawing**.

Select the Drawing Template (arrow) in the New-SOLIDWORKS Document dialog box.

Click **OK**.

The default drawing template may have settings that are used with the ISO standards. We will need to switch them to ANSI standards prior to making the drawings.

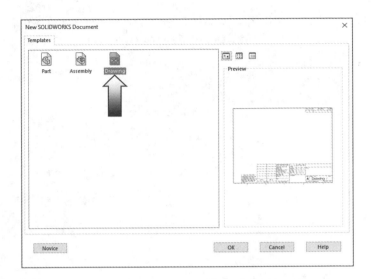

2. Selecting the ANSI standard sheet size:

Push **Escape** if you do not see the option to select the drawing sheet size, similar to the dialog below.

Right-click inside the drawing and select **Properties** from the pop-up list (arrow).

Select the **C (ANSI) Landscape** for the paper size (arrow).

Select the **Third Angle Projection** option for the Type of Projection.

Set the **Scale** to **1:1**.

Enable the **Display Sheet Format** check box and click **OK**.

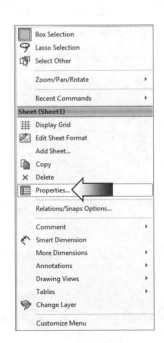

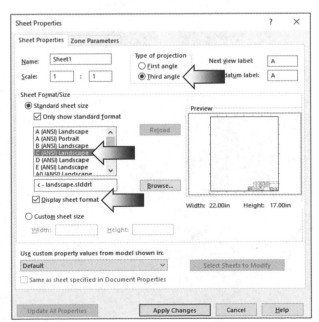

3. Switching to ANSI standards:

Click the **IPS** (or MMGS) tab at the bottom right of the screen and select **Edit Document Units** (arrow).

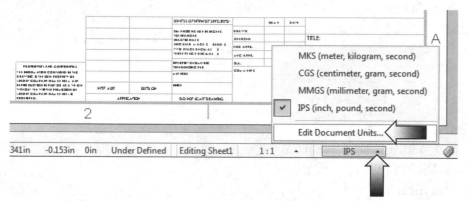

Click the **Drafting Standard** option (arrow).

Select **ANSI** from the drop-down list (arrow).

Click **OK**.

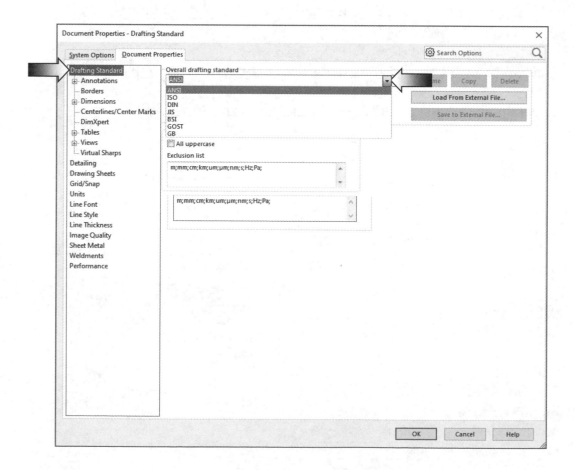

4. Using the View Palette:

Part and Assembly documents can be opened using the **View-Palette**. Drawing views and Annotations can then be dragged and dropped into the drawing.

Select the **View Palette** tab and click the **Browse** button (arrow).

Open an assembly document named: **Aluminum Enclosure Assembly.sldasm**.

Drag and drop the **Isometric Explode**d view onto the drawing as indicated below.

Click **OK**.

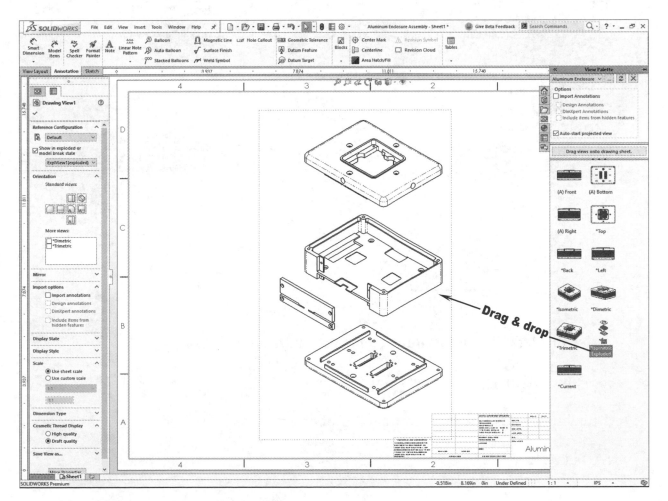

5. Inserting a Bill of Materials:

A drawing can contain an Excel-Based Bill of Materials or a Table-Based Bill of Materials, but not both. We will insert an Excel-Based BOM into the drawing.

Select the dotted border of the drawing view and click the **Annotation** tab (arrow).

Select the **Bill of Materials** option under the **Tables** drop down list (arrow).

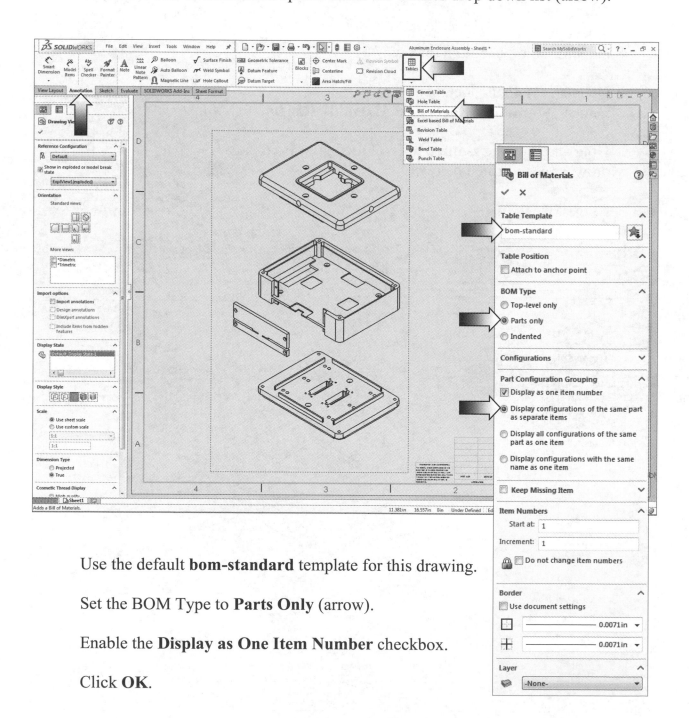

Use the default **bom-standard** template for this drawing.

Set the BOM Type to **Parts Only** (arrow).

Enable the **Display as One Item Number** checkbox.

Click **OK**.

Place the Bill of Materials table inside the drawing approximately as shown.

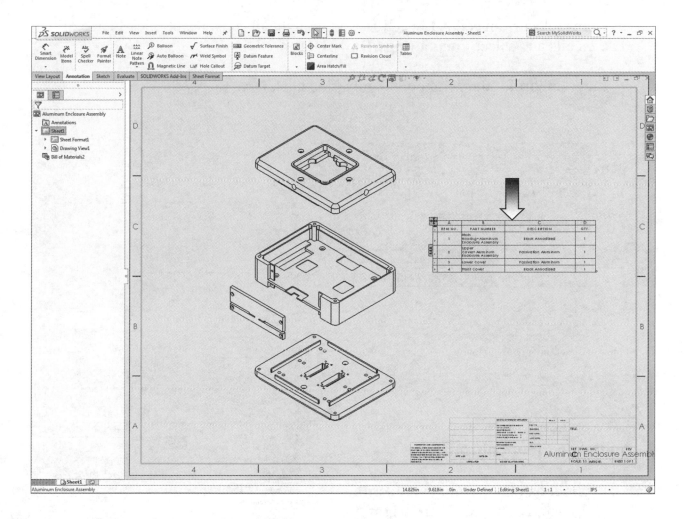

Drag the '4-way' arrow to move the BOM (arrow).

Click the **Use Document Font** button (arrow) to enable the formatting tools and modify the BOM content if needed.

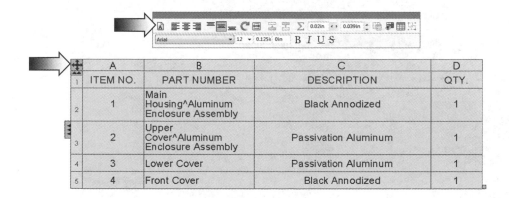

	A	B	C	D
1	ITEM NO.	PART NUMBER	DESCRIPTION	QTY.
2	1	Main Housing^Aluminum Enclosure Assembly	Black Annodized	1
3	2	Upper Cover^Aluminum Enclosure Assembly	Passivation Aluminum	1
4	3	Lower Cover	Passivation Aluminum	1
5	4	Front Cover	Black Annodized	1

6. Modifying the BOM's Row Height:

Drag-Select from cell **A1** to cell **D4** to highlight all cells. Right-click in one of the cells and select **Formatting / Row Height**.

Enter **.500in** for the height of the rows and click **OK**.

7. Adding a new column:

Right-click on the column header **D** and select **Insert / Column Right**.

A new column **E** is added to the right side of the column D.

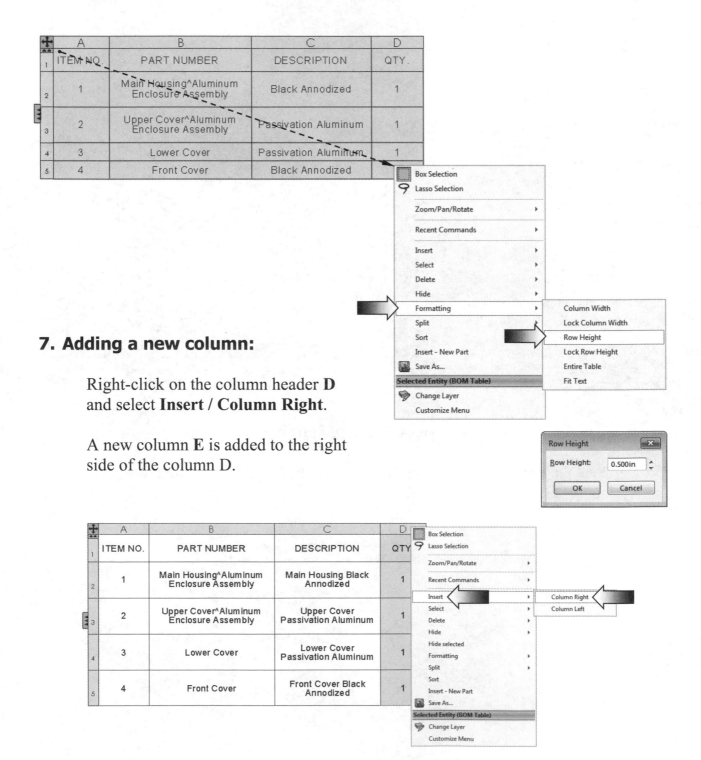

8. Customizing the new column:

Double-click on the column header **E** and select:

* Column Type:
  Custom Property

* Property Name:
  Finish

	A	B	C	
1	ITEM NO.	PART NUMBER	DESCRIPTION	QTY.
2	1	Main Housing^Aluminum Enclosure Assembly	Main Housing	1
3	2	Upper Cover^Aluminum Enclosure Assembly	Upper Cover	1
4	3	Lower Cover	Lower Cover	1
5	4	Front Cover	Front Cover	1

Column type:
CUSTOM PROPERTY
Property name:

Description
Finish
PartNo
SW-Author(Author)
SW-Comments(Comments)
SW-Configuration Name(Con
SW-Created Date(Created Dat
SW-File Name(File Name)
SW-Folder Name(Folder Name
SW-Keywords(Keywords)
SW-Last Saved By(Last Saved B
SW-Last Saved Date(Last Save
SW-Long Date(Long Date)
SW-Short Date(Short Date)
SW-Subject(Subject)
SW-Title(Title)

The Surface Finishes that were entered in the properties of each part document are displayed in the new column.

	A	B	C	D	E
1	ITEM NO.	PART NUMBER	DESCRIPTION	QTY.	Finish
2	1	Main Housing^Aluminum Enclosure Assembly	Main Housing	1	Black Annodized
3	2	Upper Cover^Aluminum Enclosure Assembly	Upper Cover	1	Passivation Aluminum
4	3	Lower Cover	Lower Cover	1	Passivation Aluminum
5	4	Front Cover	Front Cover	1	Black Annodized

Column type:
PART NUMBER
CUSTOM PROPERTY
UNIT OF MEASURE
EQUATION
ITEM NO.
PART NUMBER
COMPONENT REFERENCE
TOOLBOX PROPERTY

Double-click on the **column header B** and change the Part Number to **Custom Property**.

	A	B	C	D	E
1	ITEM NO.	BER	DESCRIPTION	QTY.	Finish
2	1	Main Housing^Aluminum Enclosure Assembly	Main Housing	1	Black Annodized
3	2	Upper Cover^Aluminum Enclosure Assembly	Upper Cover	1	Passivation Aluminum
4	3	Lower Cover	Lower Cover	1	Passivation Aluminum
5	4	Front Cover	Front Cover	1	Black Annodized

Select the **PartNo** from the **Property Name** drop down list.

The Column B is updated to display the part numbers that were previously entered at the part level.

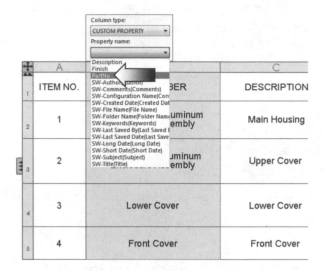

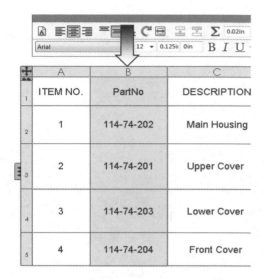

9. Changing the color of the paper:

Click **Options** on the top of the screen (arrow).

Click the **Colors** option.

Select **Drawing Paper Color** under Color-Scheme settings.

Click **Edit** and select the **White** color.

Enable the check box for **Use Specified Color For Drawings Paper Color** and click **OK**.

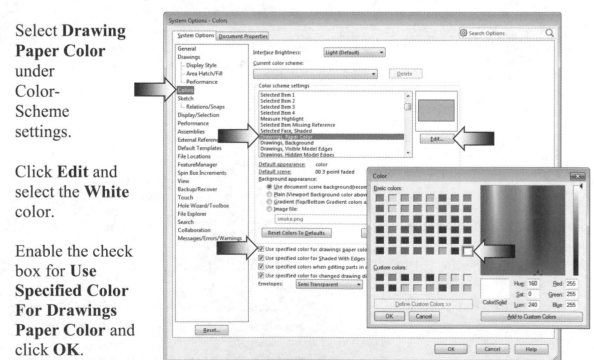

10. Adding balloons:

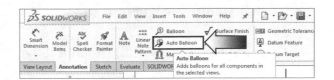

Balloons are used to identify item numbers in the bill of materials.

The balloon numbers are associated with the bill of materials as well as the order of the components displayed on the FeatureManager tree.

Select the border of the Isometric view and click **Auto Balloon** from the **Annotation** tab.

The balloons are added to each unique component.

Use the default **Circular** style.

Click **OK**.

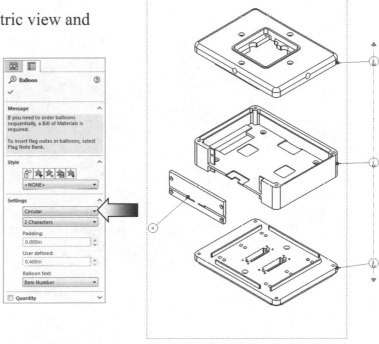

The balloons are attached to the magnetic lines to align or to move them in groups.

Drag one of the arrowheads to move the magnetic line; the balloons that attached to it are also moved.

Delete the magnetic lines and move the balloons individually if needed.

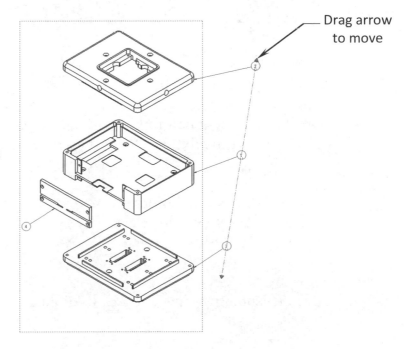

Drag arrow
to move

11. Adding an isometric view:

Expand the **View Palette** and drag/drop the **Isometric view** onto the drawing.

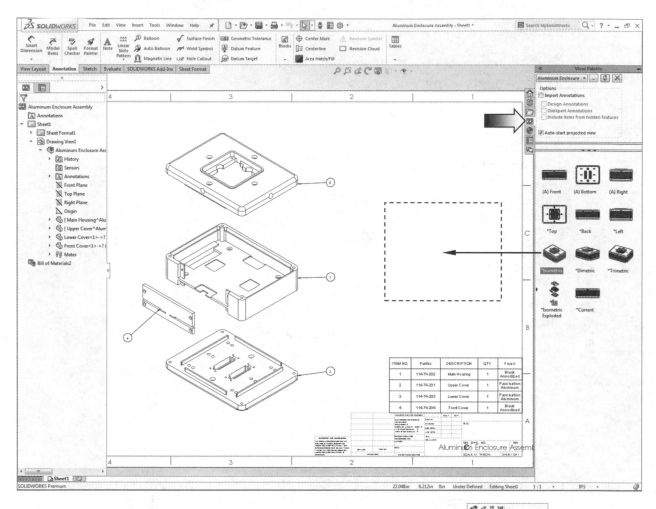

The isometric view is aligned with the exploded view by default. We will need to break the alignment before it can be moved freely.

Right-click the dotted border of the isometric view and select: **Alignment / Break Alignment**.

Move the isometric view to check its alignment.

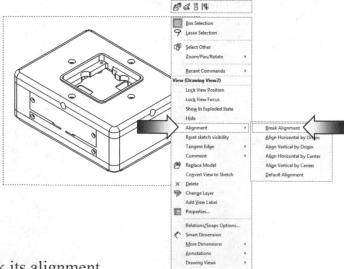

12. Changing the display of a drawing view:

Click the dotted border of the isometric view and select the **Shaded With Edges** button from the properties tree (arrow).

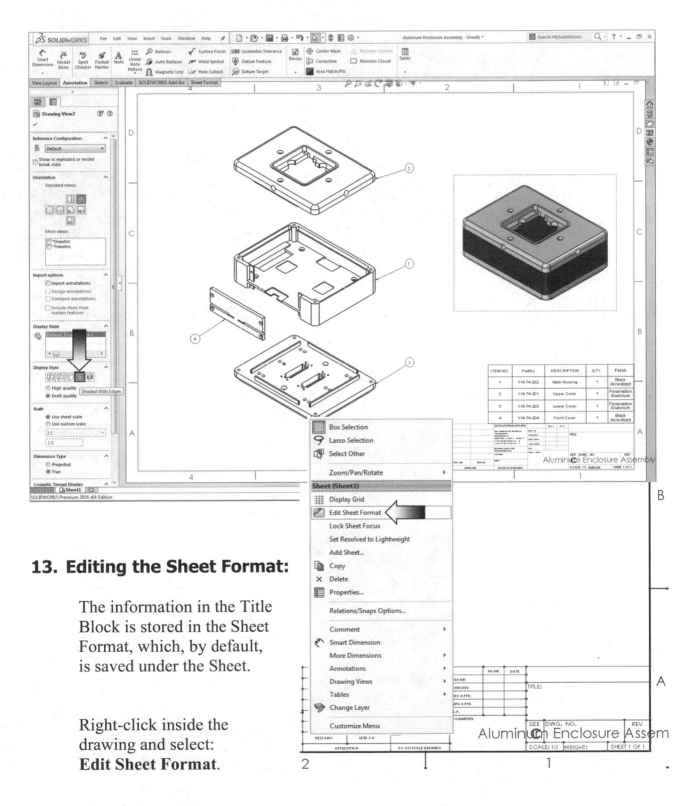

13. Editing the Sheet Format:

The information in the Title Block is stored in the Sheet Format, which, by default, is saved under the Sheet.

Right-click inside the drawing and select: **Edit Sheet Format**.

Hover the cursor over the the company name field; look for the symbol "A" (note). The property link $SPRP"COMPANYNAME" appears.

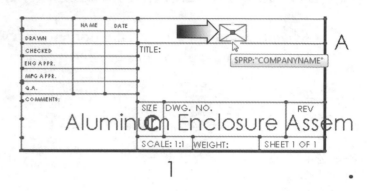

Double-click the property link **$SPRP"COMPANYNAME"** and enter the name of your company in this field.

Select the property link in the **Title** area and click the **Link To Property** button (arrow).

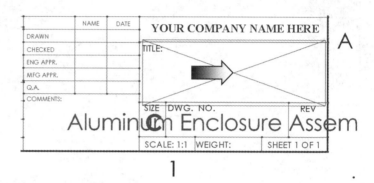

Use the default **Current Document** option (arrow).

Select the property **SW-File Name (File Name)** under the Property Name drop down list (arrow).

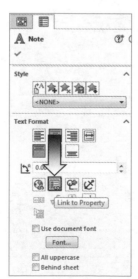
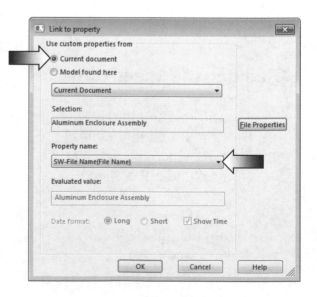

Click **OK**.

The property links to the name of the document and displays it in the title field.

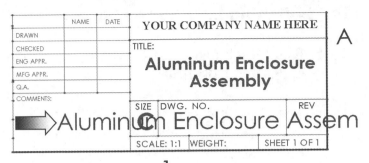

Select the **Drawing Number** (Aluminum Enclosure Assembly) and click the **Link To Property** button 📇.

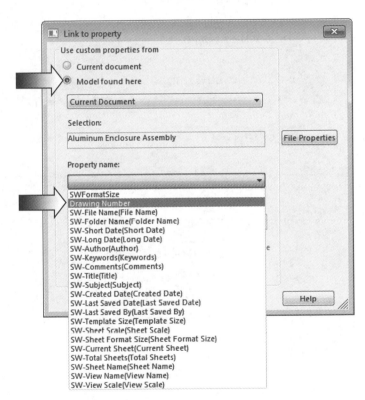

Select the **Model Found Here** option (arrow).

Under **Property Name** select **Drawing Number** from the drop down list (arrow).

Click **OK**.

The Drawing Number is linked to the property that was previously entered in the part and now is displayed in the field.

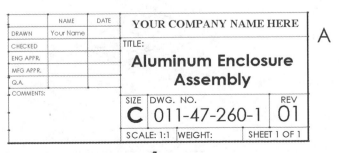

14. Switching back to the Sheet:

Right-click inside the drawing and select **Edit Sheet**.

The Sheet Layer is brought to the front and the Sheet Format layer is put to the back.

15. Saving your work:

Click **File / Save As**.

Use the default name **Aluminum Enclosure Assembly.slddrw**.

Press **Save**.

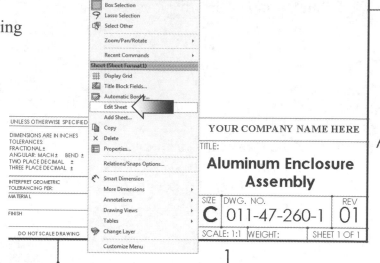

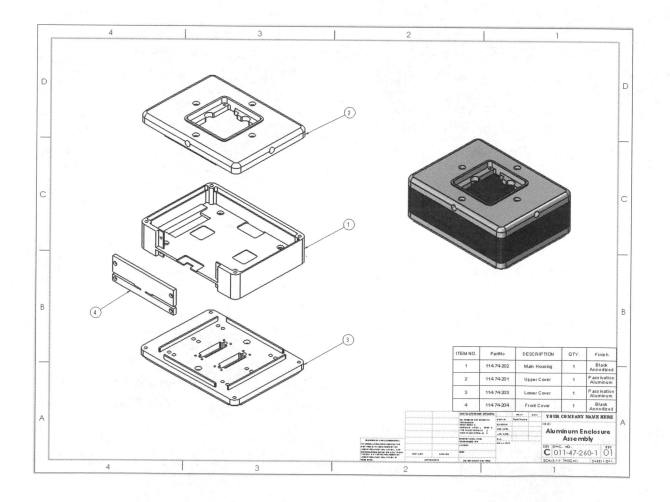

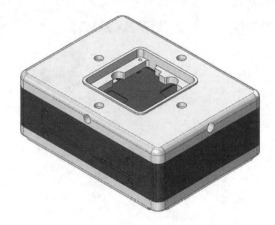

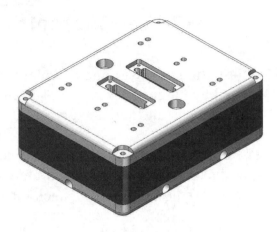

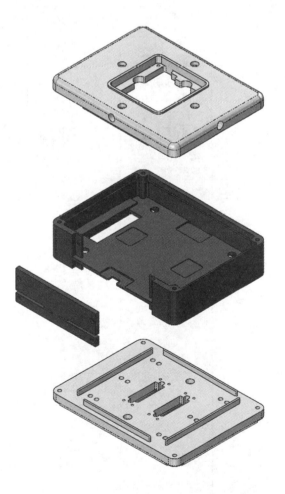

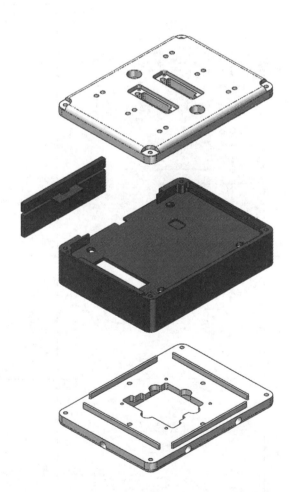

Reference Reading:
GD&T – Geometric Dimensioning & Tolerancing

Geometric Dimensioning & Tolerancing provide a vital link in the manufacturing field. It is used by engineers, designers, drafters, inspectors, machinists, and anyone involved in turning ideas into satisfactory products.

The engineer and designer can use geometric and position tolerances to communicate specifications clearly and accurately to the people who make the parts and to the inspector who will in turn check the parts against the designer's drawings.

When properly applied, the process of proper tolerancing will help eliminate or reduce problems which may occur later, and products can be produced more accurately.

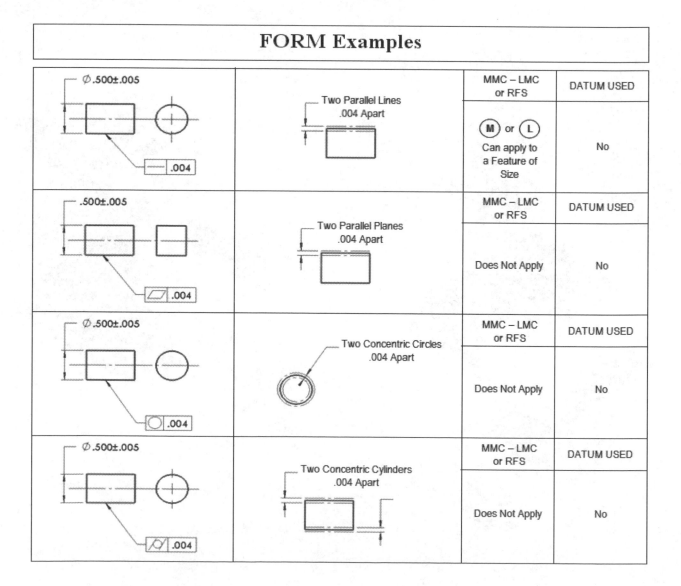

		MMC – LMC or RFS	DATUM USED
Ø.500±.005 / .004	Two Parallel Lines .004 Apart	Ⓜ or Ⓛ Can apply to a Feature of Size	No
.500±.005 / .004	Two Parallel Planes .004 Apart	Does Not Apply	No
Ø.500±.005 / .004	Two Concentric Circles .004 Apart	Does Not Apply	No
Ø.500±.005 / .004	Two Concentric Cylinders .004 Apart	Does Not Apply	No

FORM Examples

ORIENTATION Examples

		MMC – LMC or RFS	DATUM USED
//.004 A	Two Parallel Planes .004 Apart A	Ⓜ or Ⓛ Can apply to a Feature of Size	Yes
⟂.004 A A	90° A Two Parallel Planes .004 Apart	Ⓜ or Ⓛ Can apply to a Feature of Size	Yes
∠.004 A A 15°	Two Parallel Planes .004 Apart 15° A	Ⓜ or Ⓛ Can apply to a Feature of Size	Maybe Used or Not

Ⓐ .500±.005

PROFILE Examples

		MMC – LMC or RFS	DATUM USED
⌒.004 A	Two Lines .004 Apart Along True Profile	Does Not Apply	Maybe Used or Not
⌓.004 A A	Two Planes .004 Apart Along True Profile A	Does Not Apply	Maybe Used or Not

RUNOUT Examples

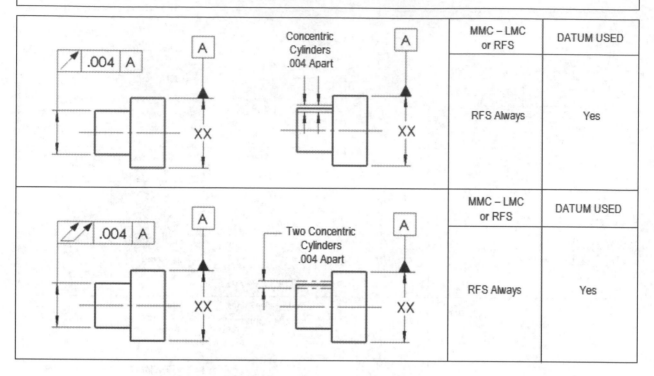

	MMC – LMC or RFS	DATUM USED
	RFS Always	Yes
	MMC – LMC or RFS	DATUM USED
	RFS Always	Yes

LOCATION Examples

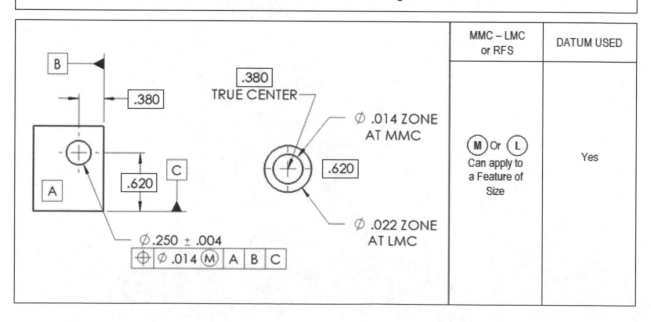

	MMC – LMC or RFS	DATUM USED
	M Or L Can apply to a Feature of Size	Yes

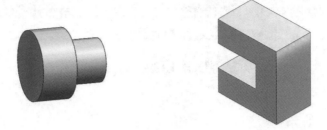

LOCATION Examples (cont.)

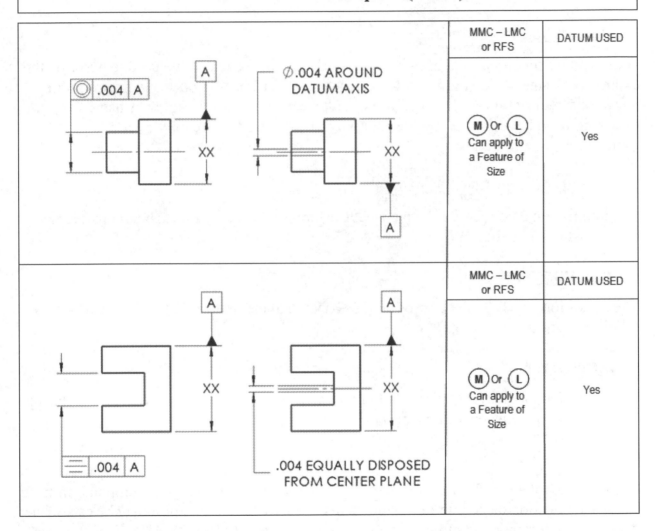

Geometric Dimensioning & Tolerancing play an important step in the advancement of design and production technology. It is a cost saver and an eliminator of confusion.

Geometric and Position Tolerances can be used at all levels. It is important to learn what they are, what they do, and what their applications are.

ASME (American Society for Mechanical Engineering)
ANSI Y14.5M (American National Standards Institute)
Symbol Descriptions

ANGULARITY:

The condition of a surface or line, which is at a specified angle (other than 90°) from the datum plane or axis.

BASIC DIMENSION: | 1.00 |

A dimension specified on a drawing as BASIC is a theoretical value used to describe the exact size, shape, or location of a feature. It is used as a basis from which permissible variations are established by tolerance on other dimensions or in notes. A basic dimension can be identified by the abbreviation BSC or more readily by boxing in the dimension.

CIRCULARITY (ROUNDNESS):

A tolerance zone bounded by two concentric circles within which each circular element of the surface must lie.

CONCENTRICITY:

The condition in which the axis of all cross-sectional elements of a feature's surface of revolution are common.

CYLINDRICITY:

The condition of a surface of revolution in which all points of the surface are equidistant from a common axis, or for a perfect cylinder.

DATUM:

A point, line, plane, cylinder, etc., assumed to be exact for purposes of computation from which the location or geometric relationship of other features of a part may be established. A datum identification symbol contains a letter (except I, O and Q) placed inside a rectangular box.

DATUM TARGET:

The datum target symbol is a circle divided into four quadrants. The letter placed in the upper left quadrant identifies its associated datum feature. The numeral placed in the lower right quadrant identifies the target; the dashed leader line indicates the target on far side.

FLATNESS: ▱

The condition of a surface having all elements in one plane. A flatness tolerance specifies a tolerance zone confined by two parallel planes within which the surface must lie.

MAXIMUM MATERIAL CONDITION: Ⓜ

The condition of a part feature when it contains the maximum amount of material.

LEAST MATERIAL CONDITION: Ⓛ

The condition of a part feature when it contains the least amount of material. The term is opposite from maximum material condition.

REGARDLESS OF FEATURE SIZE: RFS

This is the <u>default</u> condition for <u>all</u> Geometric Tolerance. This means that the feature's nominal center is to be inspected to the datum's nominal center, regardless of the size or shape of either the feature or datum.

PARALLELISM: //

The condition of a surface or axis which is equidistant at all points from a datum plane or axis.

PERPENDICULARITY: ⊥

The condition of a surface, line, or axis, which is at a right angle (90°) from a datum plane or datum axis.

PROFILE OF ANY LINE: ⌒

The condition limiting the amount of profile variation along a line element of a feature.

PROFILE OF ANY SURFACE: ⌓

Similar to profile of any line, but this condition relates to the entire surface.

PROJECTED TOLERANCE ZONE: Ⓟ

A zone applied to a hole in which a pin, stud, screw, etc. is to be inserted. It controls the perpendicularity of any hole, which controls the fastener's position; this will allow the adjoining parts to be assembled.

REGARDLESS OF FEATURE SIZE: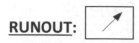

A condition in which the tolerance of form or condition must be met, regardless of where the feature is within its size tolerance.

RUNOUT:

The maximum permissible surface variation during one complete revolution of the part about the datum axis. This is usually detected with a dial indicator.

STRAIGHTNESS:

The condition in which a feature of a part must be a straight line.

SYMMETRY:

A condition wherein a part or feature has the same contour and sides of a central plane.

TRUE POSITION:

This term denotes the theoretical exact position of a feature.

FREE STATE: (F)

This symbol indicates the part must not be restricted during inspection.

References:

https://www.asme.org/products/codes-standards/
y14.5-2009-dimensioning-and-tolerancing

http://www.etinews.com/asme-y14.5-2009.html

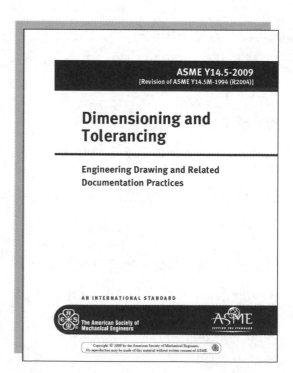

FEATURE CONTROL FRAMES

A feature control frame symbolizes the tolerance requirements for a feature of a part. It can be added to a drawing note for a feature tolerance or can be specified by running a leader line from the feature control frame directly to the feature.

The box may be attached to an extension line from the feature or it can be placed on a dimension line. A feature can have more than one feature control frame, depending on its requirements.

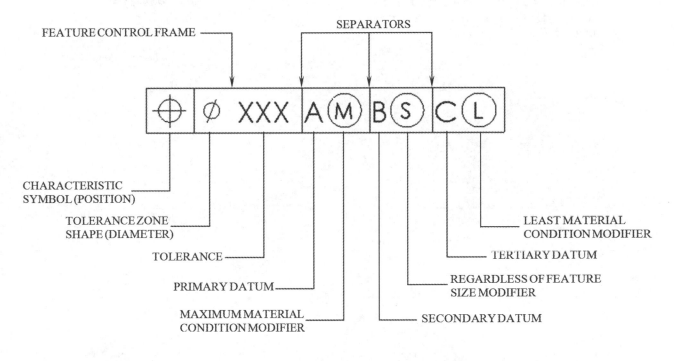

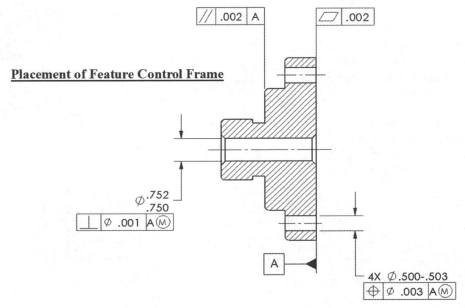

Placement of Feature Control Frame

Size at MMC – Maximum Material Condition

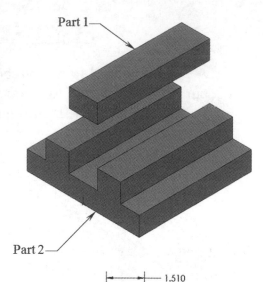

The 2 parts below are designed so that the Bar (part 1) will always fit into the slot of the Base (part 2), even when both are made at their most adverse condition of fit (MMC).

With the Bar (part 1), which is dimensioned at 1.500±.010, the MMC would be the largest within tolerance, or 1.510

With the Slot (part 2) at MMC, it would be 1.511, the smallest within tolerance.

It can safely be predicted that the Bar will always be assembled in the Slot, even if both parts are at their MMC.

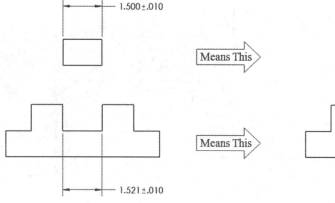

LMC – Least Material Condition

Least Material Condition is a provision that is the opposite of Maximum Material Condition. The abbreviation is LMC. Its symbol in a tolerance frame is an L within a circle Ⓛ

Least Material Condition Specifies that the tolerance Applies when feature of size is made at its largest for an internal feature – or – at its smallest for an external feature.

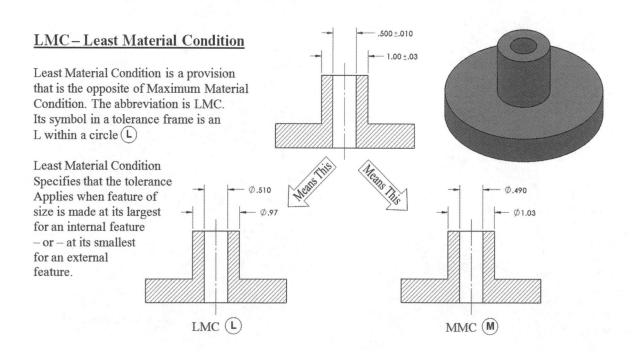

CHAPTER 16

Drawings & Detailing

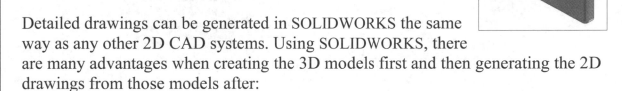

Drawings & Detailing Part-1
Front Cover Plate

Detailed drawings can be generated in SOLIDWORKS the same way as any other 2D CAD systems. Using SOLIDWORKS, there are many advantages when creating the 3D models first and then generating the 2D drawings from those models after:

* Design for the 3D models is faster than design in 2D layout.

* The drawings are created from the 3D models, so the process is more efficient.

* You can review models in 3D and check for correct geometry and design issues before generating the drawings, so the drawings are more likely to be free of design errors.

* You can insert dimensions and annotations from model sketches and features into the drawings automatically, so you do not have to create them manually in the drawings.

* Parameters and relations of models are retained in the drawings, so the drawings reflect the design intent of the model.

* Changes in models or in the drawings are reflected in their related documents, so making changes is easier, and the drawings are more accurate.

In most cases, the orthographic drawings views will be created first and the dependent views such as section views, detail views, auxiliary views, broken views, and crops views are created after.

Dimensions, datums, tolerances, notes, and other annotations are then added to specify the precision and the clarity of the features in each drawing's view.

Drawings & Detailing Part-1
Front Cover Plate

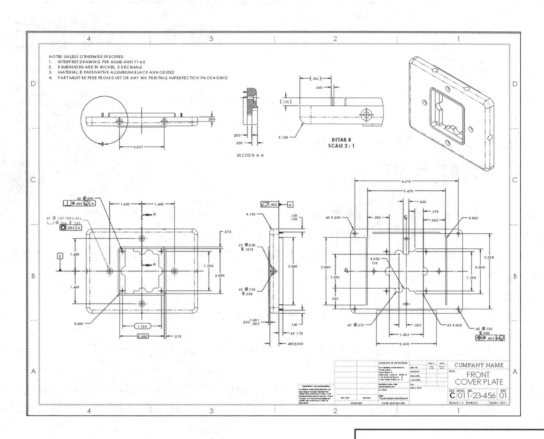

Dimensioning Standards: **ANSI**

Units: **INCHES** – 3 Decimals

Tools Needed:

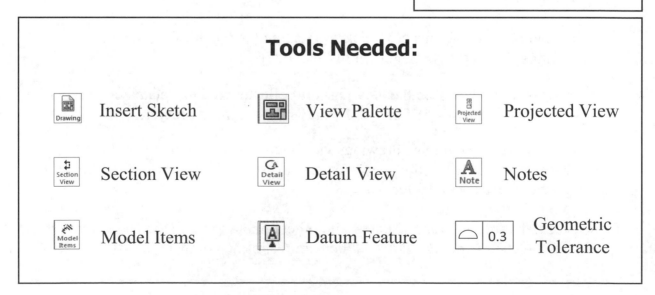

Insert Sketch

View Palette

Projected View

Section View

Detail View

Notes

Model Items

Datum Feature

Geometric Tolerance

1. Opening a part document:

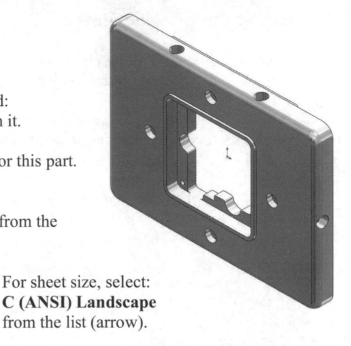

Select **File / Open**.

Browse to the part document named:
Front Cover Plate.sldprt and open it.

We will generate a detail drawing for this part.

Select **Make Drawing From Part** from the
File pull-down menu (arrow).

For sheet size, select:
C (ANSI) Landscape
from the list (arrow).

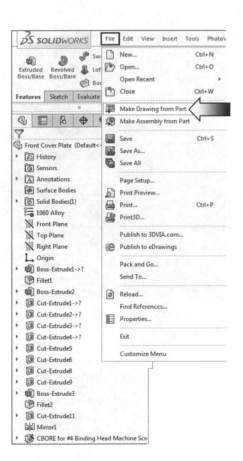

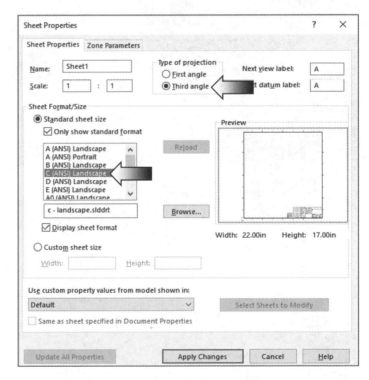

Set the following parameters:

      * Scale: **1:1**   * Type of Projection: **Third Angle**.
      * View & Datum Label: **A**   * Display Sheet Format: **Enabled**.

Click **OK**.

2. Creating the drawing views:

Expand the **View Palette** tab (arrow).
Enable only the **Auto Start Projected View** option, disable the others.

Use the View Palette to quickly insert one or more predefined views to the drawing. It contains images of standard views, annotation views, section views, and flat patterns (sheet metal parts) of the selected model.

The View Palette orientations are based on the ten standard orientations (Front, Right, Top, Back, Left, Bottom, Current, and Isometric) and any custom views in the part or assembly. After you place the views, you can fold or project views from it.

Drag and drop the **Front** view to the drawing approximately as shown.

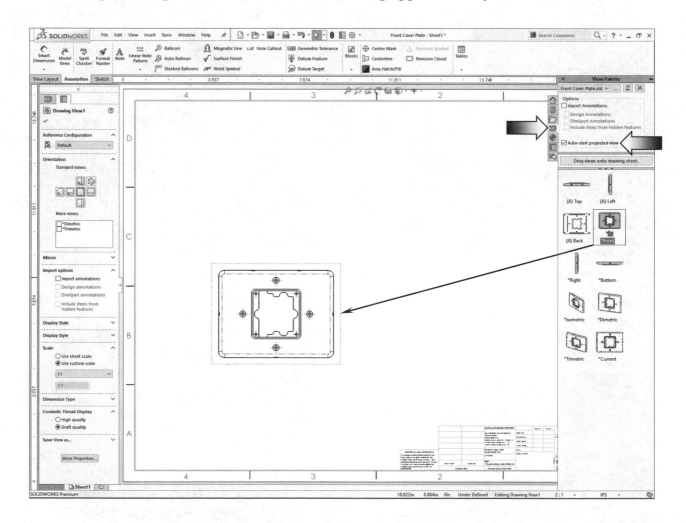

The Projected View command is activated automatically; move the mouse cursor up and down to see the previews of other drawing views.

Create the **Front** view, **Top** view, **Right** view, and **Isometric** view and place them approximately as shown below.

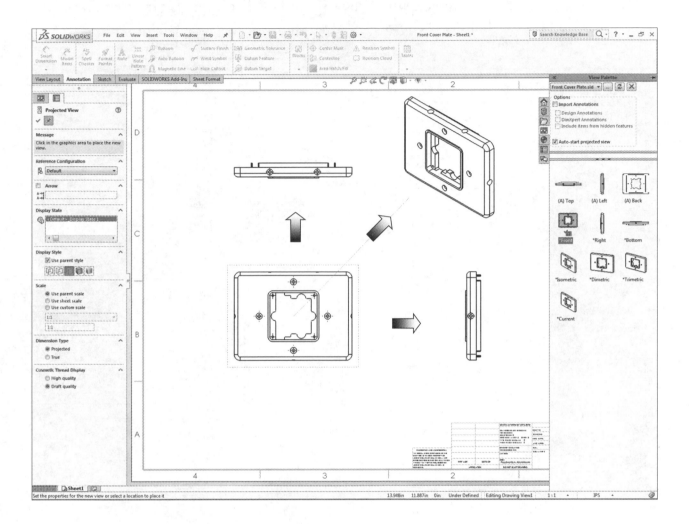

The Front, Top, and Right views are aligned along the vertical and horizontal directions by default.

The Top drawing view is not aligned with any view by default; it can be moved freely.

To break the alignment of any of the views, right-click the dotted drawing view's borders and select: **Alignment / Break Alignment**.

To re-align the views, right-click a drawing view's border and select: **Alignment / Default Alignment.**

Drawing views can also be aligned manually to one another by using their origins or by the horizontal and vertical directions.

3. Changing to Phantom line style:

The tangent edges in the drawing views should be displayed as Phantom lines instead of the default solid lines.

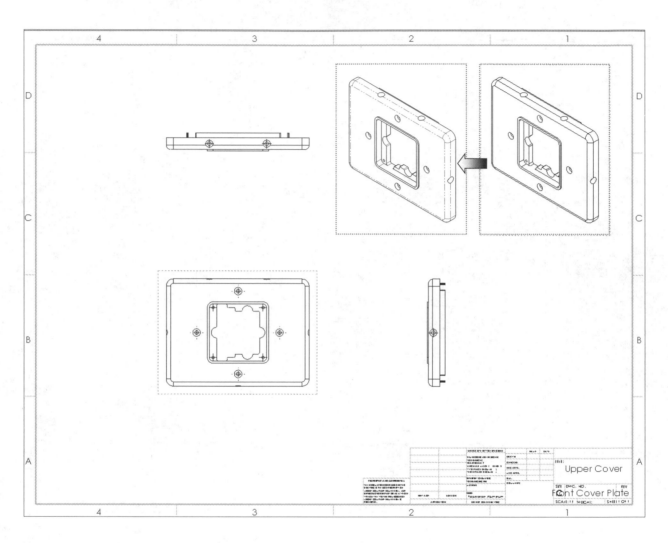

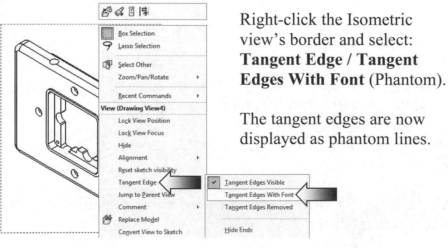

Right-click the Isometric view's border and select: **Tangent Edge / Tangent Edges With Font** (Phantom).

The tangent edges are now displayed as phantom lines.

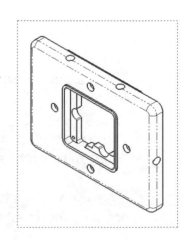

4. Creating a Partial Section view:

Switch to the **Sketch** tab and select the **Line** command. Zoom closer to the Top drawing view.

Sketch a **vertical line** from the center of the view, approximately as shown.

Section View

Create a section view to show the interior part of an object as if it were cut in half or quartered.

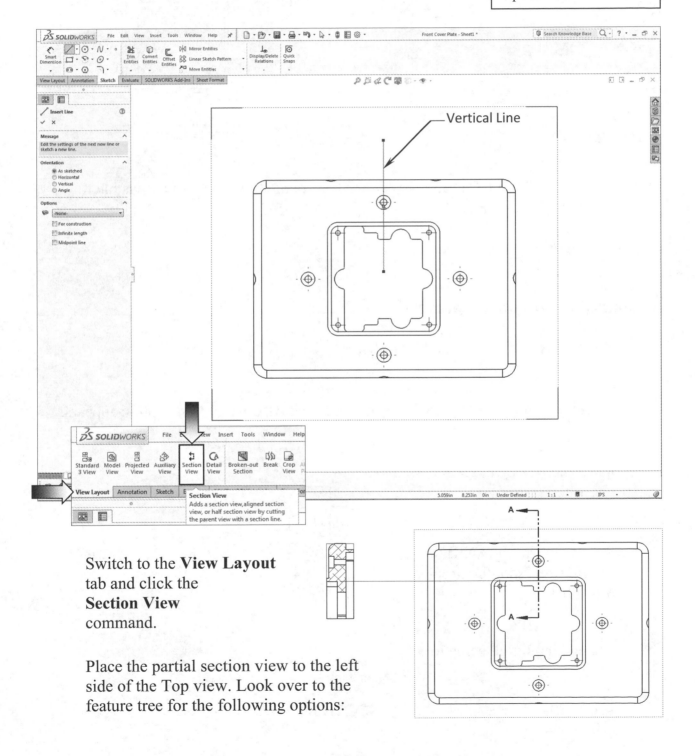

Switch to the **View Layout** tab and click the **Section View** command.

Place the partial section view to the left side of the Top view. Look over to the feature tree for the following options:

Click the **Flip Direction** button to flip the view and the section arrows.

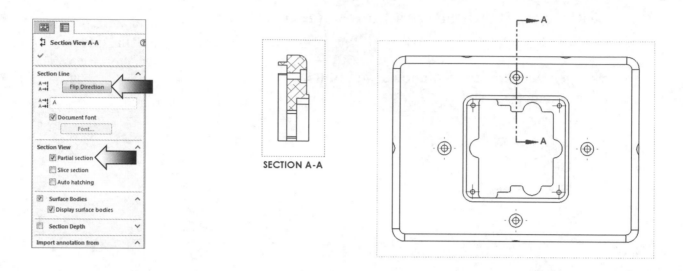

Enable the **Partial Section** checkbox. The counterbore hole is now visible for detailing.

Click **OK**.

5. Changing the hatch pattern:

Zoom closer to the partial section view.

Select <u>one of the sectioned faces</u> as noted. The Area Hatch/Fills option appears.

Clear the **Material Crosshatch** checkbox (arrow).

Select the pattern **ANSI38-Aluminum** from the drop-down list.

Change the **hatch scale** to **3**.

Click **OK**.

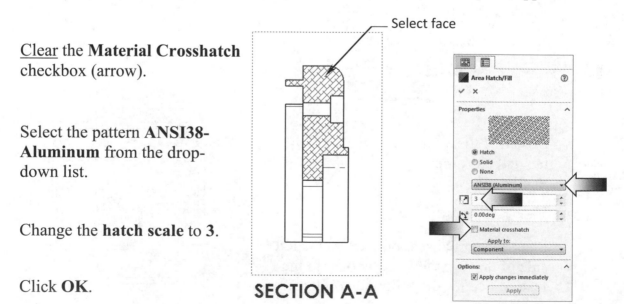

6. Creating a Detail view:

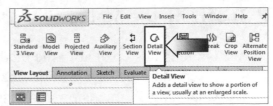

Select the **Detail View** command from the **View Layout** tab. The mouse cursor changes to the Circle command.

Zoom closer to the Right view.

Sketch a **circle** approximately as shown to capture the features on the left side of the drawing view.

> 💡 **Detail View**
>
> Create a detail view in a drawing to show a portion of a view, at an enlarged scale of 2X or more.

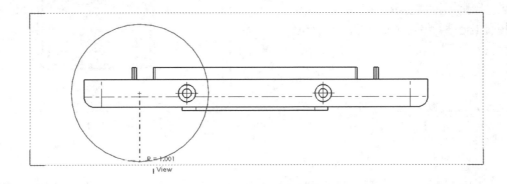

Place the detail view on the right side as shown below.

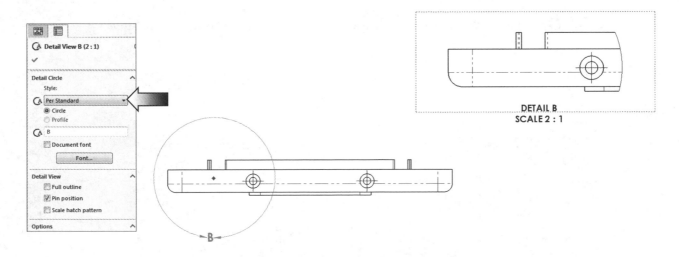

DETAIL B
SCALE 2 : 1

The default style for the detail circle has been preset to **Per Standard**.

Switch to different styles available. Change back to Per Standard when finished.

7. Inserting the model dimensions:

Dimensions created in the model get inserted to the various drawing views. Changing a dimension in the model updates the drawing, and changing an inserted dimension in a drawing changes the model.

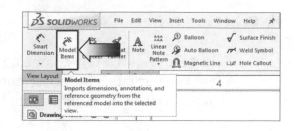

Zoom to the **Top** drawing view and click its dotted border to activate it.

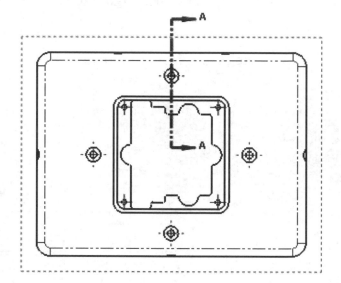

Switch to the **Annotation** tab and select the **Model Item** command (arrow).

For Source, select the option: **Entire Model**.

For Dimensions, select **Marked for Drawing** (default), **Hole Wizard Locations**, and **Hole Callouts**. Leave the other options at their default settings.

Click **OK**. The dimensions are imported but they need to be rearranged.

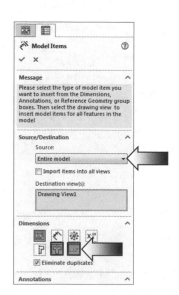

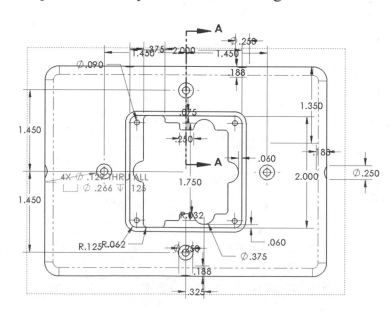

8. Cleaning up the dimensions:

Keep only the dimensions shown in the image below and delete the others.

Add the missing dimensions (circled) by using the Smart Dimension command.

The new dimensions added to the drawing will appear in **gray** color. They are called reference dimensions.

Add the number of instances in front of the dimension **Ø.090** (arrow).

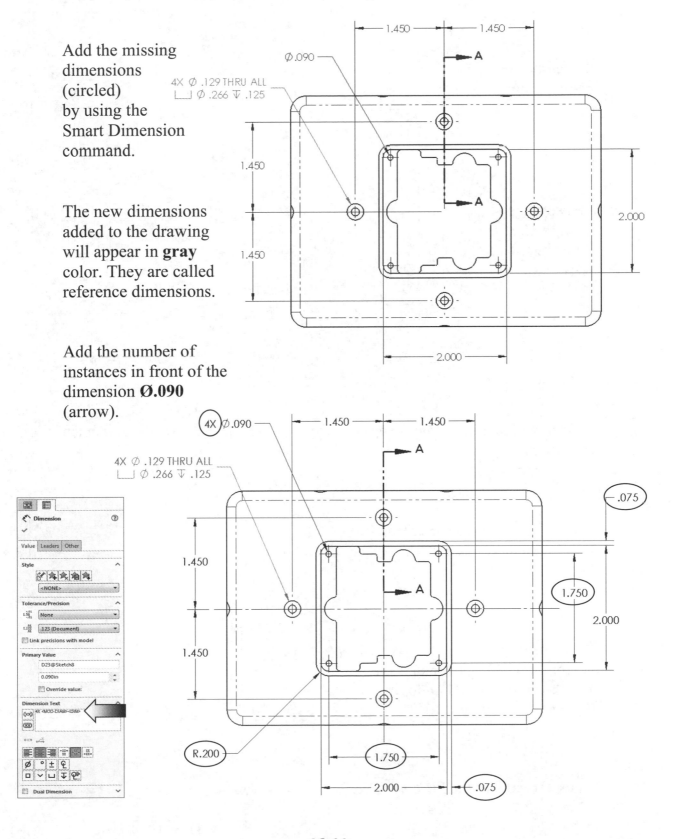

9. Inserting dimensions to another view:

Click the dotted border of the **Right** view to activate it.

Select the **Model Items** command once again.

The settings from the last time are still selected.

Click **OK**.

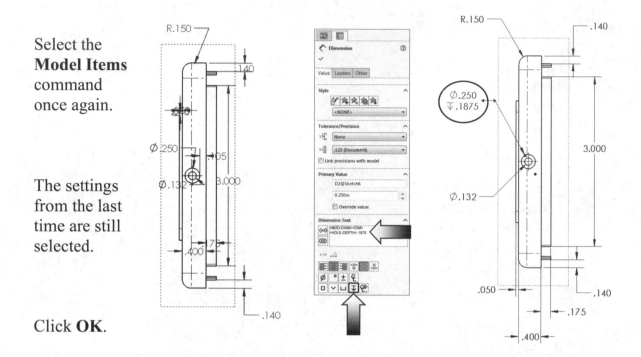

Rearrange the dimensions, keep only the ones shown in the image and delete the others.

Select the dimension **Ø.250** and add the **Depth Symbol** and the **Depth Dimension** from the PropertyManager tree (arrow).

Select the **Ø.132** dimension and add the word **THRU** under it (arrow).

Click **OK** to exit the Property tree.

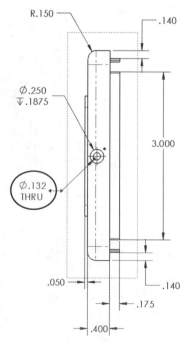

10. Breaking the view alignment:

Right-click the Partial Section
view's border and select
Break Alignment (arrow).

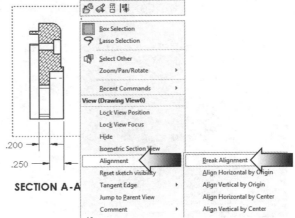

SECTION A-A

Move the partial section view by
dragging its dotted border to create
some space and add the 2 dimensions shown.

Select the **Model Items** command.

Insert the model dimensions into
the **Top** view.

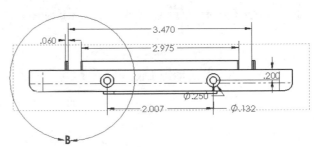

Keep only the holes' location dimensions
and delete the others.

The dimensions that
were deleted earlier
will get inserted
into other views
later on.

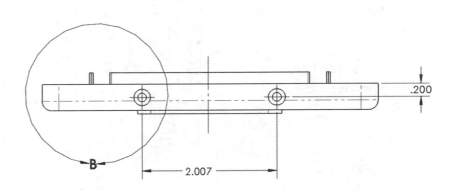

Insert the model dimensions into the **Detail View B**.

Add the reference dimensions and click the **Add Parenthesis** for each one (arrow).

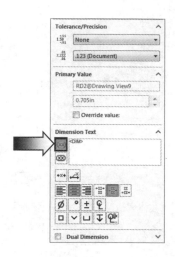

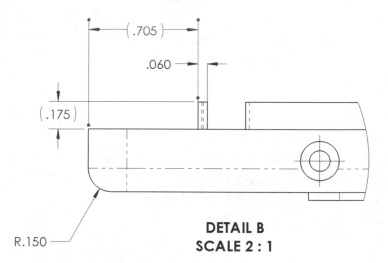

DETAIL B
SCALE 2 : 1

11. Creating a Projected View:

A Back view is needed to detail the features on the far side. There are several ways to add this view but we will try the Projected View command instead.

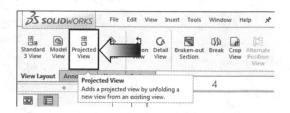

Switch back to the **View Layout** tab and click the **Projected View** command (arrow).

Move the cursor to the right to see the preview graphic of the view in the back.

Place the back view approximately as shown.

Click **OK**.

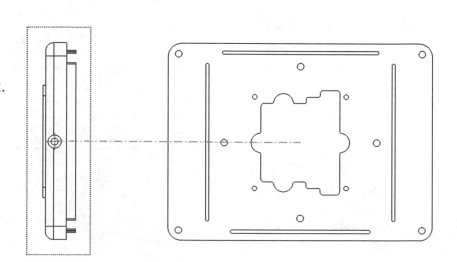

Insert the model dimensions into the back view and add the dimensions shown.

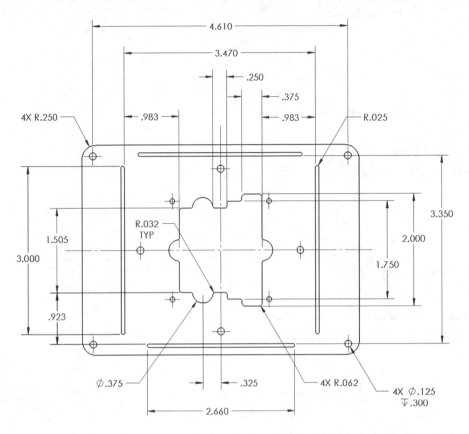

12. Adding the Centerline Symbol:

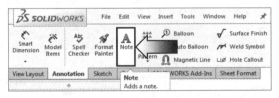

A library of symbols such as degree, depth, countersink, and so on is available. In the Dimension Value PropertyManager, click More Symbols under Dimension Text to access the library. Symbols libraries for various annotations, such as Notes, Geometric Tolerance Symbols, Surface Finish Symbols, Weld Symbols, and so on, are also available in the PropertyManager.

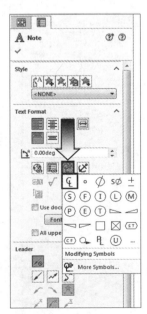

Switch to the **Annotation** tab.

Select the **Note** command; click the **Add Symbol** button (arrow).

Select the **Centerline** symbol ₵ (arrow).

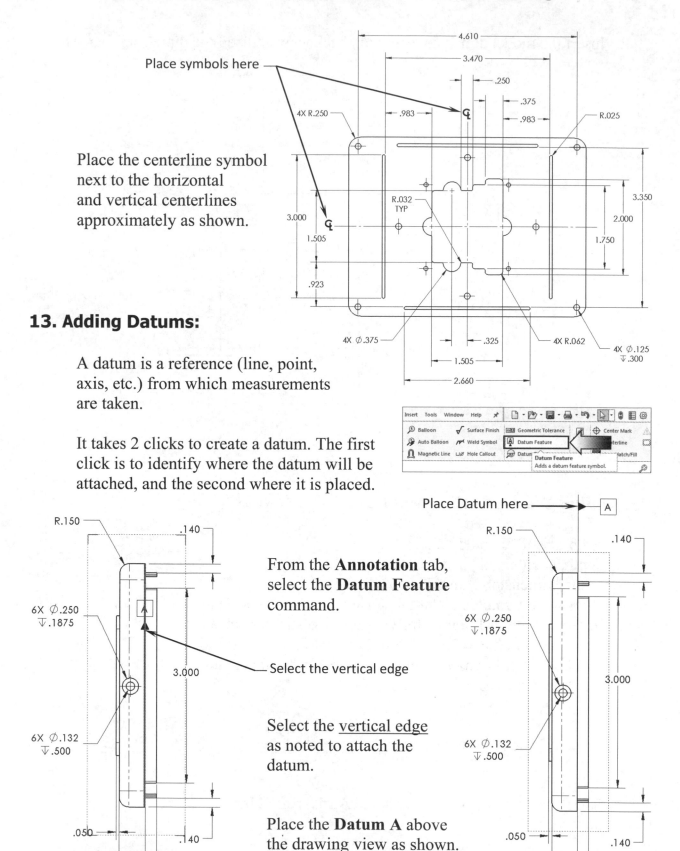

Place symbols here

Place the centerline symbol next to the horizontal and vertical centerlines approximately as shown.

13. Adding Datums:

A datum is a reference (line, point, axis, etc.) from which measurements are taken.

It takes 2 clicks to create a datum. The first click is to identify where the datum will be attached, and the second where it is placed.

Place Datum here ———→ A

From the **Annotation** tab, select the **Datum Feature** command.

Select the vertical edge

Select the <u>vertical edge</u> as noted to attach the datum.

Place the **Datum A** above the drawing view as shown. The datum can be placed on the left or right side of the line.

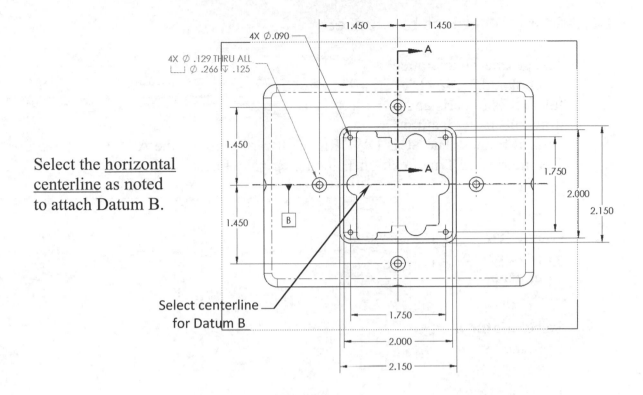

Select the <u>horizontal centerline</u> as noted to attach Datum B.

Move the cursor outward and place **Datum B** approximately as shown below.

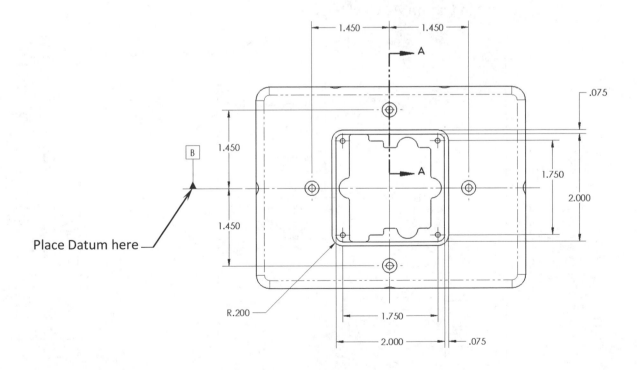

Click **OK** to exit the Datum Feature command.

14. Adding a Flatness tolerance:

The geometric tolerance symbol adds geometric tolerances to parts and drawings using feature control frames. The SOLIDWORKS software supports the ASME Y14.5-2009 Geometric and True Position Tolerancing guidelines.

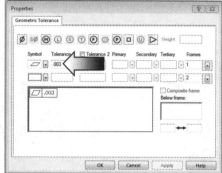

Click the **Geometric Tolerance** Command.

Select **Flatness** from the list.

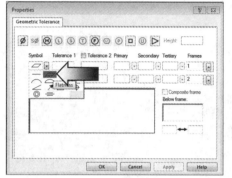

Enter **.003** under Tolerance 1.

Select the <u>vertical edge</u> as noted to attach the tolerance, and then move upward to align it next to Datum A. Click **OK**.

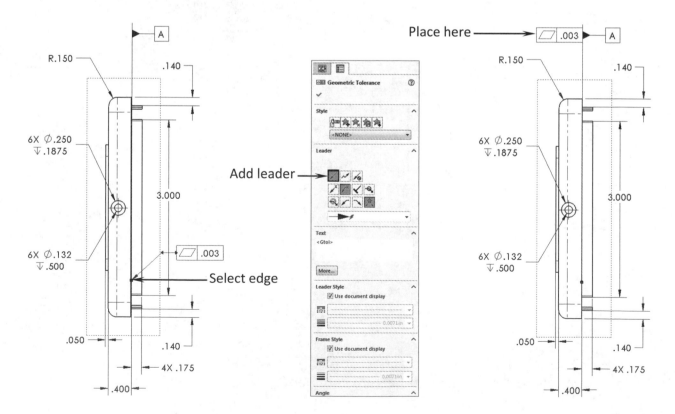

15. Adding a Perpendicular tolerance:

Perpendicularity in GD&T can mean two different things depending which reference feature is called out.

The normal form or **Surface Perpendicularity** is a tolerance that controls Perpendicularity between two 90° surfaces, or features. Surface Perpendicularity is controlled with two parallel planes acting as its tolerance zone.

Axis Perpendicularity is a tolerance that controls how perpendicular a specific axis needs to be to a datum. Axis Perpendicularity is controlled by a cylinder around a theoretically perfect parallel axis.

Select the diameter dimension of the hole (circled).

Click the **Geometric Tolerance** command.

Select the **Perpendicular** symbol from the list.

Enter **.003** under Tolerance 1.

Click the **Maximum Material Condition** symbol Ⓜ.

Enter **A** for Primary Datum.

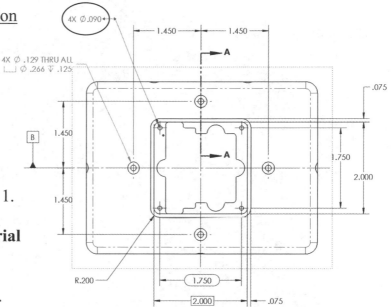

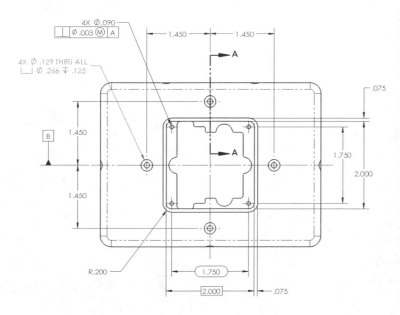

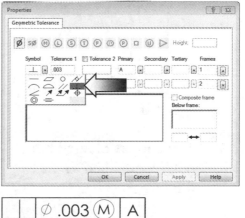

The Perpendicular control frame.

Click **OK**.

16. Copying the control frame:

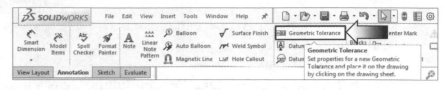

True position in terms of the axis, point, or plane defines how much variation a feature can have from a specified exact true location.

This step demonstrates how to make a copy of a geometric tolerance and attach it to another dimension.

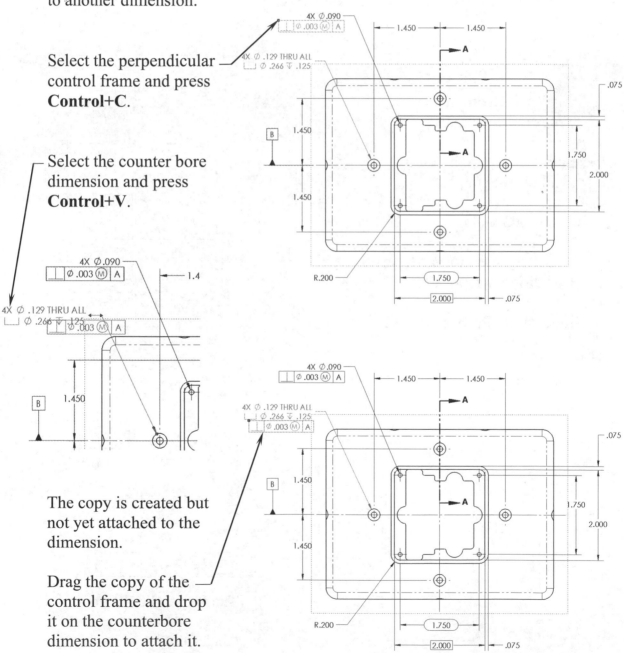

Select the perpendicular control frame and press **Control+C**.

Select the counter bore dimension and press **Control+V**.

The copy is created but not yet attached to the dimension.

Drag the copy of the control frame and drop it on the counterbore dimension to attach it.

17. Modifying the geometric tolerance:

For the counter bore feature we will change the symbol to Position, or also called True Position.

True position in terms of the axis, point, or plane defines how much variation a feature can have from a specified exact true location.

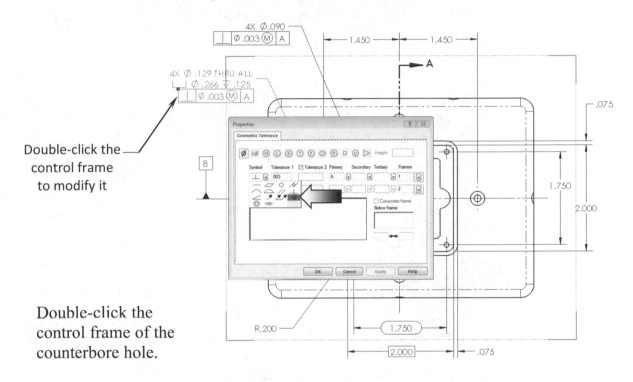

Double-click the
control frame
to modify it

Double-click the
control frame of the
counterbore hole.

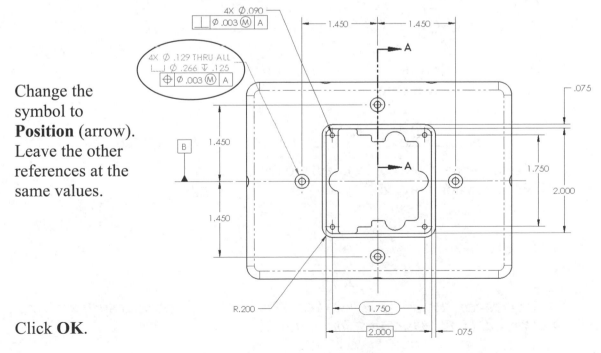

Change the
symbol to
Position (arrow).
Leave the other
references at the
same values.

Click **OK**.

18. Adding another Position tolerance:

Zoom in on the **Back** view and locate the small hole on the lower right corner.

Select the dimension **4X Ø.125** and click the **Geometric Tolerance** command.

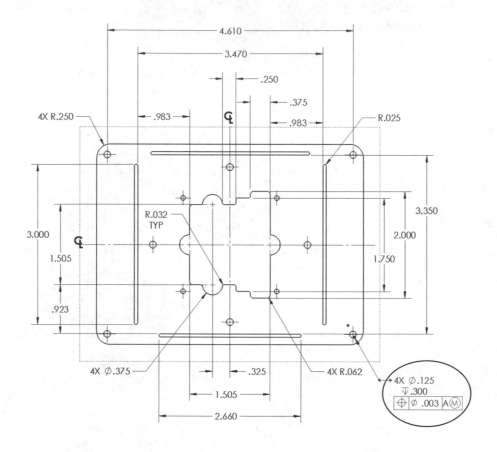

Create a True Position tolerance that looks like the one shown below.

Be sure to attach the control frame to the hole dimension (circled).

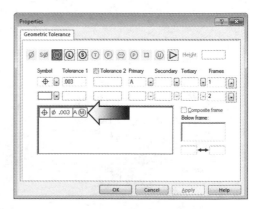

Click **OK**.

For practice purposes, we will create some other tolerances such as Symmetric, Bilateral, Limit, Basic, etc., in the next few steps.

19. Adding a Symmetric tolerance:

Zoom in on the Right view and select the **.400** dimension on the lower right side.

Expand the **Tolerance/ Precision** section and select the **Symmetric** option.

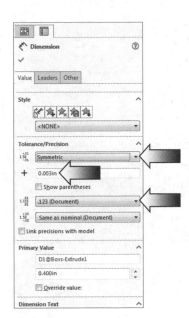

Enter **.003in** for Max-Variation.

Change the number of decimals to **3** (arrow).

Using Symmetric, only one dimension value is used (circled).

Click **OK**.

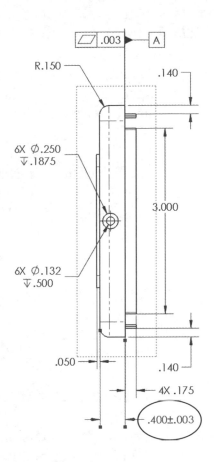

20. Adding a Bilateral tolerance:

Select the **.050** dimension in the same drawing view.

Select the **Bilateral** option under the **Tolerance/Precision** section.

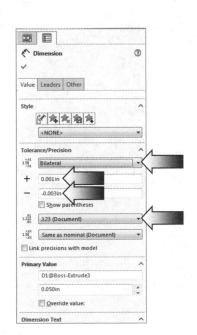

Enter **.001in** for Max-Variation

Enter **.003** for Min-Variation.

Set the number of decimals to **3 places**. Using Bilateral 2 different values are used (circled).

Click **OK**.

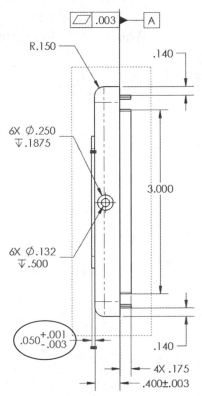

21. Adding a Limit tolerance:

Use Limit Tolerance to specify a range or the total amount a dimension can vary.

Click the dimension **.140** and select the **Limit** option.

Enter **.003** for Max Variation.

Enter **.002** for Min Variation.

Set the number of decimals to **3 places**.

Click **OK**.

22. Adding a Basic tolerance:

The Basic Dimension has a box around the dimension.
A basic dimension is a numerical value used to describe the theoretically exact size, profile, orientation, or location of a feature.

Add a **Basic** tolerance to the dimension **2.000** (circled).

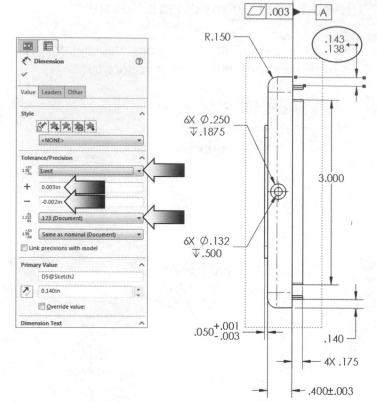

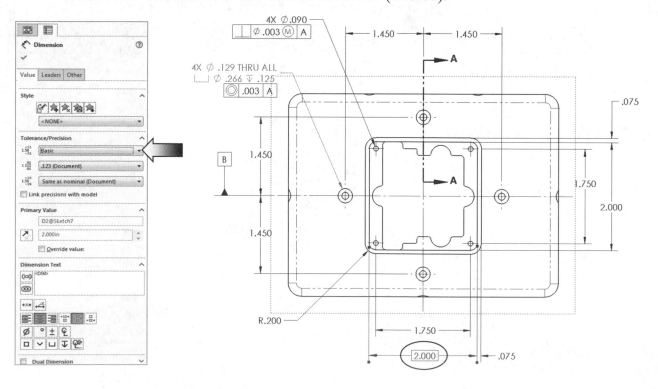

23. Adding General Notes:

Zoom to the upper left area of the drawing.

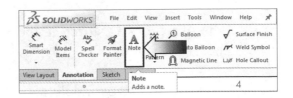

Switch to the Annotation tab and select the **Note** command (arrow).

Before typing, place the note box on the upper left corner of the drawing (arrow).

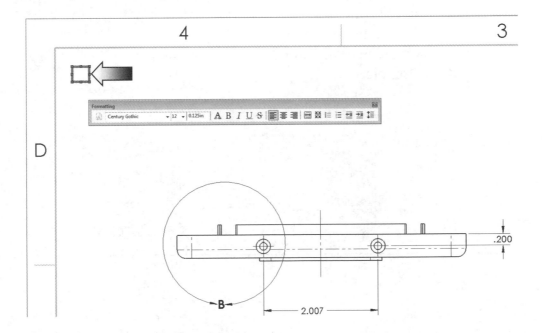

Enter the notes shown below.

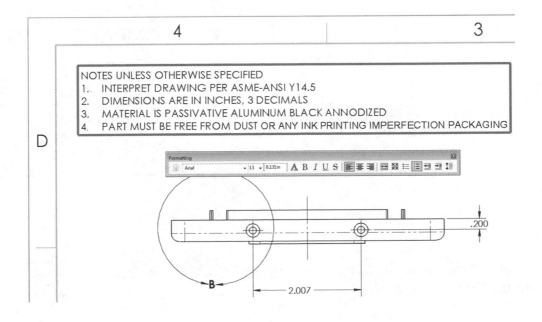

NOTES UNLESS OTHERWISE SPECIFIED
1.   INTERPRET DRAWING PER ASME-ANSI Y14.5
2.   DIMENSIONS ARE IN INCHES, 3 DECIMALS
3.   MATERIAL IS PASSIVATIVE ALUMINUM BLACK ANNODIZED
4.   PART MUST BE FREE FROM DUST OR ANY INK PRINTING IMPERFECTION PACKAGING

24. Filling out the Title Block:

The title block is stored in the Sheet Format, which is behind the Sheet layer.

You can add, move, delete, or modify any lines or text in the title block.

In the default drawing template, there is a "blank note" already placed in the center of each field. These blank notes, or any new notes, can be linked to any custom properties such as Company Name, Part Number, Description, Weight, Material, Revision, etc.

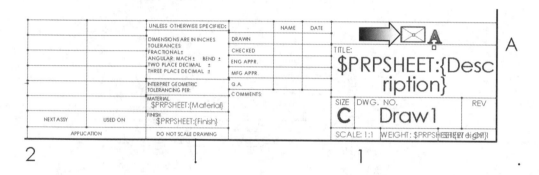

Hover the mouse cursor over the center of the Company Name field (arrow), and double-click the blank note when the note symbol appears.

Enter the name of your company or school here.

Enter the Title of the drawing, the Drawing Number, the Revision, as well as Your Name and Date in the fields as indicated.

Right-click inside the drawing and select **Edit Sheet** to return to the drawing (arrow).

25. Saving your work:

Select **File Save As**.

Enter **Front Cover Plate.slddrw** for the name of the drawing.

Click **Save**.

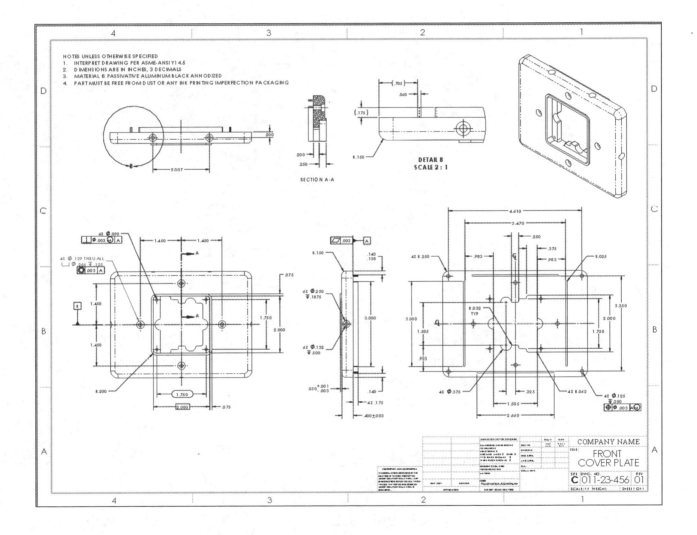

OPTIONAL:

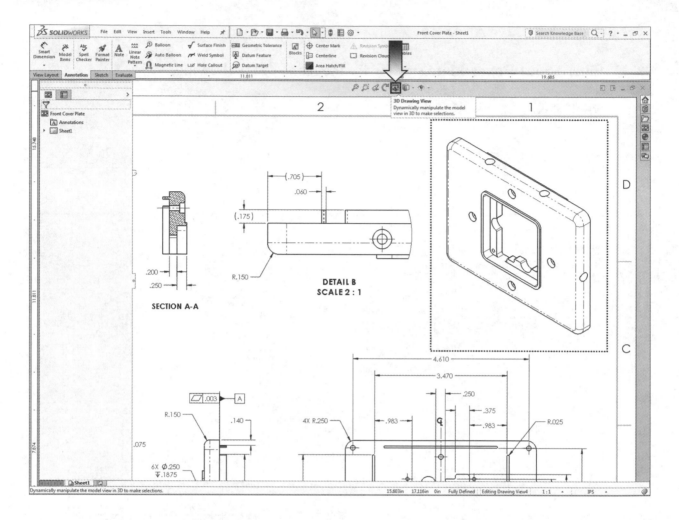

The 3D Drawing View mode lets you rotate a drawing view out of its plane so you can see the edges obscured by other entities. It is used to make geometry selection in a drawing view easier.

Click the **3D Drawing View** command from the View (Heads Up) tool bar.

The Rotate command is the default tool. Click and drag the left mouse button to rotate the view to different orientations.

The new view can be saved for future use.

Press **Esc** when finished.

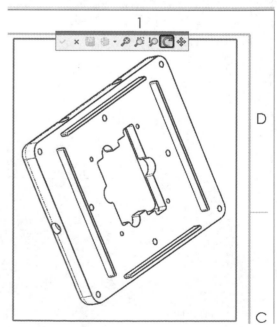

<u>Exercise:</u> Creating a drawing from a model

1. Open the model named **Elbow_Single Flange** and create the drawing shown below.
2. Dimensions are in inches, 3 decimal places.
3. Insert the dimensions from the model and add annotations where needed.

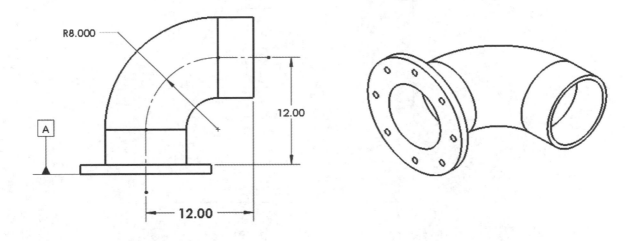

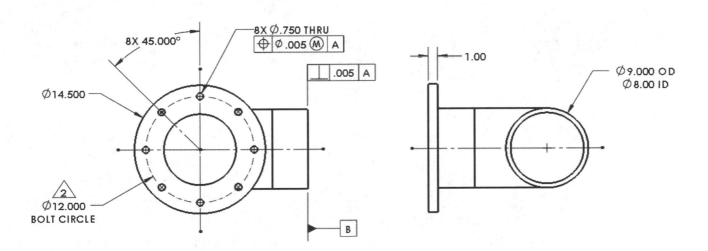

4. Save the drawing as **Elbow_Single Flange.slddrw**.

<u>Exercise</u>: Attaching a note or symbol to a dimension

1. Opening a drawing:

Open the drawing document named **Attaching Symbol to Dim.slddrw**.

2. Adding symbol to a dimension:

Select the **5.00** dimension (circled).

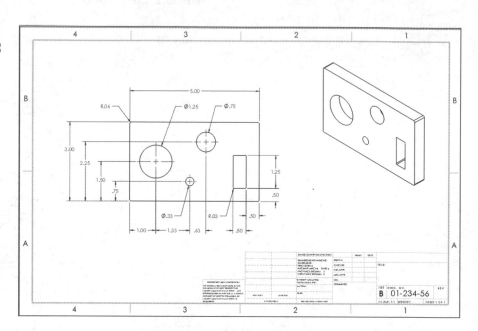

Expand the **Dimension Text** section on the property tree.

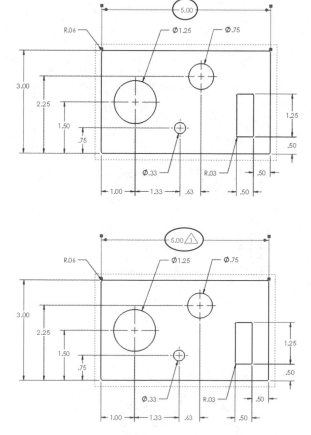

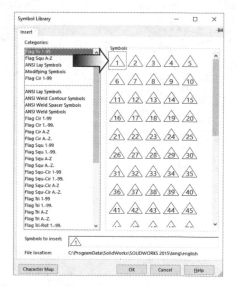

Click the **More Symbols** button (arrow).

From the Flag Tri 1-99 category, select: **Triangular 1** symbol (arrow).

Click **OK**.

Drawings & Detailing Part-2
Mounting Bracket

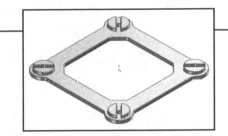

Drawings & Detailing Part-2
Mounting Bracket

When creating an engineering drawing, one of the first things to do is to create the drawing views that clearly show the details of the design.

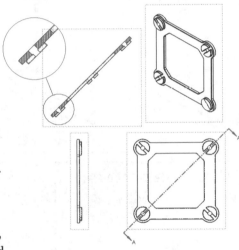

The standard drawing views such as the Front, Top, Right, and Isometric Views are usually created first, then the Section view, Detail View, Broken-Out Section views are created after to display the details that were not clear in the other drawing views.

After the drawing views are laid out, dimensions, annotations, tolerances, geometric tolerances, and notes are added to precisely control the shapes, sizes, and locations of the features.

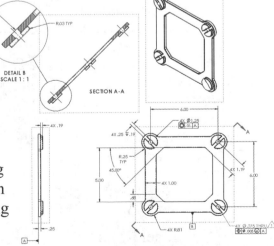

A lot of time, when modeling a part, it may seem more convenient to use fewer dimensions and more relations to constrain the geometry, but when a drawing is made from the part, those dimensions will not be available. In cases like this, reference dimensions will be used to fill in the missing dimensions. So if we have to decide between using a dimension or a relation when creating the model, go with the dimension instead.

Drawings & Detailing Part-2
Mounting Bracket

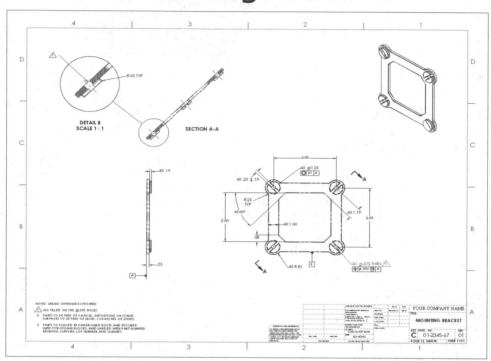

View Orientation Hot Keys:

Ctrl + 1 = Front View
Ctrl + 2 = Back View
Ctrl + 3 = Left View
Ctrl + 4 = Right View
Ctrl + 5 = Top View
Ctrl + 6 = Bottom View
Ctrl + 7 = Isometric View
Ctrl + 8 = Normal To Selection

Dimensioning Standards: **ANSI**
Units: **INCHES** – 3 Decimals

Tools Needed:

Section View	Area Hatch/Fill	Centerline
Detail View	Hole Callout	Geometric Tolerance
Model Items	Datum Feature	Note

1. Opening a part document:

Click **File / Open**.

Browse to the Training
Folder and open the part
document named:
Mounting Bracket.sldprt

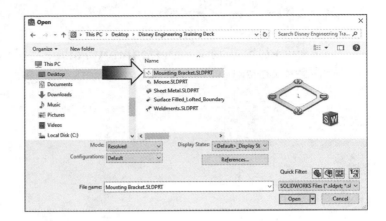

2. Making drawing from part:

Select **File / Make Drawing from Part** (arrow).

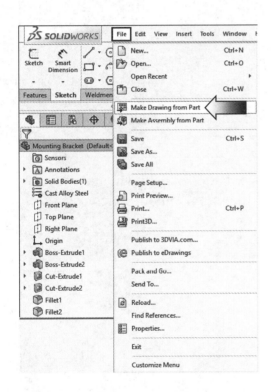

In the Sheet Properties, select the following:

* Scale: **1:2**

* Type of Projection: **Third Angle**

* Sheet Size: **C (ANSI) Landscape**

* Display Sheet Format: **Enabled**

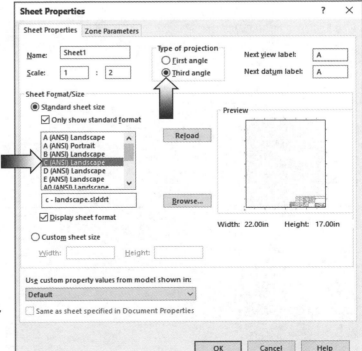

Click **OK**.

NOTE: _If you have your company's
format already created and
want to use it with this lesson.
Click the Browse button,
locate it and open it._

3. Using the View Palette:

Expand the **View Palette** (arrow) and enable only the **Auto-Start Projected View** checkbox and clear all others (for this lesson only).

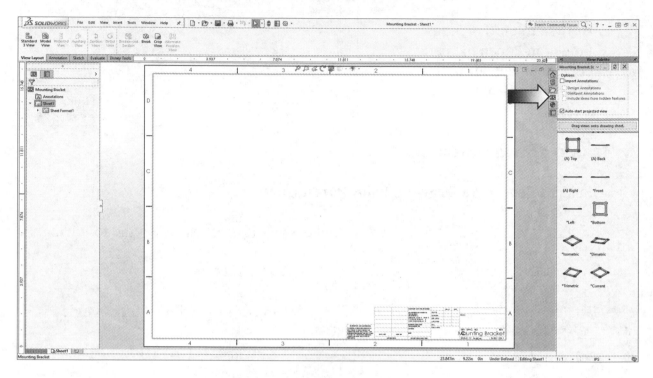

Drag the **Top** view and drop on the drawing approximately as shown, and move the cursor to the left and place the side view, and then the Isometric view.

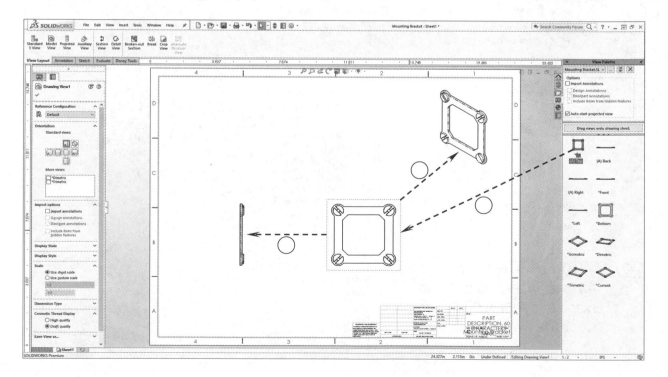

4. Changing the Line Style:

The default line style for all object outlines is solid line, which sometimes is a little confusing when the tangent edges are too close to the object lines.

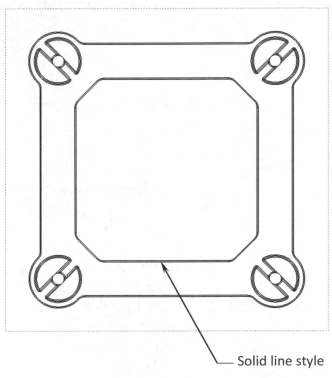

For the purpose of this exercise, we will change the tangent edges to phantom line style (with Font), to differentiate the object lines from the edges of the fillets.

Right-click on the dotted border of the top view and select: **Tangent Edge**, **Tangent Edges With Font** (arrow).

Solid line style

Phantom line style

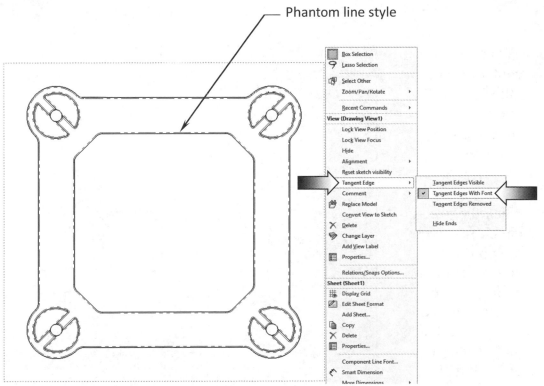

5. Creating a Section View:

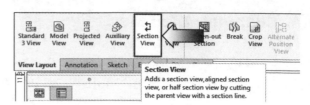

Click the **View Layout** tool tab.

Select the **Section View** command ⤢ .

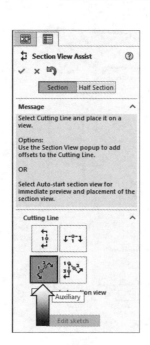

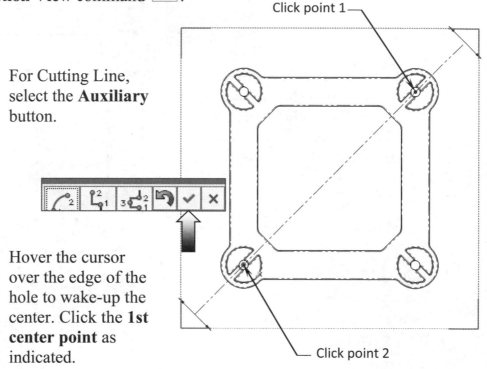

For Cutting Line, select the **Auxiliary** button.

Hover the cursor over the edge of the hole to wake-up the center. Click the **1st center point** as indicated.

Click point 1

Click point 2

"Wake-up" the center of the other hole and click the **2nd center point** as noted.

Click the **green check mark** on the pop-up confirmation dialog box to accept the line placement. Place the view on the upper left side.

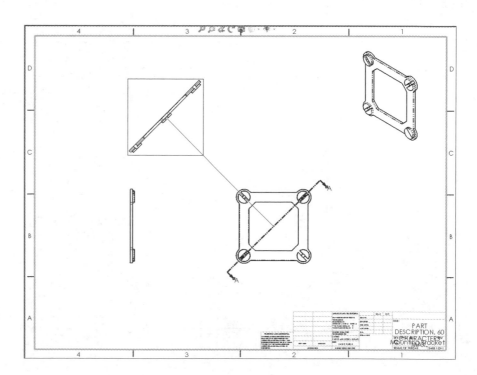

6. Modifying the crosshatch properties:

The default cross hatch style is quite confusing, they are not only parallel to the object outlines but also too far apart.

We will change it to make the cutting surfaces more visible, even when viewing it at a distance.

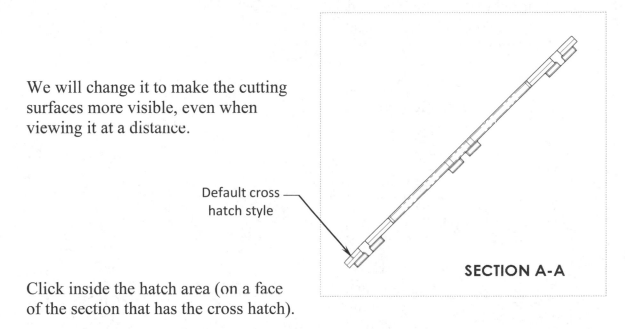

Default cross hatch style

SECTION A-A

Click inside the hatch area (on a face of the section that has the cross hatch).

The **Area Hatch/Fill** properties appear on the Properties tree.

ANIS31 cross hatch style

Change the Hatch-Pattern to **ANSI31** (ANSI BrickStone).

Change the Scale to **4** and the Angle to **45**.

Click **OK**.

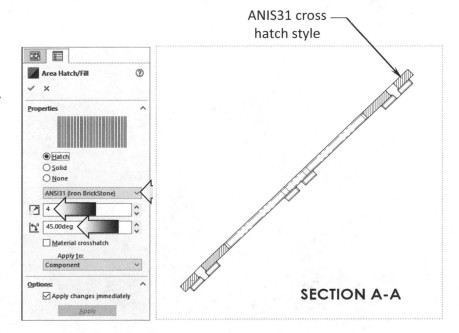

SECTION A-A

7. Adding a Detail View:

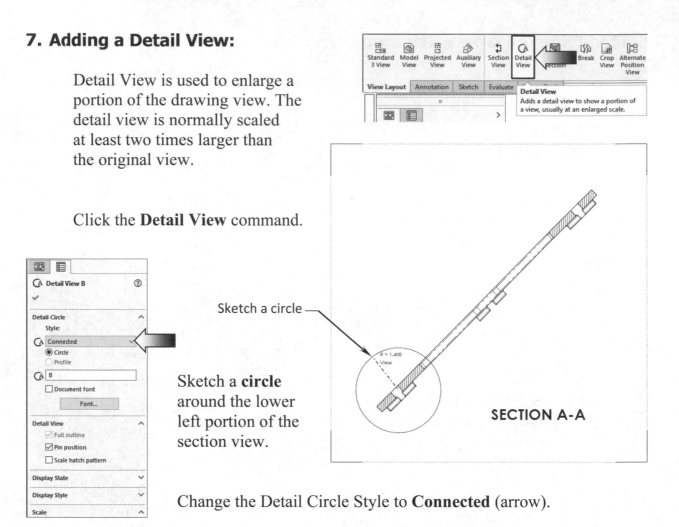

Detail View is used to enlarge a portion of the drawing view. The detail view is normally scaled at least two times larger than the original view.

Click the **Detail View** command.

Sketch a circle

Sketch a **circle** around the lower left portion of the section view.

SECTION A-A

Change the Detail Circle Style to **Connected** (arrow).

Click **OK**.

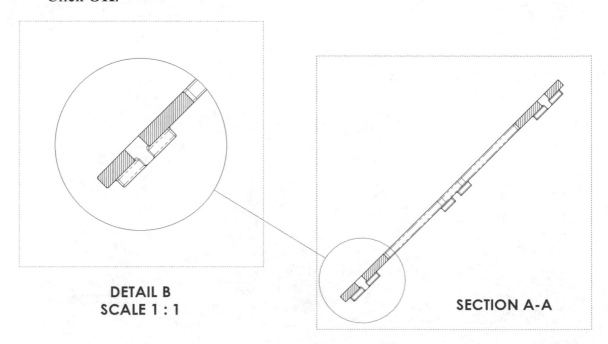

DETAIL B
SCALE 1 : 1

SECTION A-A

8. Modifying the Section Line Style:

Sometimes the thick section line can make other smaller features hard to see. To over come this, we will change the section line to an Alternative Without Connector style.

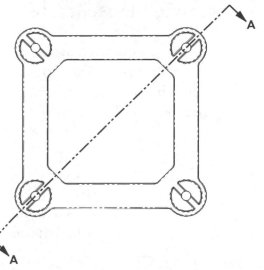

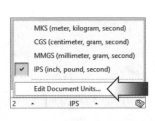

Select **Edit-Document Units** at the lower right corner of the screen.

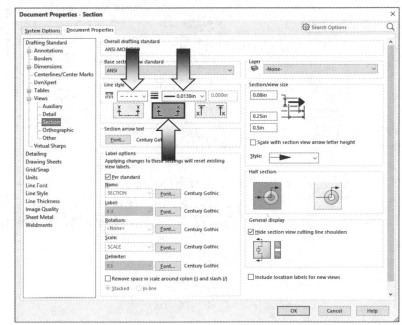

Expand the **Views** options and click **Section**.

Select the following:

    * Line Style: **Phantom**

    * Line Thickness: **0.0138in**.

    * Section Arrows: **Alternative Without Connector**.

Click **OK**.

The center portion of the Section Line is removed; only the 2 ends are visible.

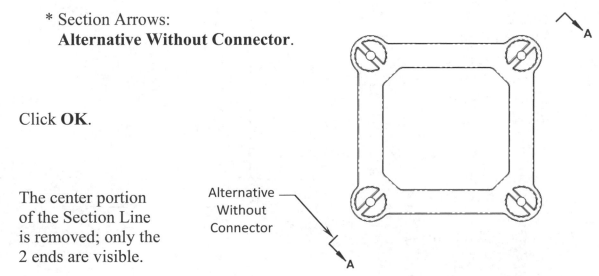

Alternative Without Connector

9. Inserting the Model Dimensions:

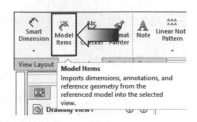

Dimensions are usually specified when designing a part, then they are exported from the model into the drawing views. Changing a dimension in one document changes it in any associated documents.

Switch to the **Annotation** tab and select the **Model Items** command.

Change the **Source** to **Entire Model**.

For Dimensions, enable: **Marked** and **Not Marked for Drawing**, **Hole Wizard Locations**, and **Hole Callout options**.

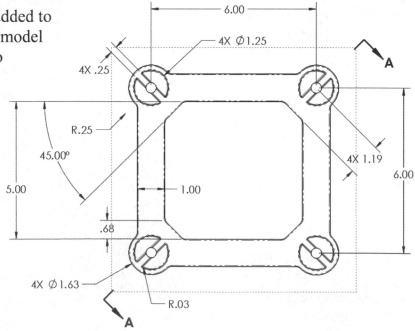

Dimensions are often added to several sketches in the model using the same plane to define the features. This causes them to look cluttered and quite confusing when exporting to a drawing.

Rearrange the dimensions similar to the image shown on the right, to make them easier to read.

10. Arranging dimension handles:

Since we cannot see the entire circle, the dimension R.25 looks like it was attaching to the wrong side of the arc. See example below.

To attach the handle point to the opposite side of the arc: Select the **R.25** dimension, click the **Leaders** tab and enable the checkbox **Dimension to Inside of Arc** (arrow).

Click the "round handle" on the arrowhead to flip it to the opposite side.

Click **OK**.

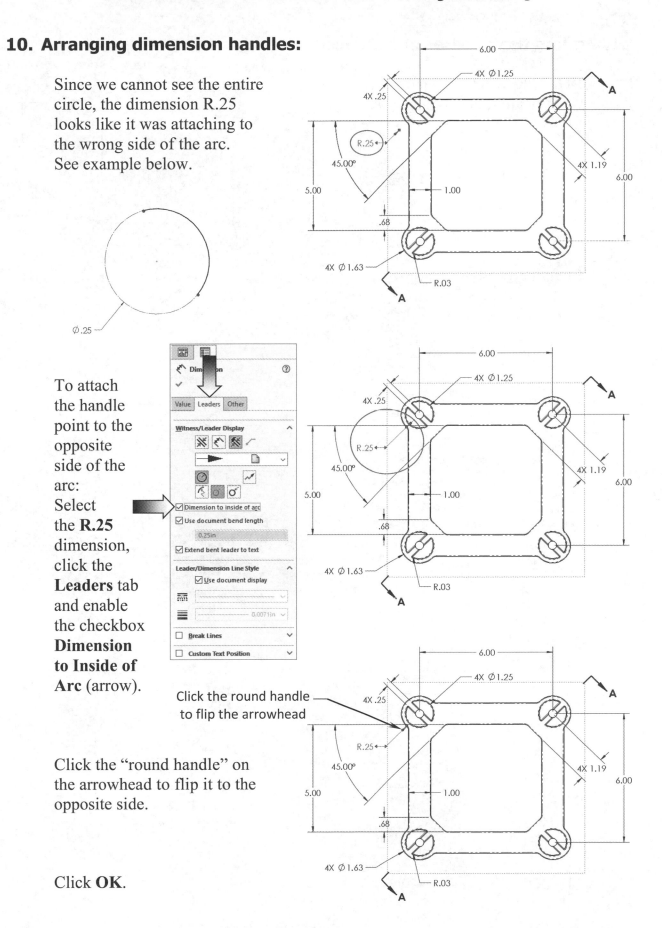

11. Adding the number of instances:

Delete the **R.03** dimension, we will add it in another view.

Select the dimension **Ø1.63** (circled).

Click the **Leaders** tab (arrow).

Select the **Radius** button (arrow).

The dimension **Ø1.63** is changed to **R.81**.

Switch to the **Value** tab.

Enter **4X** in front of the R<DIM> (arrow).

Click **OK**.

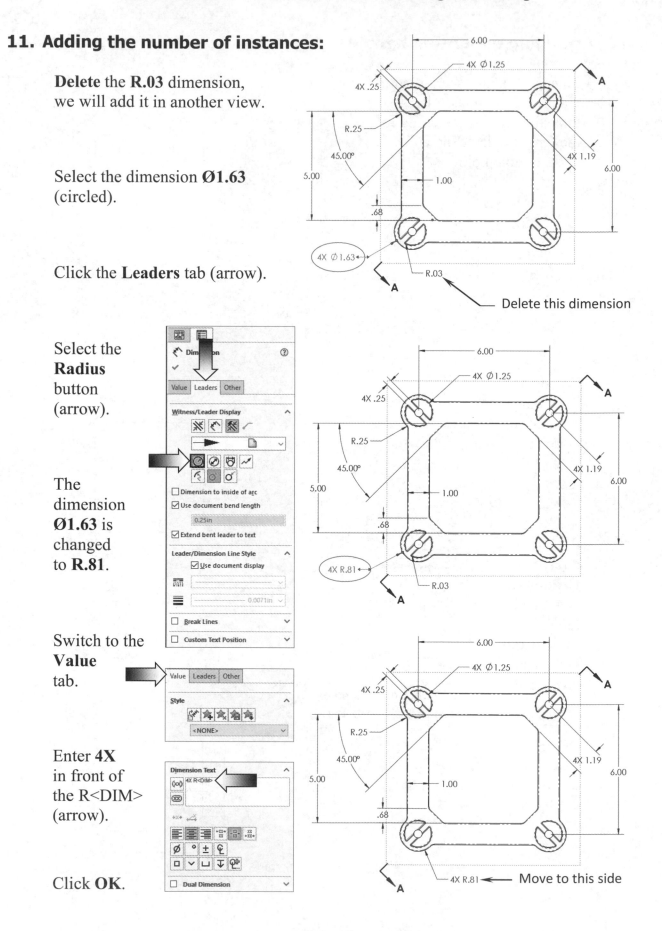

Delete this dimension

Move to this side

12. Using the Hole Callout:

Select the **Hole Callout** from the **Annotation** tab.

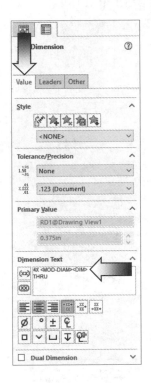

Click the **circular edge** of one of the 4 holes (circled) and place the dimension.

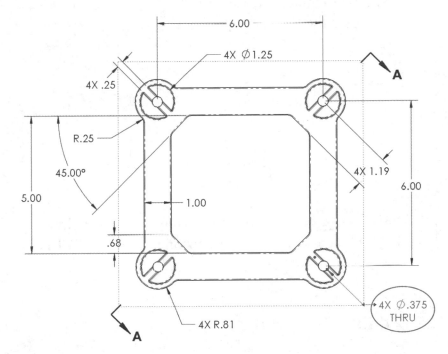

Click the **Value** tab (arrow).

Add the **4X** in front of the dimension as shown.

Zoom in on the **Left** view.

Delete the **R.03** dimension. It is not clear where the dimension is attached. We will add it to the Detail View.

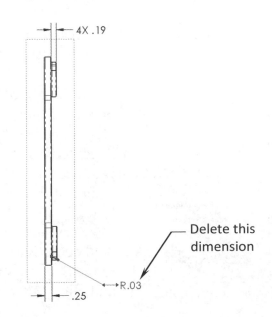

13. Adding reference dimensions:

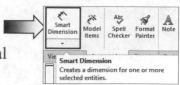

Reference dimensions cannot be used to change the actual geometry, but when the feature is changed in the model, the reference dimension will be updated accordingly.

Zoom closer to the Detail View. Select the **Smart Dimension** command.

Select the edge shown to attach the radius dimension.

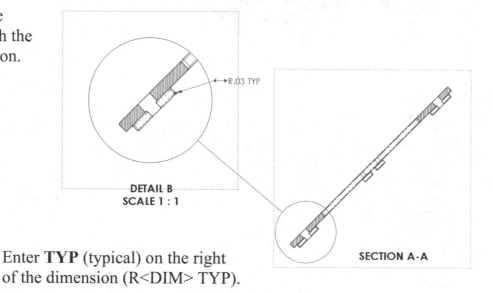

Enter **TYP** (typical) on the right of the dimension (R<DIM> TYP).

Click **OK**.

14. Adding Centerlines:

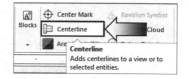

Click the **Centerline** command on the **Annotation** tab.

Enable the **Select View** checkbox (arrow). This option will insert the centerlines automatically to the selected view.

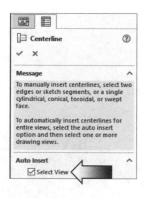

Click the dotted border of the Detail View. A centerline is added to the center of the hole.

Add centerlines to the Section View and click **OK** when finished.

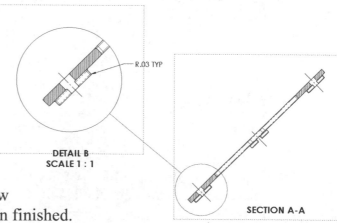

15. Adding Center Marks:

Select the **Center Mark** command. Zoom in on the Top view.

Select the **circular edge** of the 4 holes as noted.

Uncheck the **Use Document Default** and the **Extended Lines** checkboxes.

Enter **.080in** for Mark size, click **OK**.

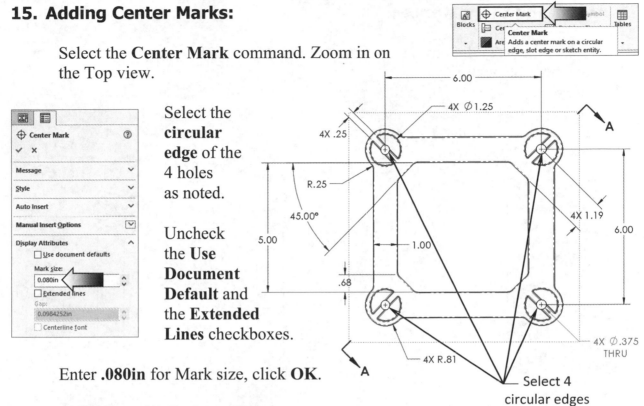

16. Modifying the hole callout:

The **.25** dimension is missing the depth callout.

Select the **.25** dimension (circled). Enter **4X** in front of the <DIM>.

Click to the right of the <DIM>, select the **Depth** symbol and enter **.19** next to it.

Click **OK**.

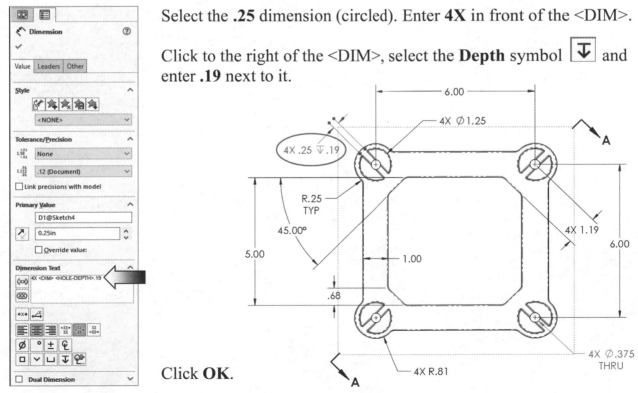

17. Modifying the instances:

Select the **R.25** dimension. Click to the right of the <DIM> and press enter to go down to the next line.

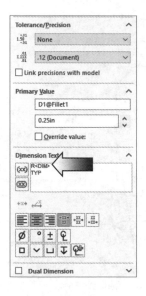

Enter **TYP** (typical) (arrow).

Click **OK**.

Select the dimension **1.00** (circled).

Add the **4X** in front of the <DIM>.

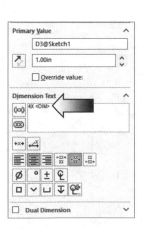

Click **OK**.

18. Adding Datum Features:

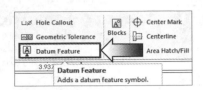

A Datum is a point of reference: Use a point, a line, a surface, or an axis as a basis for measurement or calculation of feature's precision.

Select the **Datum Feature** command from the **Annotation** tool tab.

The default datum symbol should be selected already by default (arrow).

For Datum A, select the edge on the far left as indicated.

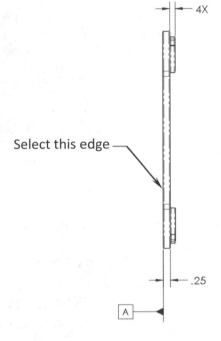

Select this edge

Place the **Datum A** below the drawing view as shown.

For Datum B, select the bottom edge as noted.

Place the **Datum B** approximately as shown.

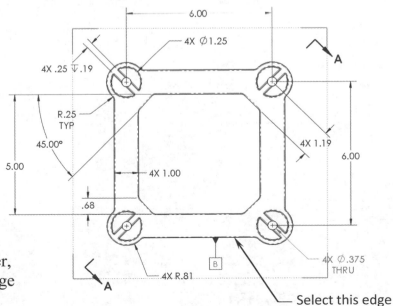

To change the Datum letter, click the symbol and change it from the Properties tree.

Select this edge

19. Adding Geometric Tolerances:

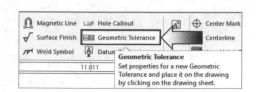

Geometric Dimensioning & Tolerancing or GD&T is a mechanical engineering language, which represents a way to define the size, location, orientation, and form of a part feature.

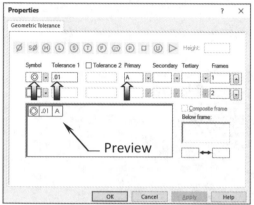

Select the dimension **4X Ø1.25**.

Click the **Geometric Tolerance** command.

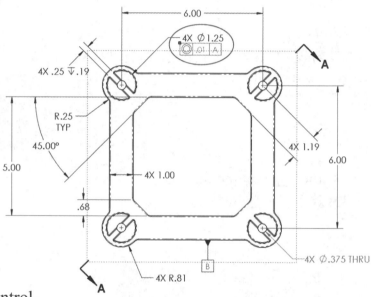

For Symbol, select: **Concentric** (arrow).

For Tolerance, enter **.01**

For Primary Datum, enter **A**

Click **OK**. The Geometric Control-Frame is attached to the dimension automatically.

Repeat the last step and add the **Position** tolerance to the dimension **4X Ø.275 THRU**.

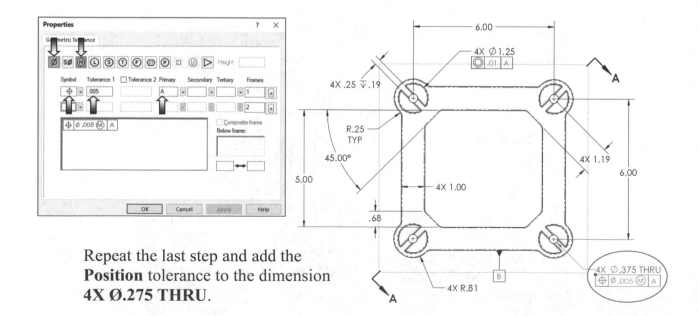

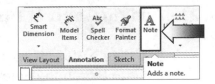

20. Adding Flag Notes:

Flag Notes are used to communicate and detail the characteristics of a part's features in a drawing. This is common to the vocabulary of people who work with engineering drawings in the manufacture and inspection of parts and assemblies.

Select the **Note** command.

Click the **Add Symbol** button (arrow).

Select **More Symbols...** (arrow).

Select **Flag Tri 1-99** (arrow).

Select the symbol **Triangle 1** from the list.

Select the **left edge** of the Detail View to attach the Flag Note.

Click **OK**.

Select the **Bent Leader** option if needed (arrow).

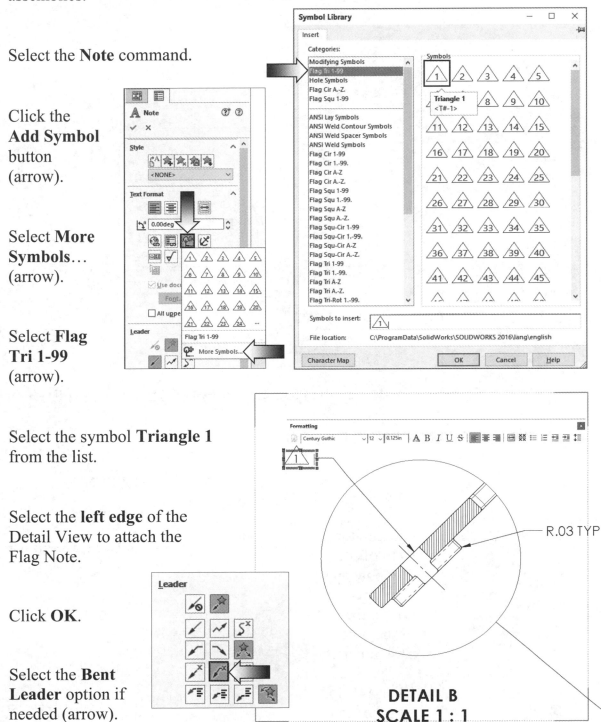

21. Adding General Notes:

A Flag Note can be linked to General Notes simply by selecting the symbol while typing the notes.

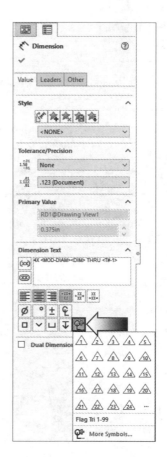

Select the **Note** Command, place the blank note box on the lower left side of the drawing

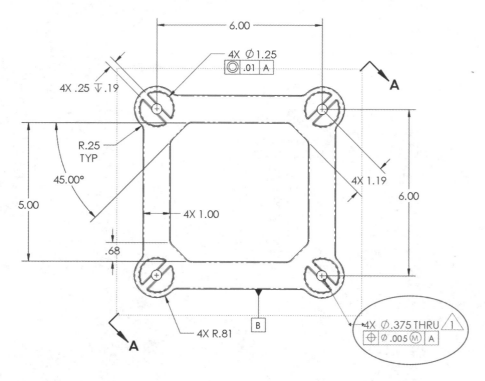

Enter: **NOTES: UNLESS OTHERWISE SPECIFIED**.

Click the **Add Symbols** button.

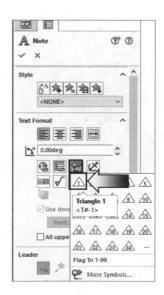

Select the flag note **Tri-1**.

Click to the right of the flag note and continue with typing the rest of the notes shown here.

Click **OK** to exit the Note.

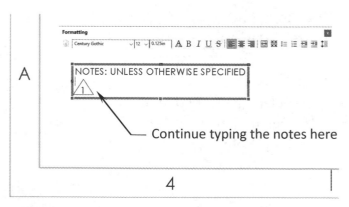

Continue typing the notes here

NOTES: UNLESS OTHERWISE SPECIFIED

⚠1 NO FILLETS ON THE ⌀.375 HOLES

2 PARTS TO BE FREE OF CRACKS, DISTORTIONS OR STAINS. SURFACES TO BE FREE OF NICKS, SCRATCHES OR DENTS.

3 PARTS TO PLACED IN CARDBOARD BOXES AND SECURED WITH STYROFOAM BLOCKS, AND LABELED WITH PART NUMBER REVISION, SUPPLIER, LOT NUMBER AND QUANITY.

22. Filling out the Title Block:

By default, the Sheet Format layer is placed in the back of the Sheet layer.

The Sheet layer is used to create the drawing, and the Sheet Format layer contains the Title Block information.

Right-click inside the drawing and select: **Edit Sheet Format**.

Zoom in on the Title Block area and hover the cursor over the center of each field until the symbol **A** appears, then double-click the empty note to modify it.

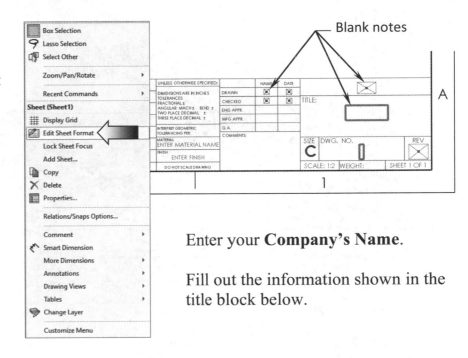

Blank notes

Enter your **Company's Name**.

Fill out the information shown in the title block below.

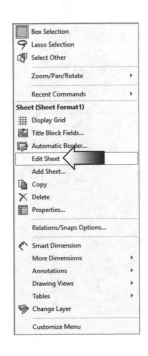

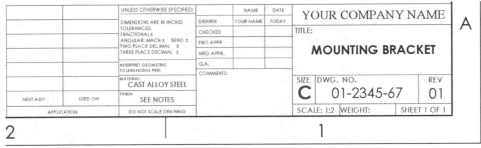

For Material, enter **Cast Alloy Steel**.

For Finish, enter **See Notes**.

Change the font size if needed to fit each field.

Click **OK** to exit the Note command.

23. Saving your work:

Select **File / Save As**.

Enter **Mounting Bracket.slddrw** for file name.

Click **Save**.

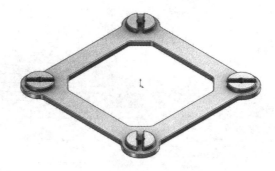

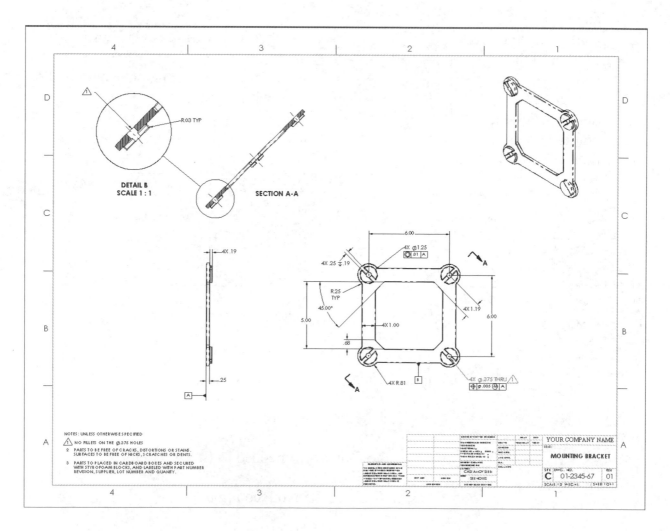

CHAPTER 17

SOLIDWORKS MBD

SOLIDWORKS MBD
Introduction Model Based Definition

SOLIDWORKS MBD (Model Based Definition) is an add-in application that lets you create models without the need for drawings giving you an integrated manufacturing solution for the SOLIDWORKS software. MBD operates within the SOLIDWORKS environment with its own CommandManager.

SOLIDWORKS MBD helps you define, organize, and publish 3D product and manufacturing information (PMI), including 3D model data in industry standard file formats such as Geometric Dimensioning and Tolerancing (GD&T) and 3Ddimensions and annotations.

In a Part or Assembly document SOLIDWORKS MBD offers the following:

* 2D, 3D views and exploded views.

* Supports all native SOLIDWORKS 3D part and assembly data, such as configurations, display states, constraints, and PMI.

* Creates output files such as 3D PDF and eDrawings.

* Supports tables, reference dimensions, DimXpert dimensions, annotation views, 3D views, and custom properties.

SOLIDWORKS MBD guides the manufacturing process directly in 3D:

* By avoiding the unnecessary and costly revision of 2D drawings, SOLIDWORKS MBD streamlines production, cuts cycle time, improves communication with the supply chain, and reduces manufacturing errors.

* By providing help to meet industry standards such as MIL-STD-31000 Rev-A and ASME 14.41.

SOLIDWORKS MBD
Introduction to Model Based Definition

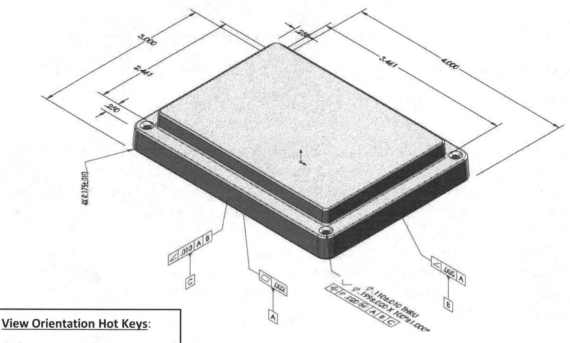

View Orientation Hot Keys:

Ctrl + 1 = Front View
Ctrl + 2 = Back View
Ctrl + 3 = Left View
Ctrl + 4 = Right View
Ctrl + 5 = Top View
Ctrl + 6 = Bottom View
Ctrl + 7 = Isometric View
Ctrl + 8 = Normal To Selection

Dimensioning Standards: **ANSI**
Units: **INCHES** – 3 Decimals

Tools Needed:

	Auto Dimension Scheme		Datum
	Location Dimension		Size Dimension
	Geometric Tolerance		Show Tolerance Status

Prismatic Parts

1. Opening a part document:

Click **File** / **Open**.

Browse to the Training Files folder
and open a part document named
Top Cover.sldprt.

2. Enabling SOLIDWORKS MBD:

Right-click any of the tool tabs (Features,
Sketch, etc. and select **MBD** (arrow).

The process of creating interactive 3D PDF
or eDrawing files requires some planning
and preparation within the Part or Assembly
followed by the use of specific MDB tools.

The appropriate views are created first, and
DimXpert tools are used to create the datums,
dimensions, geometric tolerances, and other
annotations.

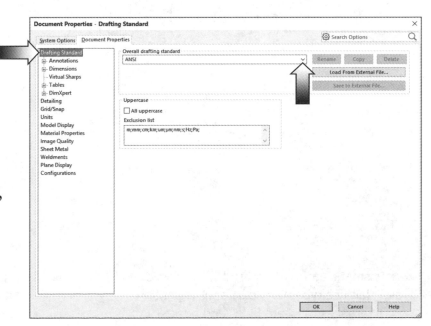

Lastly, publish to
a 3D PDF or an
eDrawing file.

3. Setting the options:

Click **Tools, Options,
Document Properties,
Drafting Standards,
Overall Drafting
Standard**.

Select **ANSI** (arrow).

Expand the **DimXpert** option.

Click the **Geometric Tolerance** option.

Enable the checkboxes:

Apply MMC to datum features of size, and

Create Basic Dimension

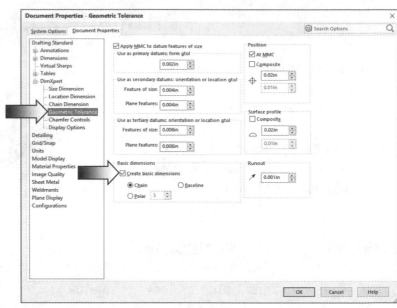

Set the tolerance values to match the ones in the dialog box.

Click the **Display Options** (arrow).

Enable the checkboxes:

Eliminate Duplicates

Show Instance Count

Click **OK**.

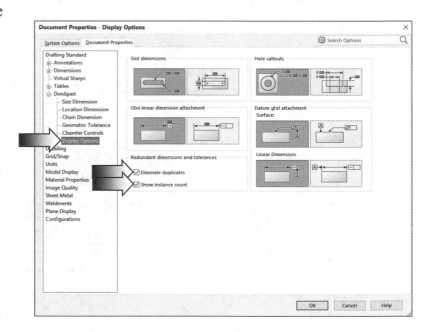

Prismatic parts

For Prismatic parts, when used with Geometric as Tolerance type, DimXpert applies <u>position tolerances</u> to locate holes and bosses.

Turn parts

For Turn parts, when used with Geometric as Tolerance type, DimXpert applies <u>circular runout tolerances</u> to locate holes and bosses.

The **Auto Dimension Scheme** tool 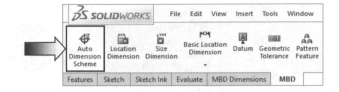 automatically applies dimensions and tolerances to the manufacturing features of a part.

4. Adding datums:

Right-click one of the tool tabs and enable the following:

     * **MBD Dimensions** tab
     * **MBD** tab

Click **Auto Dimension Scheme** on the **MBD** tool tab (arrow).

For Part Type, select **Prismatic**.

For Tolerance Type, select **Geometric**.

For Pattern Dimensioning, select **Linear**.

For Primary Datum, select the **bottom face** of the part.

For Secondary Datum, select the **right face**.

For Tertiary Datum, select the **front face**.

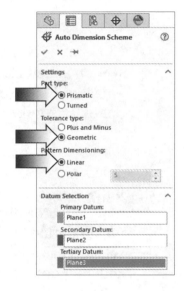

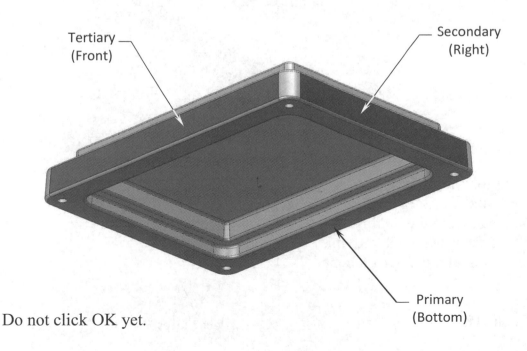

Do not click OK yet.

5. Using Scope:

The options in Scope control which features DimXpert considers for dimensioning and tolerancing.

* All Feature: Applies dimensions and tolerances to the entire part. DimXpert considers all previously defined features, including those not listed under Feature Filters, as well as those features listed under Feature Filters.

* The objective is to dimension and tolerance the selected features.

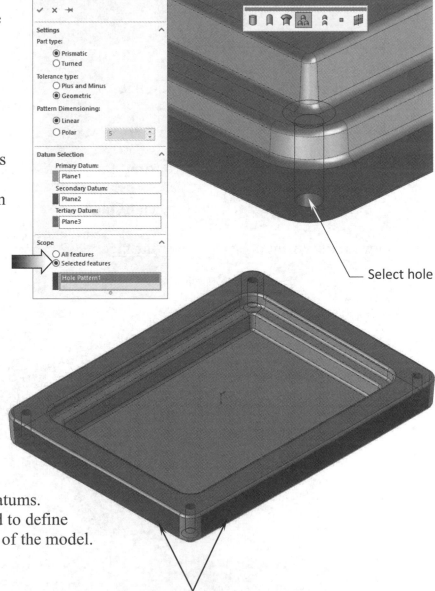

Click in the **Scope** section and click the **Selected Features** option (arrow).

Select one of the **holes**. The selection pop-up dialog allows you to define the selection as a Pattern and all 4 holes are selected.

Select hole

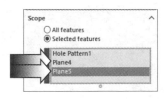

Additionally, select the **2 planar faces** opposite from the datums. These faces are used to define the width and depth of the model.

Click **OK**.

Select 2 faces

Switch to the DimXpertManager tree (arrow) to see the tolerances and annotations.

Mapping Features: DimXpert features map directly to the annotations in the model.

* Datum A comes with a Flatness tolerance.
* Datum B comes with an Angularity tolerance.
* Datum C comes also with an Angularity tolerance.

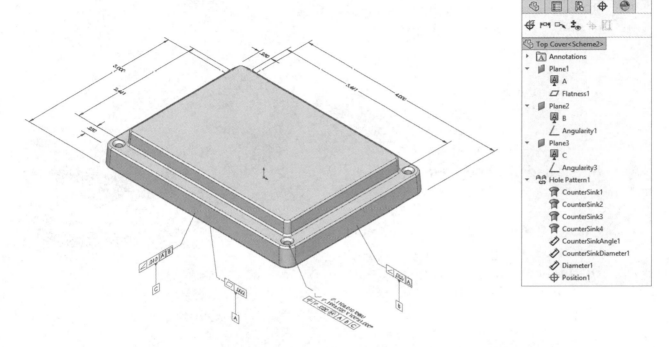

6. Showing the constraint status:

Click **Show Tolerance Status**.

The **Green** faces appearing in the model indicate Fully Constrained geometry.

The original color of the part remains on those faces that were not selected in the tolerance scheme.

Click-off Show Tolerance Status.

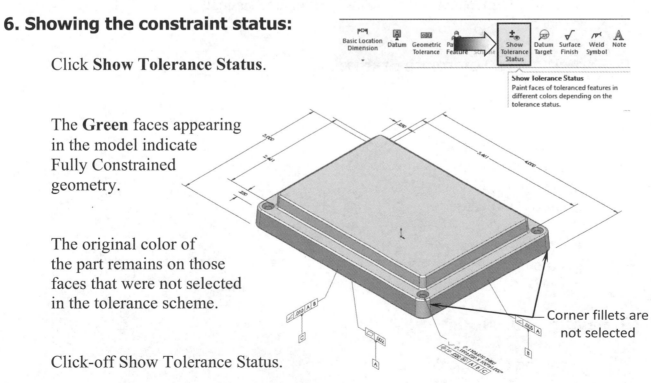

Corner fillets are not selected

7. Adding the size dimensions:

To further constrain the geometry the Size and Location dimensions are added.

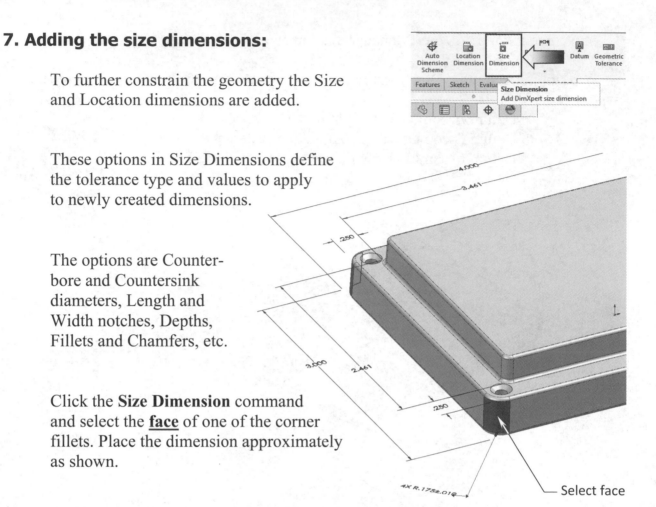

These options in Size Dimensions define the tolerance type and values to apply to newly created dimensions.

The options are Counterbore and Countersink diameters, Length and Width notches, Depths, Fillets and Chamfers, etc.

Click the **Size Dimension** command and select the **face** of one of the corner fillets. Place the dimension approximately as shown.

Select face

The Size Dimensions adds the 4X callout automatically.

8. Showing the Tolerance Status:

Click **Show Tolerance Status**.

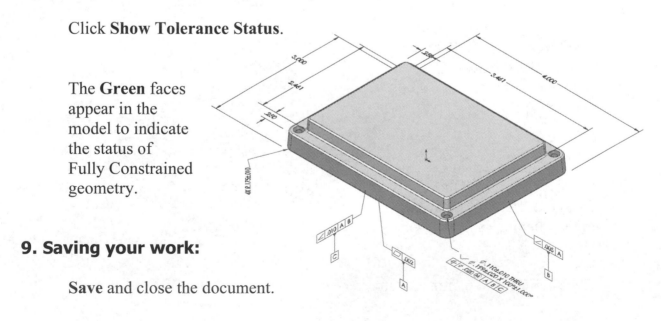

The **Green** faces appear in the model to indicate the status of Fully Constrained geometry.

9. Saving your work:

Save and close the document.

Exercise: Adding the Size and Location Dimensions

1. Opening a part document:

Click **File / Open**.

Browse to the Training Files folder and open a part document named **Mounting Bracket.sldprt**.

2. Adding the Datum reference:

Change to the **MBD** tab.

Click **Datum** (arrow).

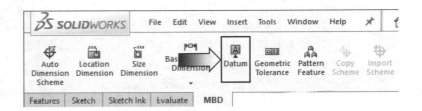

For **Primary Datum**, select the <u>bottom face</u> of the model.

For **Secondary Datum**, select the <u>back face</u> of the model.

For **Tertiary Datum**, select the <u>side face</u> as shown.

Click **OK**.

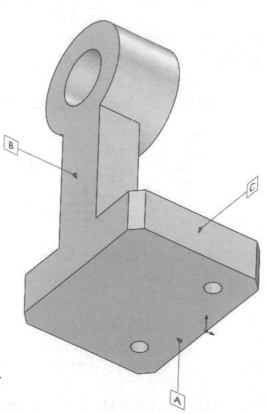

Double click a datum to edit its Label, Leader Style, Text, Line Style, Line Thickness, etc.

3. Adding the Size Dimensions:

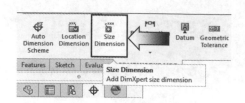

The Size Dimension is used to define the tolerance type and values to apply to newly created size dimensions.

Click the **Size Dimension** command (arrow).

Select the <u>edge</u> of the upper **Hole**.
Click the **Hole** option in the
Selection Pop-up dialog (arrow).
Place the dimension as shown.

Select either the <u>edge</u> or the <u>inner</u>
<u>face</u> of the **Counterbore** hole.
Click the **Pattern** option in the
Selection Pop-up dialog (arrow).
Both counterbore holes are selected.
Place the dimension to the side
of the hole.

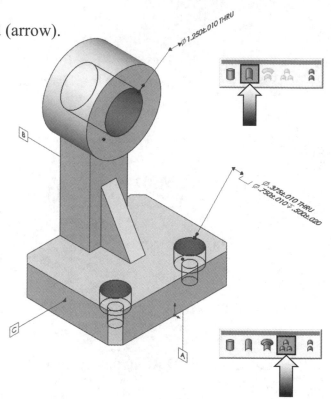

4. Showing the Tolerance Status:

Click the **Show Tolerance Status** command.

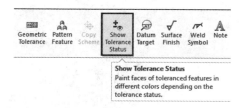

Show Tolerance Status
Paint faces of toleranced features in
different colors depending on the
tolerance status.

DimXpert displays feature faces in
default colors representing one of the
three states:

    * Under constrained = **Yellow**
    * Fully constrained = **Green**
    * Over constrained = **Red**

The Location Dimensions are added next.

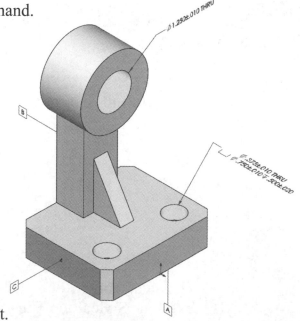

5. Adding the Location Dimensions:

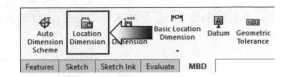

Set Location Dimension options to define the tolerance type and values to apply to newly created linear and angular dimensions defined between two features. These options apply to dimensions created with the Location Dimension and Auto Dimension Scheme tools. These options do not affect pre-existing features, dimensions, or tolerances.

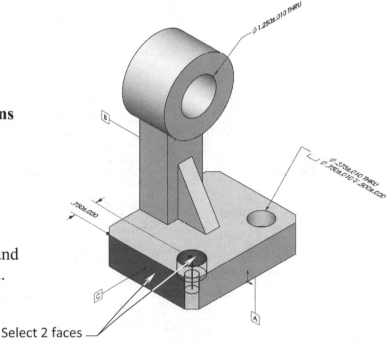

Select the **Location Dimensions** tool (arrow).

Click the <u>left face</u> of the base and <u>inside face</u> of the hole as noted.

Select 2 faces

Place the dimension on the back of the model approximately as shown.

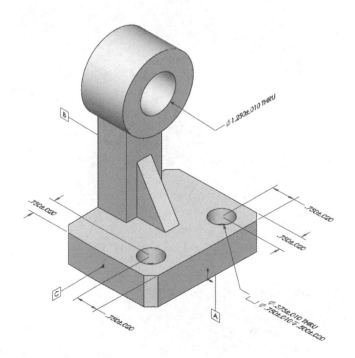

Continue with adding the location dimensions to locate the two counterbore holes.

Click **OK**.

6. Adding the Size Dimensions to other features:

Click the **Size Dimension** command again.

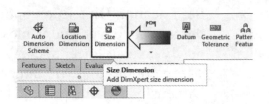

It is a little easier to add dimensions to one feature at a time so that duplicate dimensions can be eliminated.

Add the Size Dimensions to the base, the vertical column, the rib, and the circular boss, similar to the ones shown in the image.

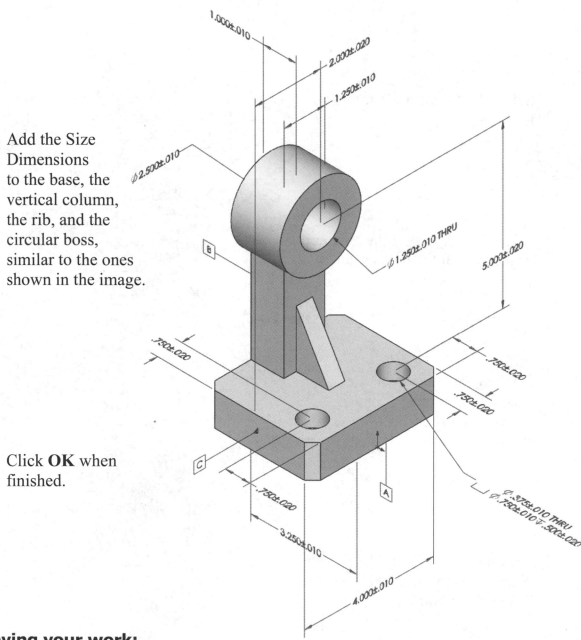

Click **OK** when finished.

7. Saving your work:

Save and close the document.

Turned Parts

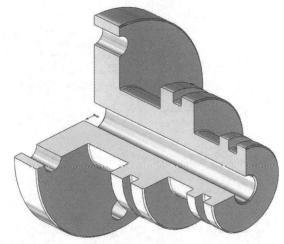

1. Opening a part document:

Click **File / Open**.

Browse to the Training Files folder
and open a part document named
Turned Part.sldprt.

The Turn option in the Auto Dimension
Scheme is used to customize the annotations
and dimensions for a cylindrical type of part
(normally cut on an engine lathe) with diameters
instead of lengths.

2. Adding Datums:

Click **Auto Dimension Scheme** .

For Part Type, select the **Turned** option.

For Tolerance Type, select
Geometric.

For Pattern Dimensioning,
select **Linear**.

For Primary Datum, select the <u>face</u>
on the left side as noted.

For Secondary Datum, select
the <u>hole</u> in the center of the model.

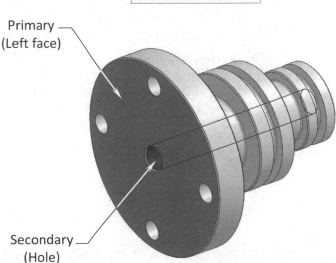

Primary — (Left face)

Secondary — (Hole)

For **Scope**, click the **Selected Features** option (arrow).

Select hole

Select all <u>18 faces</u> of the model, except for the chamfers.

Also select one of the <u>holes</u>; all 4 holes should be highlighted at the same time.

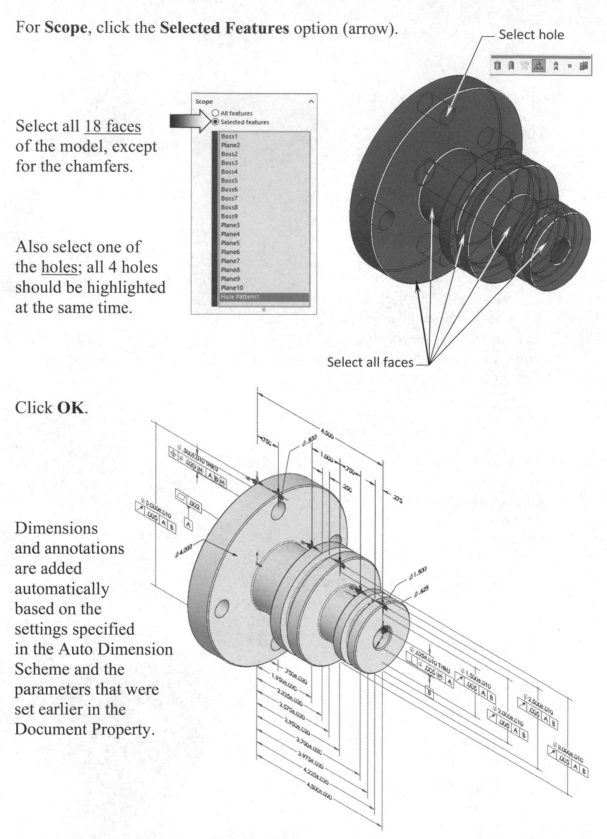

Select all faces

Click **OK**.

Dimensions and annotations are added automatically based on the settings specified in the Auto Dimension Scheme and the parameters that were set earlier in the Document Property.

Rearrange the dimensions and annotations so that they do not overlap one another and are easy to read.

3. Displaying the Annotation tree:

By default the Dimxpert-Manager is displaying the **Feature Based** tree.

Right-click the <u>name</u> of the part and select **Tree Display, Show Annotation Based Tree** (arrows).

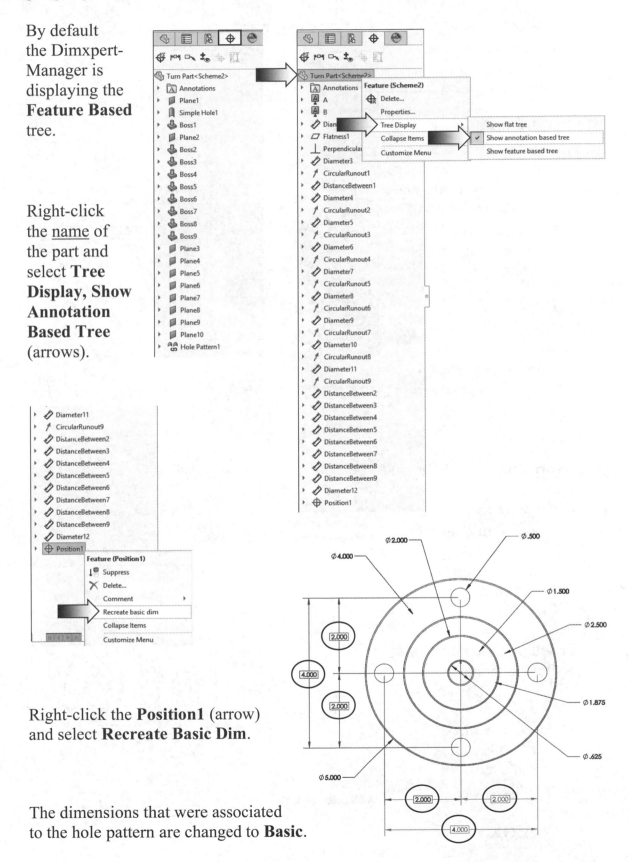

Right-click the **Position1** (arrow) and select **Recreate Basic Dim**.

The dimensions that were associated to the hole pattern are changed to **Basic**.

4. Deleting the dimension scheme:

If needed, individual dimensions can be deleted one at a time or all at once by using the Delete-All-Tolerances option.

We are going to try out the Plus/Minus and the Polar options, so the current dimension scheme must be deleted.

Click **Tools**, **DimXpert**, **Delete All Tolerances** (arrow). Click **Yes** to confirm the delete.

5. Using Plus and Minus:

Click **Auto-Dimension Scheme**.

Select **Turned**, **Plus and Minus**, **Linear**.

Select the same **Datum A** and **B** as in step 2, page 13.

For Scope, click Selected Features and select one of the <u>holes</u>.

Click **OK**.

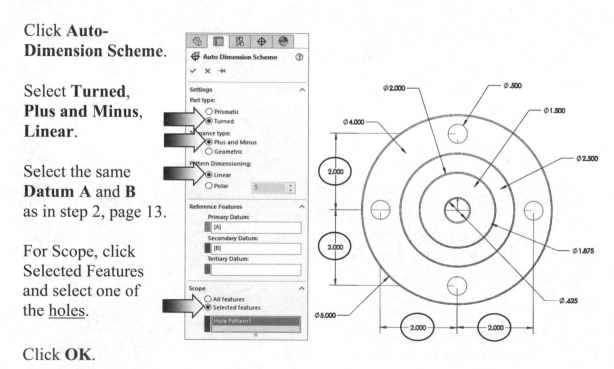

6. Deleting scheme:

Click **Tools**,
DimXpert, **Delete
All Tolerances**
(arrow).

Click **Yes** to confirm
the delete.

We will now try out
the Polar option.

7. Using Polar:

Click **Auto-
Dimension Scheme**.

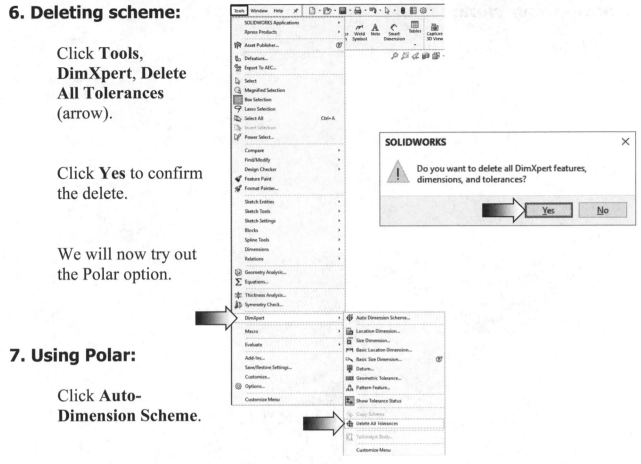

Select **Turned**, **Plus and Minus**,
Polar.

Select the same
references for
Datum A and **B**
as in the
previous step.

For Scope,
click **Selected
Features** and
select one of the
holes.

Click **OK** and
examine the results.

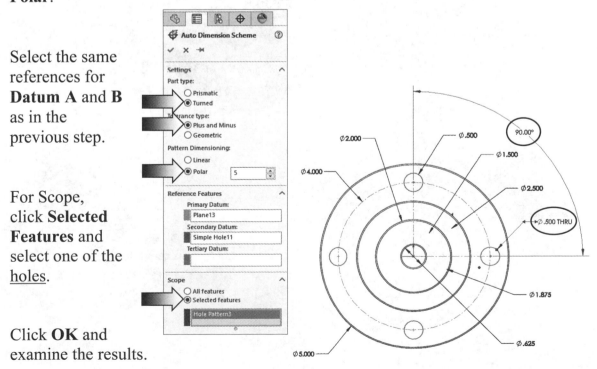

8. Saving your work:

Save your work and close all documents.

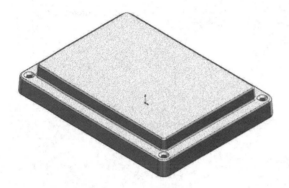

Capturing 3D Views

1. Opening a part document:

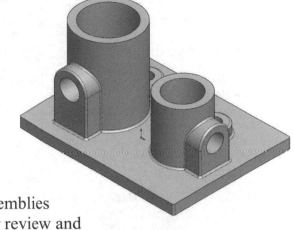

Click **File / Open**.

Browse to the Training Files folder
and open a part document named
3D Views.sldprt.

You can create 3D views of parts and assemblies
that contain the model settings needed for review and
manufacturing.

The command Capture 3D Views are used to capture Product-Manufacturing-
Information (PMI). The 3D views are used to create a 3D PDF or an eDrawing
file. Use the 3D Views tab at the bottom left of the screen to publish the 3D Views.

2. Rearranging dimensions:

Right-click **Annotation** folder and select **Show DimXpert Annotations** (arrow).

For clarity and readability, rearrange the dimensions and tolerances so that they
are more visible and easier to read.

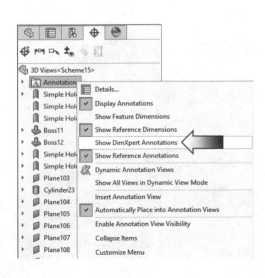

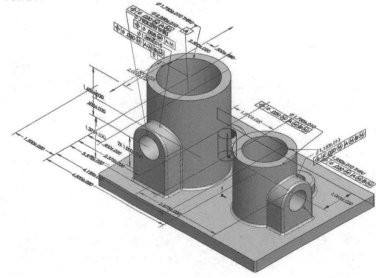

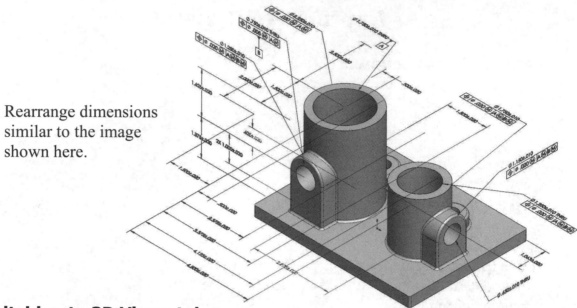

Rearrange dimensions
similar to the image
shown here.

3. Switching to 3D Views tab:

Select the **3D Views** tab at the bottom left corner
of the screen (arrow).

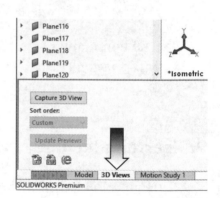

When you capture a 3D view, you are capturing
the current image and position of the model in
the graphics area at that particular moment such
as the active configuration, exploded and model
break views, annotations, display states, zoom
level, view orientation, and section views. This is a
useful way to include or exclude 3D PMI in a 3D view for better organization.

Change to the **Isometric View**
(Control+7). This will be the first
3D view captured with all DimXpert
annotations visible.

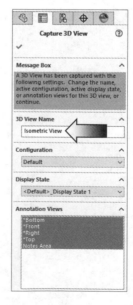

Click **Capture 3D View** Capture 3D View ;
the 3D Views tab opens.

For 3D View Name, enter
Isometric View.

Select <u>all</u> Annotation Views (default)

Click **OK**. A preview of the Isometric
View appears in the 3D Views pane.

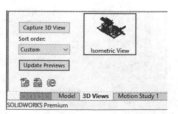

4. Capturing the Orthographic views:

Change to the **Front** orientation (Control+1).

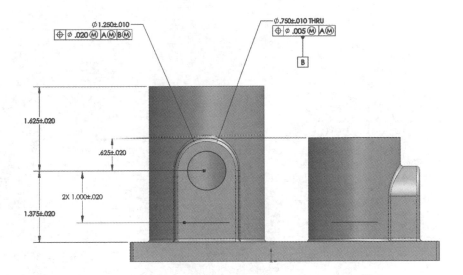

Rearrange the dimensions to look similar to the image above.

Click **Capture 3D View** Capture 3D View .

For 3D View Name, enter **Front View**.

Click **OK**.

The Front View appears in the 3D Views pane.

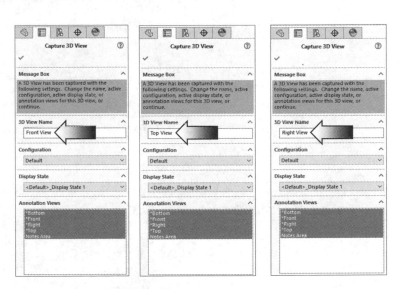

Repeat the same step and capture the **Top** and the **Right** views.

If you need to make any changes simply right click the preview image and select **Recapture View**.

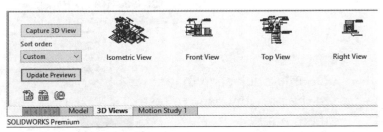

The section view is next.

5. Creating a Section view:

Click the **Section View** command on the View Heads-Up toolbar.

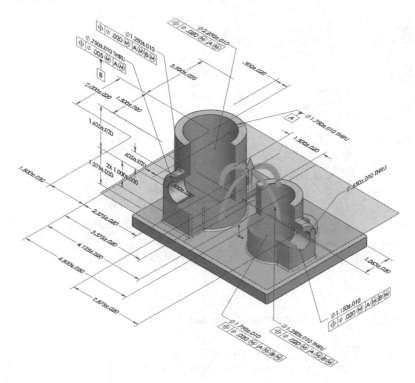

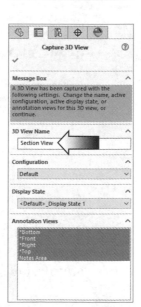

For Section Method, select **Zonal** (arrow).

For Section 1, select the **Front** plane and enter **0.000in** for Offset Distance.

For Section 2, select the **Top** plane and enter **1.250in** for Offset Distance.

For Section 3, select the **Right** plane and enter **-1.250in** for Offset Distance and click **OK**.

Select **Capture 3D View** | Capture 3D View |.

For 3D View Name, enter **Section View**.

Click **OK**.

Double-click each 3D view in the preview pane to test them out.

Publishing PDF and eDrawing Files

(Use the same part document from the previous lesson for this exercise.)

1. Publishing to 3D PDF:

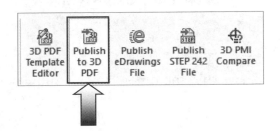

Publishing to 3D PDF or eDrawing is the last step in using SOLIDWORKS MBD.

Publishing takes a few steps to complete. The user has to define the template, the selection of the views and output details that include text and property.

Click **Publish to 3D PDF** (either on the MBD toolbar or at the lower left corner).

For Theme, select **Template_Approval** (arrow). This template comes with 2 sheets by default.

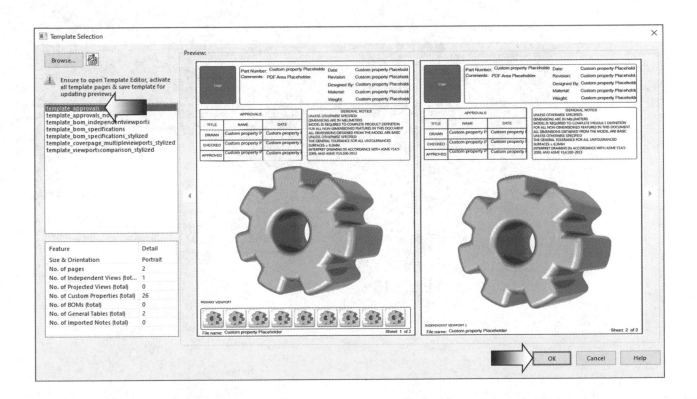

Click **OK**.

The **Publish to 3D PDF** screen appears displaying the details of the 2 sheets.

Click inside the box under Primary & Thumbnail View and select the **Front View**, **Top View**, and **Right View** in the bottom pane.

The 1st view in the list (the Isometric View) will be used as the Primary View, and the other views will be added to the Thumbnail Carousel in the PDF file.

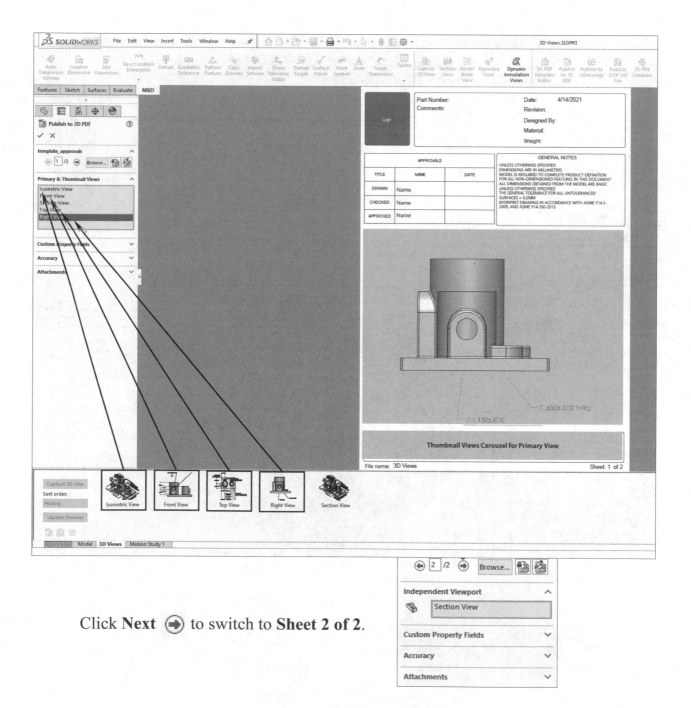

Click **Next** to switch to **Sheet 2 of 2**.

Expand the **Independent Viewport** section, click in the box and then select the
Section View from the 3D Views tab in the bottom pane.

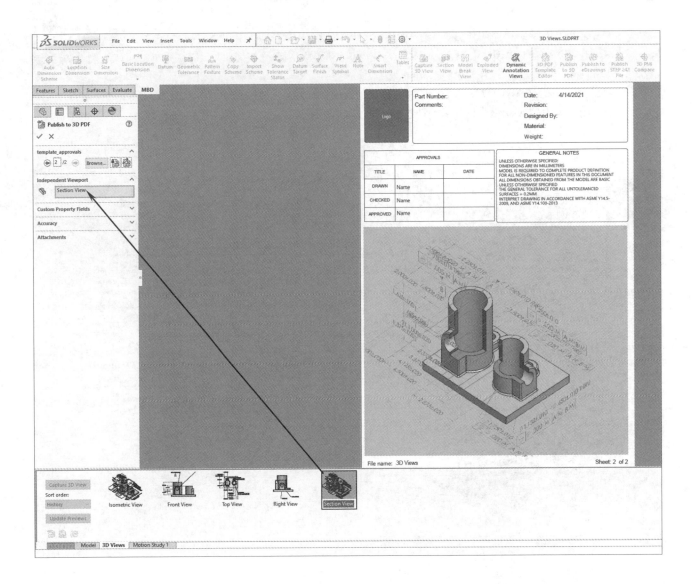

The Section View will be displayed in sheet 2 of the PDF file.

Click **OK**.

The Publish to 3D PDF PropertyManager reads the specify details of the active
model and saves them in a 3D PDF file.

2. Viewing the 3D PDF file:

Open the PDF file manually if the PDF application is not launched automatically.

If prompted, expand the **Options** section at the top right side of the screen and select: **Trust this document always**.

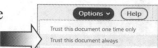

Select the **Isometric View** from the Thumbnail Carousel to activate it. Click any other view to display it in the graphics area.

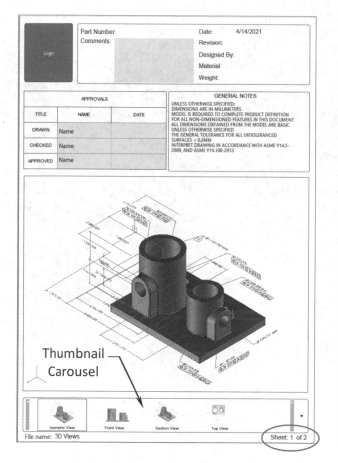

Thumbnail Carousel

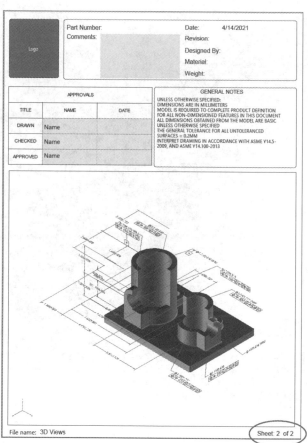

* Hover the mouse cursor over the model area and roll the scroll-wheel up and down to zoom in and out.
* Press and hold the scroll-wheel to rotate the view.
* Press and hold the control key + the left mouse button to pan.
* Press and hold the shift key + the left mouse button to zoom in/out.

3. Saving the PDF document:

Save, **close** and **return** to the previous SOLIDWORKS document.

4. Publishing to eDrawing:

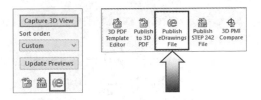

The 3D Views data can also be published
to an eDrawing file.
eDrawing provides a lot more manipulation
options than 3D PDF and it is more familiar to most SOLIDWORKS users.

Click **Publish eDrawing File** (arrow).
After SOLIDWORKS MBD runs through the 3D Views saved in the lower pane
the eDrawing application is launched.

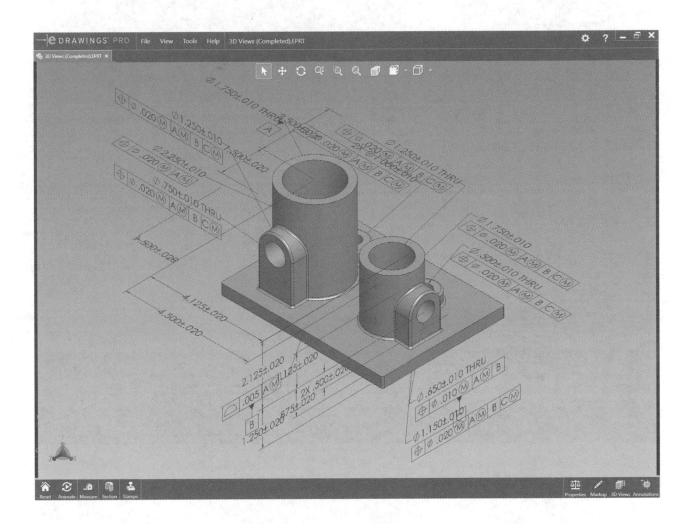

The dimensions and tolerances are displayed by default.

In eDrawing you can view and animate, measure, extract mass properties, create
markups, and add stamps before sending to others.

5. Changing to other 3D Views:

If the eDrawing is saved with 3D Views, the 3D Views command appears at the bottom right corner.

Click the **3D View** button (arrow). The saved views are displayed at the bottom.

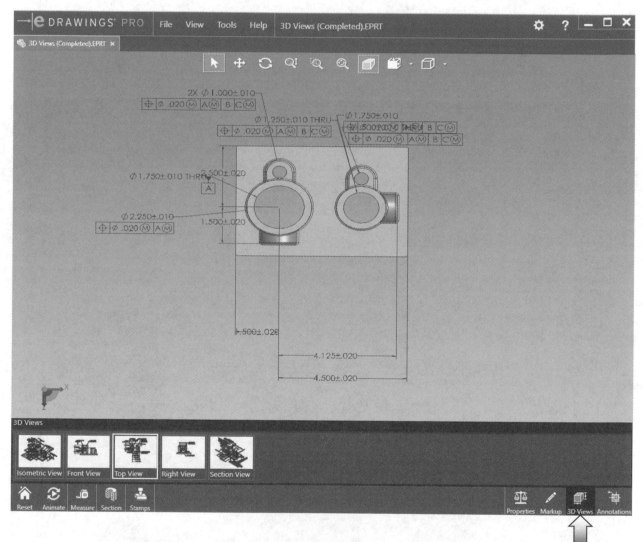

Double-click the **Top View** thumbnail graphic to activate the view.

Double-click the other 3D Views to test them out. Also try out other options like **Animate**, **Measure**, **Section**, **Stamps**, **Markup**, and **Annotations**.

Click **Reset** to return to the default Isometric View.

6. Extracting the mass properties:

Information from the model such as Material, Mass, and Volume can be retrieved by accessing the Mass Properties option.

Click **Mass Properties** (arrow).

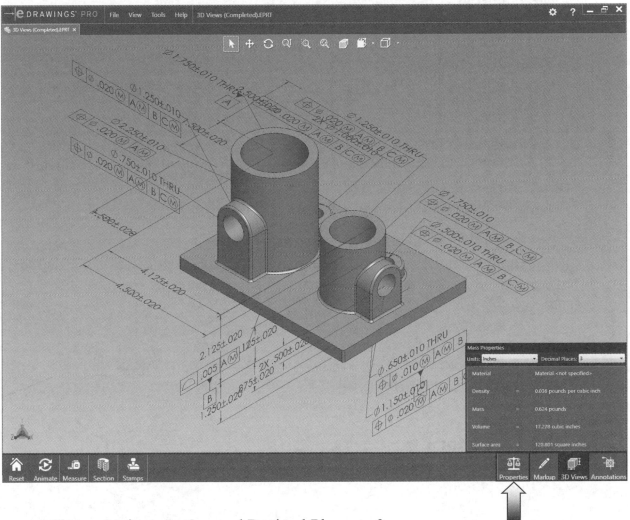

Change Units to **Inches** and Decimal Places to **3**.

Information regarding the model is displayed in a dialog at the lower right side.

7. Saving your work:

Save your drawing and close all documents.

3D PDF Template Editor

The 3D PDF Template Editor is an independent application that runs in its own window. The basic components of a template include  the company's logo, user's comments, custom properties, text, and table place-holder, thumbnails, and model areas for displaying 3D views.

The 3D PDF Templates are stored in the following directory: C:\Program Files\SOLIDWORKS Corp\SOLIDWORKS\Data\Theme folder. The supplied templates are:

* Template_Approvals
* Template_BOM_IndependentViewports
* Template_BOM_Specification_Stylized
* Template_ViewportsComparison_Stylized

* Template_Approval_Notes
* Template_BOM_Specifications
* Template_CoverPage_Multiple
Viewports_Stylized

New templates can be created from scratch, but it is easier to modify an existing template and then save it with a new name. The new template can have its own page size and orientation.

For this exercise we will modify an existing template to include new images, text, background color, and then save it as a custom template for future use.

1. Opening a part document:

Click **File / Open**.

Browse to the Training Files folder and open a part document named **Part for Template.sldprt**.

This part document has five 3D-Views previously saved: the Isometric, Front, Top, Right, and Section Views.

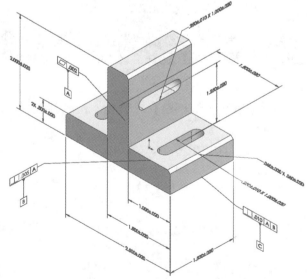

2. Launching the Template Editor application:

Click the **3D PDF Template Editor** command (arrow).

Minimize the SOLIDWORKS application.

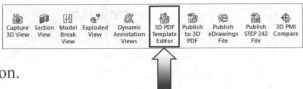

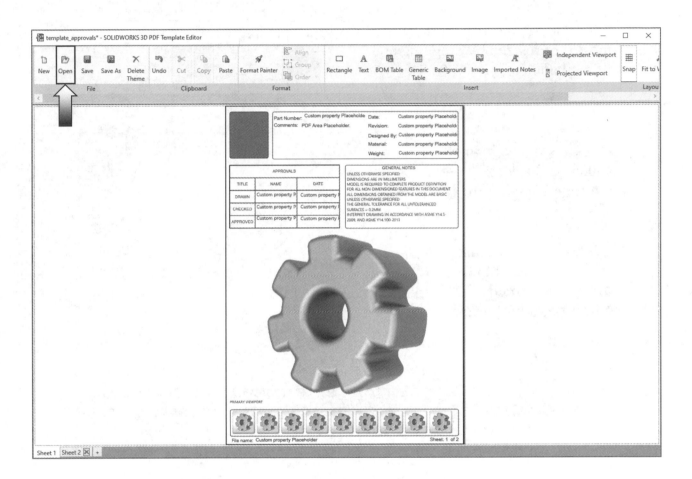

The 3D PDF Template Editor application appears in its own window.

Click **Open** (arrow).

Scroll down the list and select the template named **Template_Approvals_Notes** (arrow) and click **Open**.

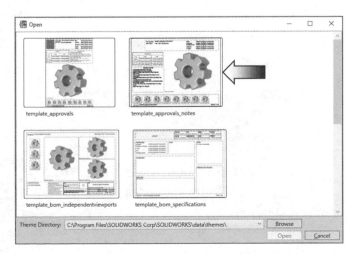

3. Inserting a logo:

The company logo must be created and saved using one of the Windows supported formats such as Jpeg, Bmp, Png, Gif, Tif, etc.

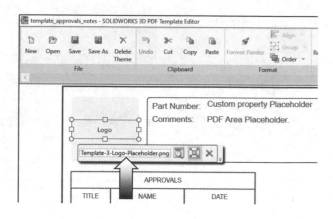

After the logo is inserted it can be moved or resized by dragging one of the handles.

Click inside the logo square and select: **Change Image** (arrow).

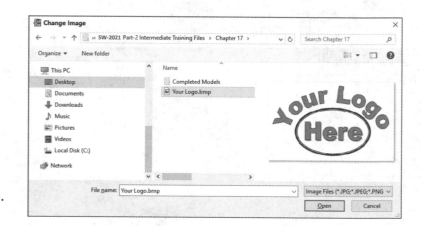

Browse to the Training folder and select the image file named: **Your Logo.bmp** (arrow).

NOTE: *When replacing an image file it may become distorted or out of proportion. Click* **Reset Original Size** *to resize it without distortion. Click* **Lock Aspect Ratio** *and then drag the handles to resize.*

Click and drag from the inside of the image to move it.

Drag the handles to resize the image.

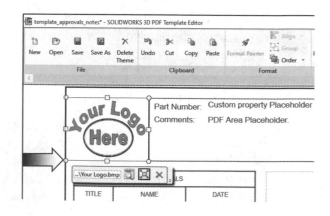

Deselect the Lock Aspect Ratio to resize the logo unproportionally, if needed.

4. Changing the background:

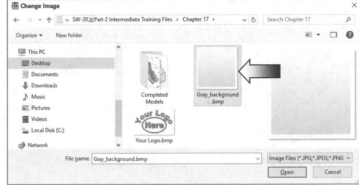

The background image must also be saved under one of Windows supported formats. For this exercise, we will use a bitmap file (bmp) to replace the current background.

Click **Background** on the **Insert** section (arrow).

Locate the bitmap file named **Gray_Background .bmp** from the same training files folder and open it.

The background color is changed to gray gradient.

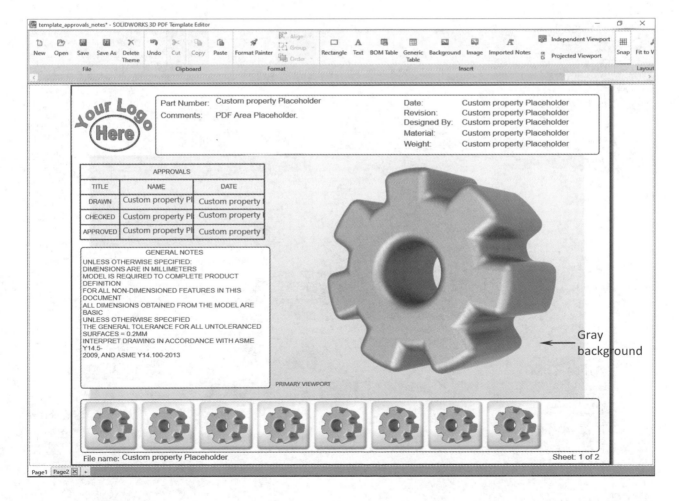

5. Adding text:

All of the supplied templates have multiple text types available:

* **Template Text Filed** is a text-string.
  It is used for Titles and Headings where it will not get changed.

* **Text Filed** is a text box with user defined text.
  It is filled in by the publisher of the PDF.

* Custom property Field is a text box that is tied to a **Custom-Property** of the Part or Assembly.
  The properties used in the standard templates are:

  • SW-Part Number   • SW-Short Date
  • Revision   • SW Material

* **PDF Form Field** is a text box with user defined text.
  It is filled-in by the user of the PDF.

These text types are distributed throughout the template as indicated below.

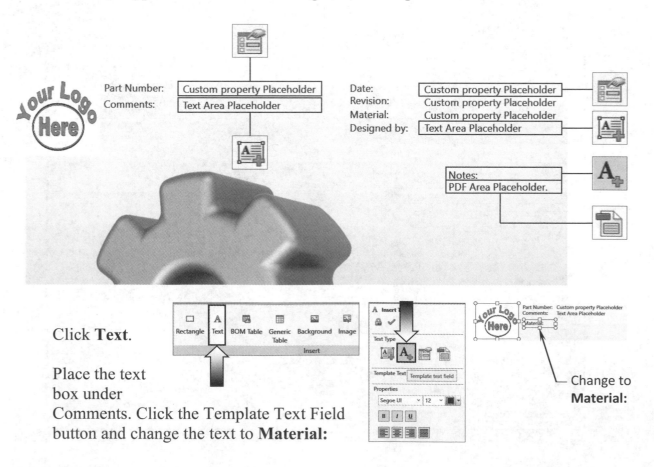

Click **Text**.

Place the text box under Comments. Click the Template Text Field button and change the text to **Material:**

Drag the text box and position it
<u>below</u> **Comments**, under the
Text Area Placeholder.

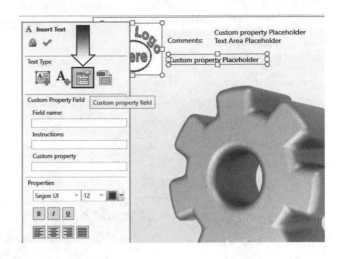

Select the new Material <u>text box</u> and
click the **Custom Property Field**
button (arrow). Click **OK**.

The custom property **Material**
is assigned to this text
box. The material of the
part will populate when
published to PDF.

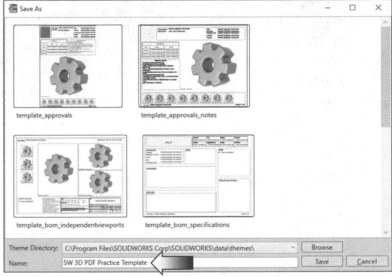

6. Saving the template:

Select **File / Save As**.

Enter **SW 3D PDF
Practice Template**
for the file name.
(Save in the Chapter-17
training files folder.)

Press **Save**.

<u>NOTE:</u> *The template can be edited after testing it out to
adjust the text position or to add new text.*

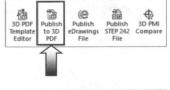

7. Publish to 3D PDF:

<u>Exit</u> the 3D PDF Template Editor and return to the
SOLIDWORKS application. (Use the **Part for Template**.)

Click **Publish to 3D PDF** command on the MBD toolbar.

Click **Browse**. Go to the Training Folder, Chapter 17
and open the template: **SW 3D PDF Practice Template**.

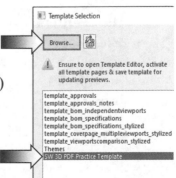

The Publish to 3D PDF window appears in the graphics area.

Expand the section: Primary & Thumbnail Views and select the **Isometric View, Front View, Top View, and Right View** from the 3D Views area in the bottom pane.

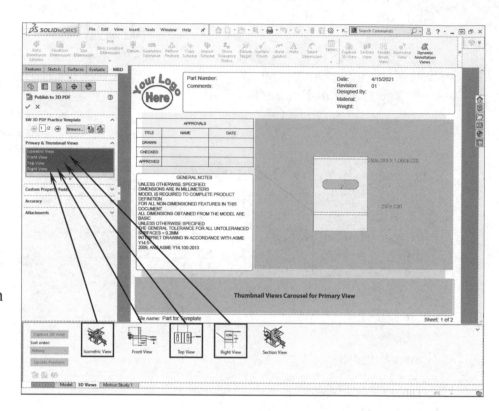

Click **Next** to switch to **Sheet 2**.

Click inside the box: **Independent Viewport** and select the **Section View** from the 3D Views, in the bottom pane.

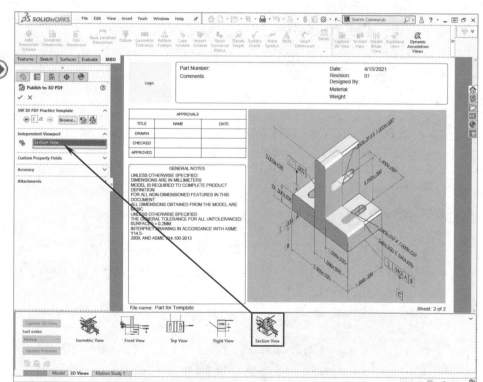

Click **OK** to save to PDF.

For file name, enter **3D PDF Test-Publish**.

Press **Save**.

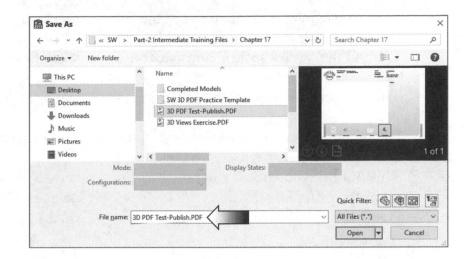

SOLIDWORKS MBD runs through each 3D view that was created and saved in the 3D View window and launches the Acrobat Reader program.

NOTE: _If 3D PDF does not open the template on its own, launch the Acrobat Reader separately and then open the 3d PDF Test-Publish document from there._

If prompted, select: **Trust this document always** (arrow).

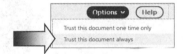

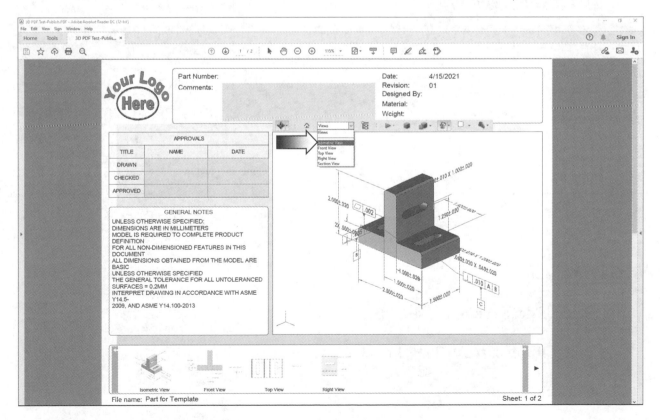

Click the **Model Tree** icon to see the information available (arrow).

Expand the **Views** section and select **Section View** (arrow).

The **Custom Properties** are filled-in automatically (arrow).

Change to different views and zoom, rotate to test them out.

This 3D PDF document can now be sent to others to present the product and manufacturing information (PMI). This PMI includes 3D Views, Dimensions. Geometric Dimensioning and Tolerancing (GD&T) data, and annotations.

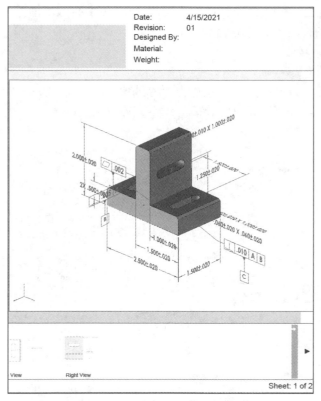

Sheet 1of 2

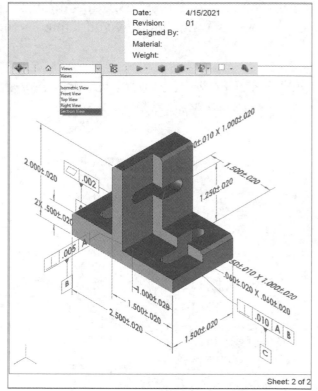

Sheet 2 of 2

8. Saving your work:

Select **File, Save As**.

Enter: **Publish to 3D PDF** for the file name.

Click **Save**.

CHAPTER 18

Sheet Metal Parts

Sheet Metal Parts
Hard Drive Enclosure

Sheet metal parts are generally used as enclosures for components or to provide support to other components.

A sheet metal part can be created on its own without any references to the components it will enclose, or it can be designed in the context of an assembly that contains the enclosed components, or it can also be designed within another part document in a multibody environment.

There are several approaches to designing a sheet metal part:

1. Design as sheet metal part from the very beginning: When you create a part initially out of sheet metal you can use the **Base Flange** tool with an open or closed profile sketch. The Edge Flange tool will add the other flanges to the base. This eliminates extra steps because you create a part as sheet metal from the initial design stage.

Base Flange   Edge Flanges   Flat Pattern

2. If you build a solid using non-sheet metal features, you need more features: **Base Extrude**, **Shell**, **Rip**, and **Insert Bends**. However, there are instances when it is preferable to build a part and then insert sheet metal bends. For example, conical bends are not supported by sheet metal features such as **Base Flange** and **Edge Flange**. Therefore, you must build the part using extrusions, revolves, and so on, and then add bends to the conical part.

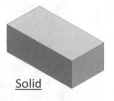   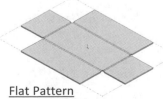

Solid   Shell   Rip Edges   Flat Pattern

Sheet Metal Parts
Hard Drive Enclosure

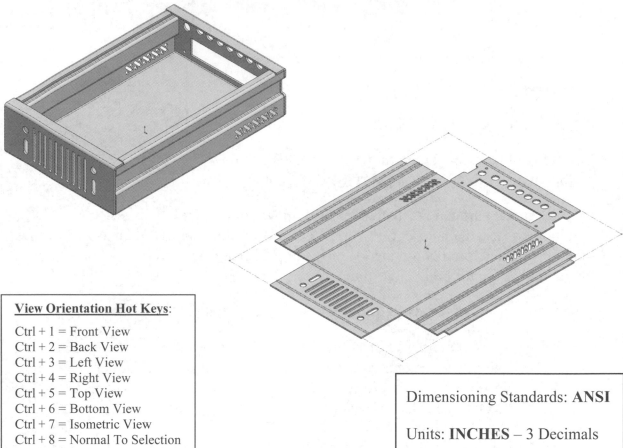

View Orientation Hot Keys:

Ctrl + 1 = Front View
Ctrl + 2 = Back View
Ctrl + 3 = Left View
Ctrl + 4 = Right View
Ctrl + 5 = Top View
Ctrl + 6 = Bottom View
Ctrl + 7 = Isometric View
Ctrl + 8 = Normal To Selection

Dimensioning Standards: **ANSI**

Units: **INCHES** – 3 Decimals

Tools Needed:

Insert Sketch	Line	Rectangle
Centerline	Convert Entities	Base Flange
Edge Flange	Extruded Cut	Library Features
Flat Pattern	Mirror Feature	Cut List

1. Starting a new part document:

Click **File / New / Part**.

From the bottom right corner, set the
Units to **IPS**.

Click **Edit Document Units** and select
ANSI under the Drafting Standard option;
also set the number of decimals to 3 places.

Open a **new sketch** on the Front plane.

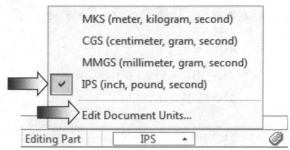

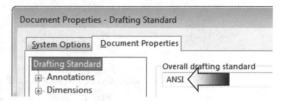

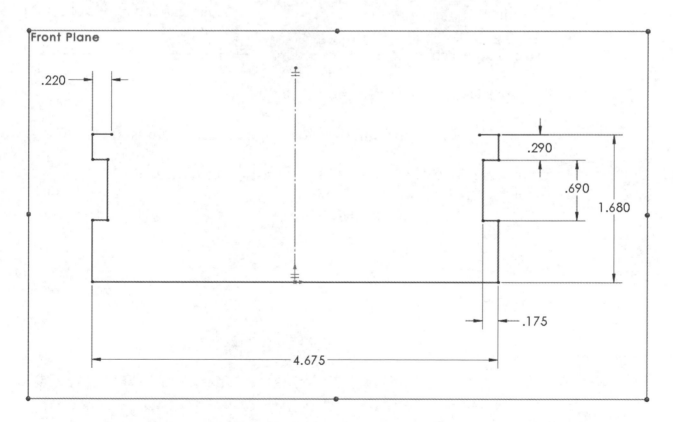

Sketch the profile shown above; start with the vertical centerline ✎ .

Either use the Mirror tool or the Symmetric relations to keep the sketch entities
constrained about the vertical centerline.

Add the dimensions shown to fully define the sketch.

2. Creating the Base Flange :

If the Sheet Metal tab is not visible, right-click one of the tool tabs and enable the Sheet-Metal checkbox.

Switch to the **Sheet Metal** tab and click the **Base Flange** command.

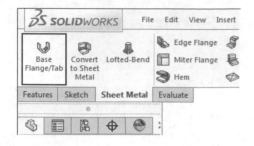

Set the following parameters:

* Direction 1: **Mid Plane**

* Depth: **7.280in**

* Use Gauge Table: **Enabled**

* Sample Table: **Steel** (Steel Air Bending)

* Thickness: **16 Gauge** (added to the outside of the sketch)

* Bend Radius: **.030in** (Override Radius enabled)

Click **OK**.

> 💡 **Base Flange**
>
> A base flange is the first feature in a new sheet metal part. It is also considered as the parent feature of a sheet metal part.

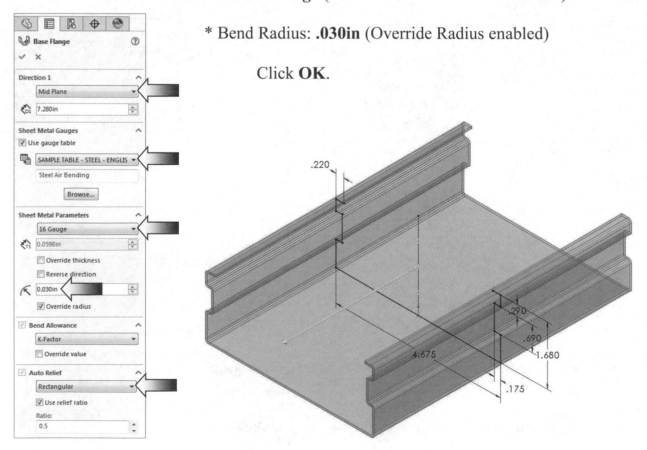

3. Creating a cut feature:

Select the <u>upper face</u> as noted and open a **new sketch**.

Sketch a **Corner-Rectangle** 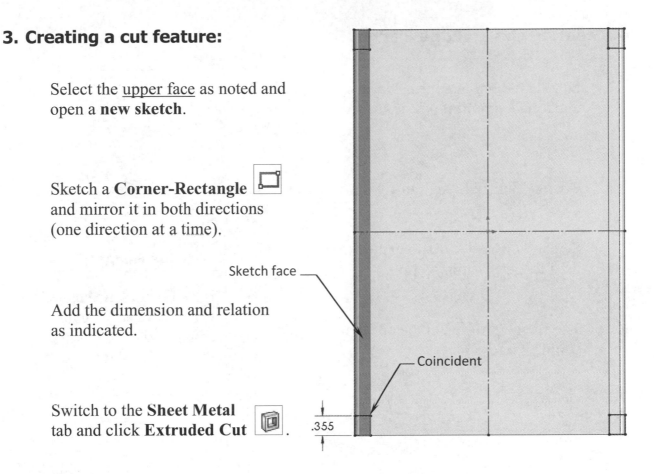 and mirror it in both directions (one direction at a time).

Add the dimension and relation as indicated.

Switch to the **Sheet Metal** tab and click **Extruded Cut**.

Change Direction 1 to **Up To Vertex** and <u>select the vertex</u> as noted in the image below.

Click **OK**.

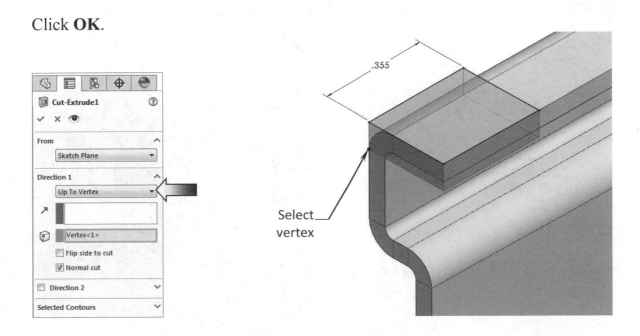

4. Creating the 1st Edge Flange :

Select the <u>bottom edge</u> as noted and click the **Edge Flange** command.

Change Flange Length to **Up To Vertex** and select the <u>upper vertex</u> as indicated.

Select the **Bend Outside** button (arrow) from the Flange Position section.

The flange profile will be edited in the next couple of steps.

> ### Edge Flanges
>
> Edge Flanges are created from the linear or curved edges of the model; its thickness is automatically linked to the thickness of the sheet metal part.

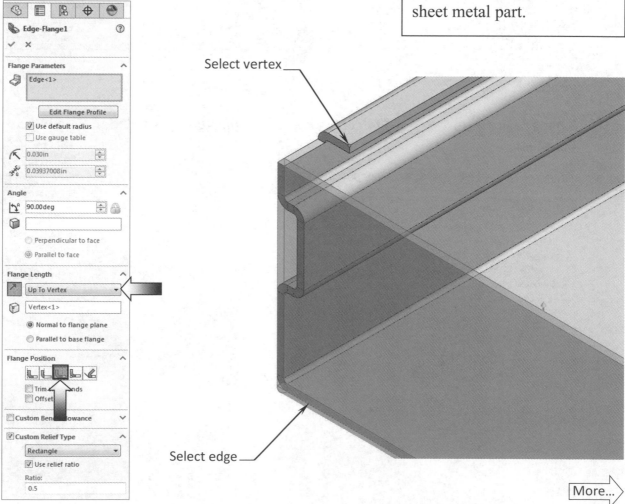

Select vertex

Select edge

More...

Click **Edit Flange Profile**
to edit the sketch (arrow).

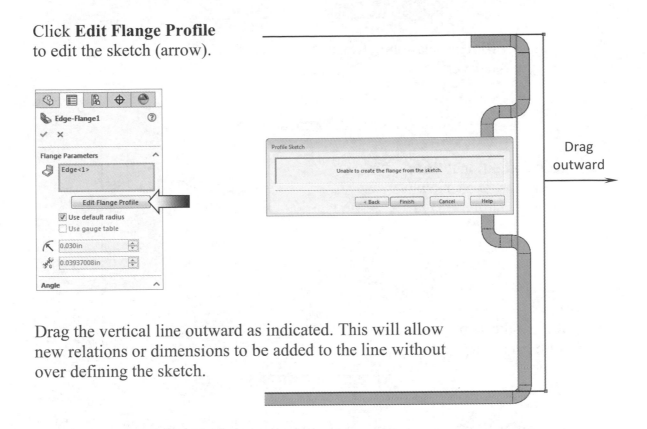

Drag the vertical line outward as indicated. This will allow
new relations or dimensions to be added to the line without
over defining the sketch.

Hold the **Control** key and select the 7 edges as indicated.

Click the **Convert
Entities** command

.

The 7 selected
edges are converted
to 2D sketch
entities.

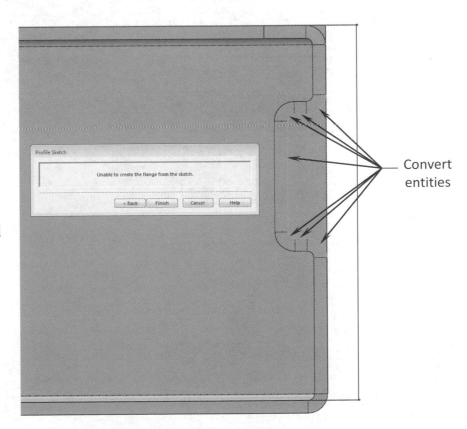

The next step is to
join them into one
continuous profile.

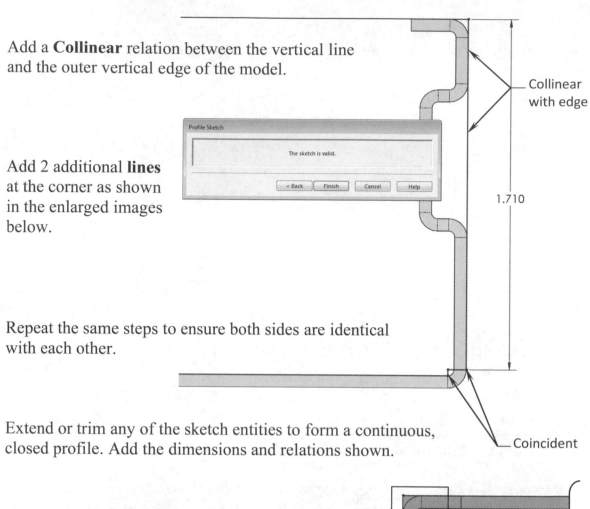

Add a **Collinear** relation between the vertical line and the outer vertical edge of the model.

Collinear with edge

Add 2 additional **lines** at the corner as shown in the enlarged images below.

1.710

Repeat the same steps to ensure both sides are identical with each other.

Extend or trim any of the sketch entities to form a continuous, closed profile. Add the dimensions and relations shown.

Coincident

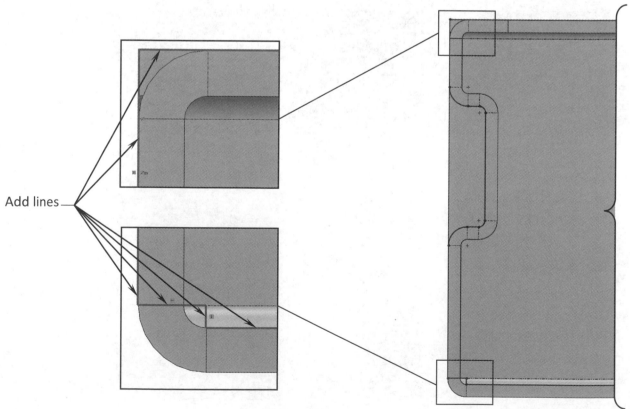

Add lines

Click the **Finish** button
on the Profile Sketch dialog
to exit the edit sketch mode.

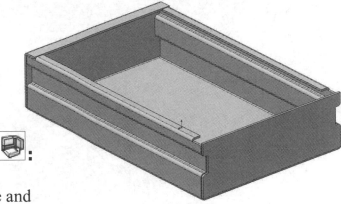

5. Creating the 2nd Edge Flange 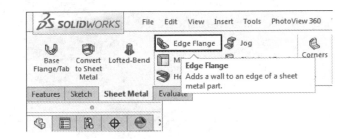:

Select the <u>outer edge</u> of the flange and
click the **Edge Flange** command once again.

Move the cursor <u>inward</u> and click the mouse, then enter **.410in** for Flange Length.

Select the **Material Inside** button
(arrow).

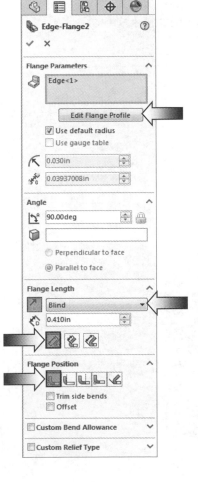

Click the **Edit Flange Profile** button (arrow).

The Flange Profile and length will be edited in
the next couple of steps.

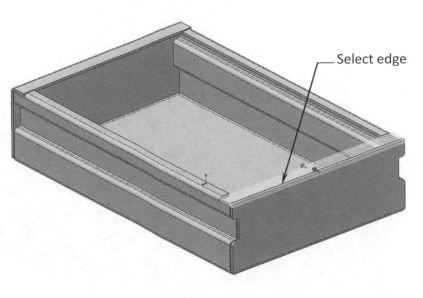

Select edge

Add the dimension **.410"** as shown.

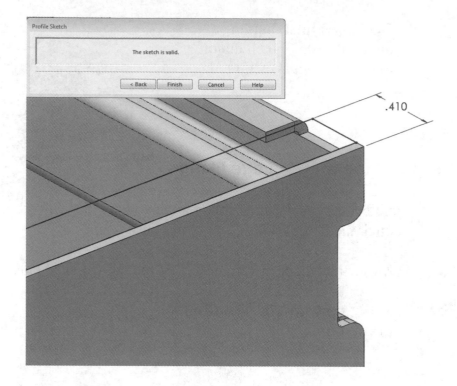

The sketch should be fully defined at this point.

Click the **Finish** button Finish to exit the edit mode.

Examine the resulted flange and compare it with the image shown.

It is good practice to save your work every once in a while. Save the first half of your work as **SM Hard Drive Enclosure.sldprt**.

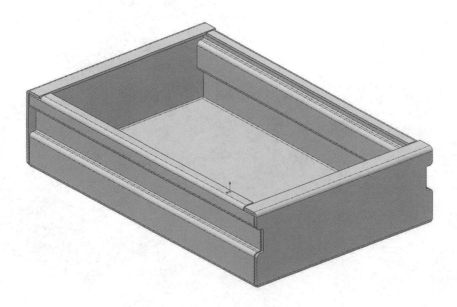

6. Copying and Pasting a sketch:

Keep your model open while opening another document.

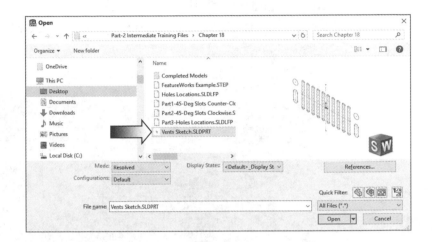

Select **File / Open**.

Browse to the training folder and open the document named **Vents Sketch**.

Select the **Sketch1** from the Feature tree and press **Control + C**.

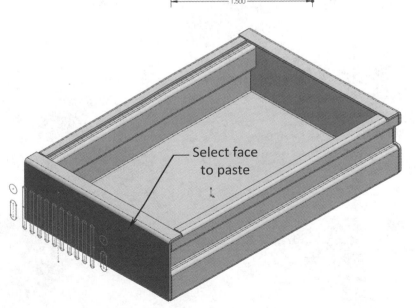

Press **Control + Tab** to switch back to the sheet metal part.

Select the <u>face</u> as noted and press **Control +V**.

Select face to paste

A copy of the selected sketch appears on the front face of the model. The next step is to center it on the face.

Select the Vents Sketch and click **Edit Sketch** 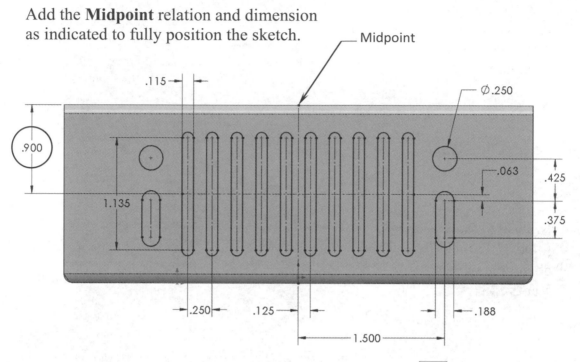.

Add the **Midpoint** relation and dimension
as indicated to fully position the sketch.

Switch to the **Sheet Metal** tab and select **Extruded Cut** .

Keep the Direction 1 at **Blind** (default).

Enable the **Link To Thickness** check box.

The **Normal Cut** check box is selected by default.

Click **OK**.

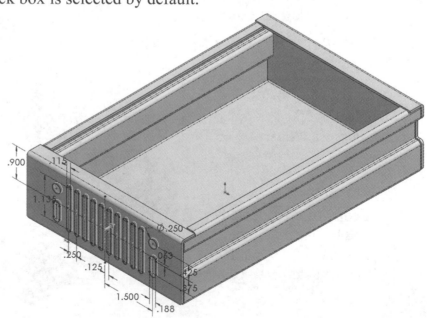

7. Creating more cuts:

The next cut feature will also be copied from an external sketch, similar to the last step.

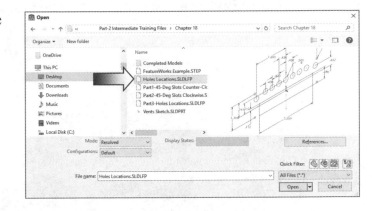

Select **File / Open**.

Browse to the training folder and open a document named **Holes Locations.sldlfp** (*Solid Library Feature Part*).

Select the **Sketch1** from the Feature tree and press **Control + C**.

Press **Control + Tab** to switch back to the sheet metal part.

Select the <u>face</u> as noted and press **Control + V**.

A copy of the selected sketch appears on the back face of the model. The next step is to center it on the face.

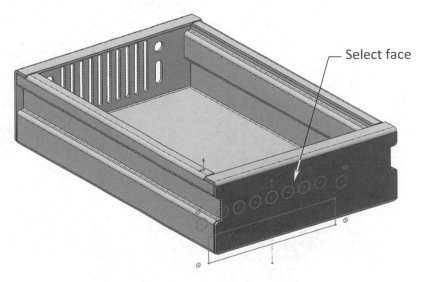

Select face

Select the Hole Locations sketch from the Feature tree and click **Edit Sketch** .

Add the **Midpoint** relation and dimensions as indicated (circled) to fully position the sketch.

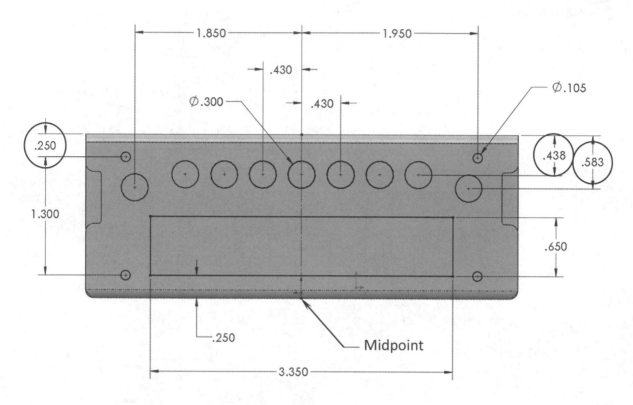

Switch to the **Sheet Metal** tab and click **Extruded Cut** .

Leave Direction 1 at **Blind** (default).

Enable the **Link-To Thickness** and the **Normal Cut** check boxes.

Click **OK**.

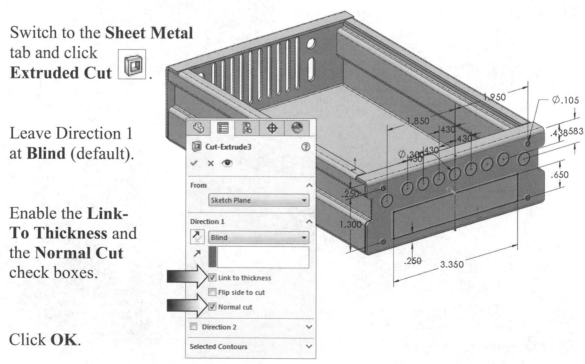

8. Adding a Library Feature Cut :

Any features created previously and saved as a Library Feature Part can be "re-used" by dragging and dropping from the file Explorer window.

Select the **File Explorer** tab (arrow), browse to the training folder and expand the Sheet Metal Parts folder.

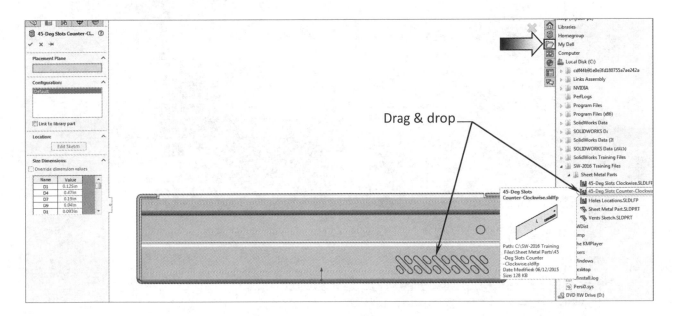

Drag and drop the library feature named **45-Deg Slots Counter-Clockwise** on the <u>right side</u> of the model as shown (Control + 4 = Right side view).

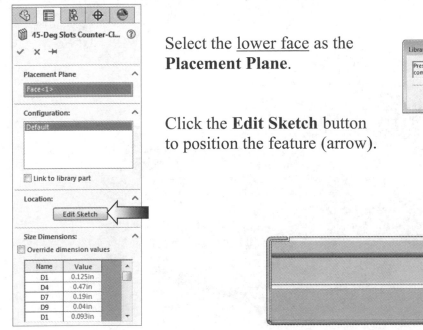

Select the <u>lower face</u> as the **Placement Plane**.

Click the **Edit Sketch** button to position the feature (arrow).

Library Feature Profile

Press Back to go back to editing the feature settings, Finish to complete the feature, or Cancel to abort the sketch edit.

< Back | Finish | Cancel | Help

Placement Plane

Add the 2 location dimensions (circled) to
fully define the sketch.

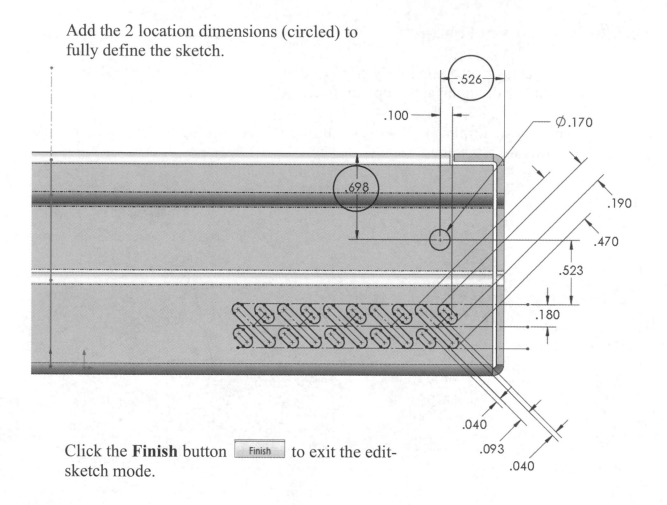

Click the **Finish** button Finish to exit the edit-
sketch mode.

The cut feature will be mirrored to the opposite side in the next step.

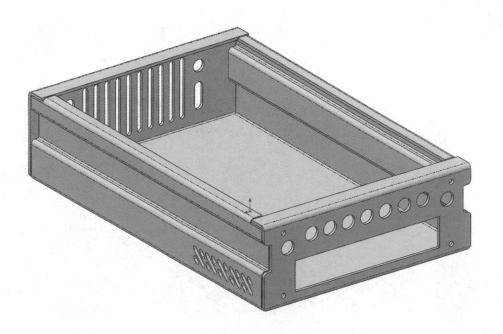

9. Mirroring a Feature:

Once applied to the part, the library features can be copied, patterned, or mirrored just like any other features.

Switch to the **Features** tab and click the **Mirror** command .

For Mirror Face/Plane, select the **Right** plane from the FeatureManager tree.

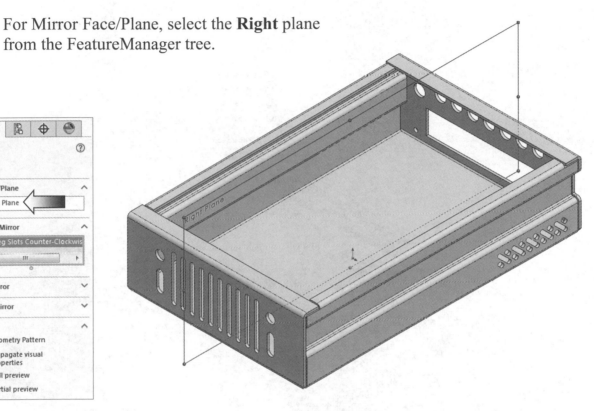

For Features to Mirror, select the **45-Deg Slots** also from the FeatureManager tree.

The preview graphics show the selected feature is being mirrored to the opposite side.

Click **OK**.

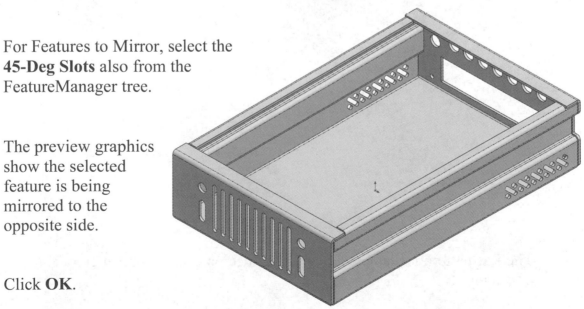

10. Toggling the Flat Display:

A flat-pattern can now be "superimposed" on top of the fold-pattern. This unique option allows the user to preview the flat pattern while comparing the features between the flat and fold patterns.

Right-click a face of the sheet metal part and select **Toggle Flat Display**.

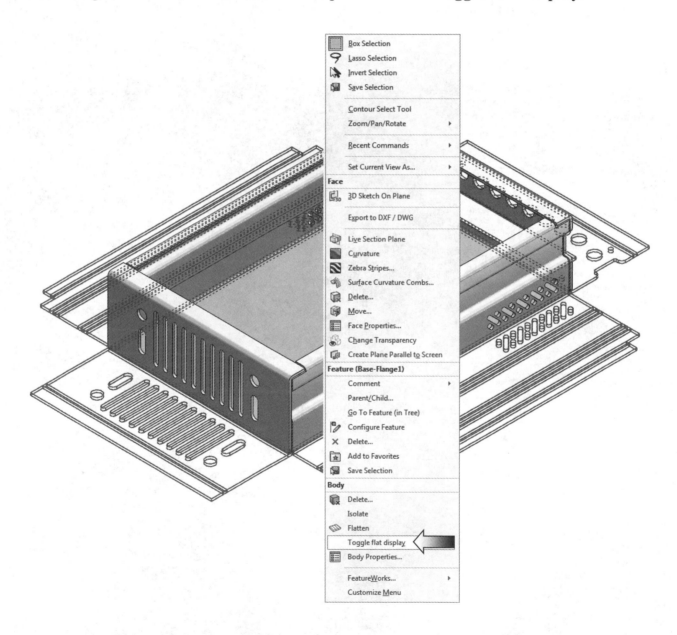

The flat-pattern is displayed over the folded part, in the preset color (blue, green, or orange).

11. Creating the Flat Pattern:

An actual Flat pattern can be created at any time. All features within the sheet metal part will be flattened, and a bounding box is added to show the overall size of the flat sheet metal.

Change to the Sheet Metal tab and click the **Flatten** command .

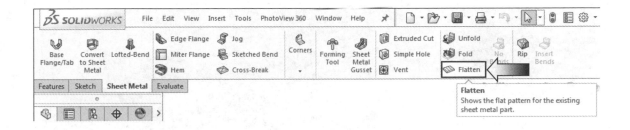

The sheet metal part is flattened, and all of its features are unfolded with it.

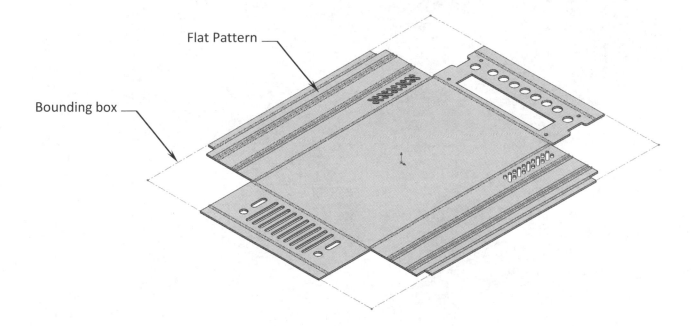

Flat Pattern

Bounding box

A "dotted box" is added around the part to show the **bounding box area** of the flat sheet. Further information like thickness, bend allowance, bend radius, etc. can be retrieved by accessing the properties of the sheet.

12. Accessing the Sheet Properties:

Expand the **Cut List** from the FeatureManager tree and right-click on the **Sheet** item and select **Properties** (arrow). *Note: The Cut List must be updated.*

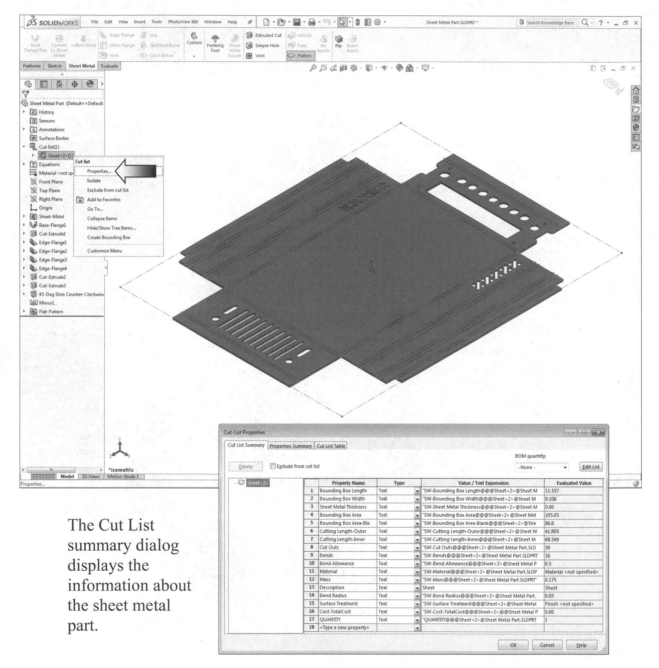

The Cut List summary dialog displays the information about the sheet metal part.

13. Saving your work:

Save your work as **SM Hard Drive Enclosure.sldprt** and replace the one saved earlier.

Sheet Metal Parts
Multibody Design

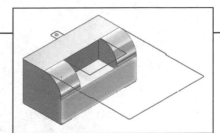

Sheet Metal Parts
Multibody Design

Sheet metal parts are generally used as enclosures for components or to provide support to other components. A sheet metal part can be designed on its own without any references to the parts it will enclose, or it can be designed in the context of an assembly that contains the enclosed components, or a sheet metal part can be designed within another part document in a multibody environment.

This handout will discuss the method of designing a sheet metal part in a multibody environment.

SOLIDWORKS multibody part functionality allows you to work with several bodies in one part to create complex sheet metal designs.

Multibody sheet metal parts can consist of multiple sheet metal bodies or a combination of sheet metal and other bodies such as weldment bodies.

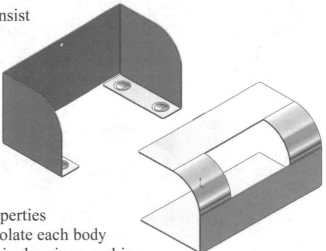

Each body has its own sheet metal and material definition, flat pattern, and custom properties.

You can use these body-related properties in BOMs and drawings. You can isolate each body and display each body individually in drawings, making it possible to interact with the part as if it is an assembly.

Sheet Metal Parts
Multibody Design

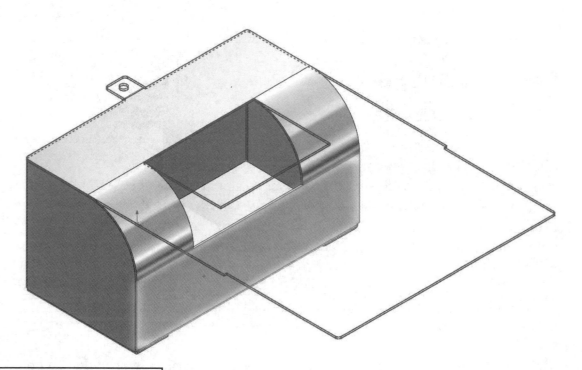

View Orientation Hot Keys:

Ctrl + 1 = Front View
Ctrl + 2 = Back View
Ctrl + 3 = Left View
Ctrl + 4 = Right View
Ctrl + 5 = Top View
Ctrl + 6 = Bottom View
Ctrl + 7 = Isometric View
Ctrl + 8 = Normal To Selection

Dimensioning Standards: **ANSI**
Units: **INCHES** – 3 Decimals

Tools Needed:

 Base Flange

 Edge Flange

 Forming Tool

 Mirror Feature

 Break Corner

 Extruded Cut

1. Starting a new part document:

Click **File / New**.

Select the **Part** template and click **OK**.

Set the Drafting Standard to **ANSI**.

Set the Units to **IPS**, **3 decimal places**.

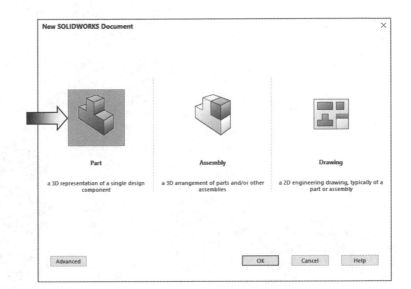

2. Creating the main sketch:

Open a **new sketch** on the <u>Top</u> plane.

Sketch a **horizontal centerline** starting from the origin.

Add the relation and dimensions shown to fully define the sketch.

Right-click on one of the tool tabs (Features, Sketch, Evaluate, etc.) and enable the **Sheet-Metal** tool bar.

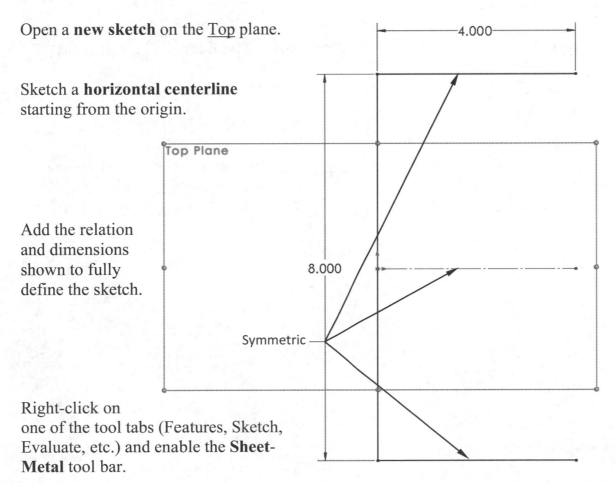

3. Creating the 1st body:

Click the **Base Flange** command .

Expand the Direction 1 section and select the **Blind** type.

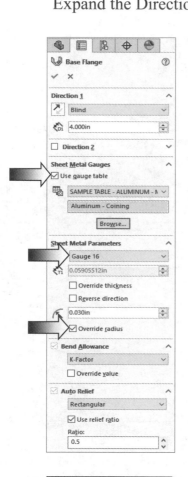

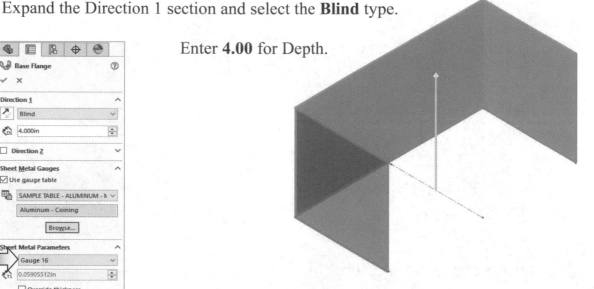

Enter **4.00** for Depth.

Enable the **Use Gauge Table** checkbox and select:
Sample Table - Aluminum – English Units.

For Sheet Metal Parameters, select **16 Gauge**.

Click the **Override Radius** checkbox and enter: **.030in**

Use the default **K-Factor** for Bend Allowance.

For Auto Relief, select the **Rectangular** type.

Use the relief Ratio of **0.5**

Click **OK**.

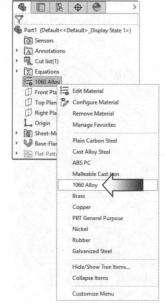

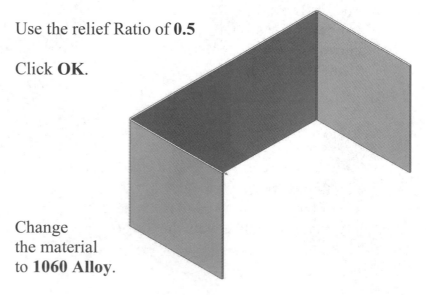

Change
the material
to **1060 Alloy**.

4. Adding fillets:

Click **Corners**, **Break Corners**.

Select the **Fillet** option (arrow).

Enter **1.750in** for radius size.

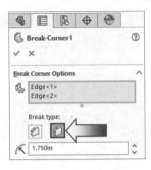

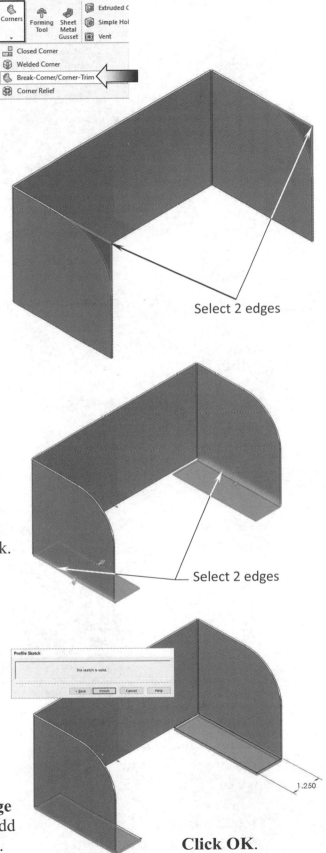

Select the **2 edges** indicated.

Click **OK**.

5. Adding the Edge Flanges:

Click **Edge Flange** .

Select the **1st edge** on the left side and drag inward and click.

Select the **2nd** edge as noted.

For Flange Position, select the option: **Bend Outside**.

Enable the **Offset** checkbox and enter **.030in**.

Click the **Edit Flange Profile** button and add the **1.250** dimension.

Click OK.

Press **Control + 4** to inspect the .030in. offset distance from the right side.

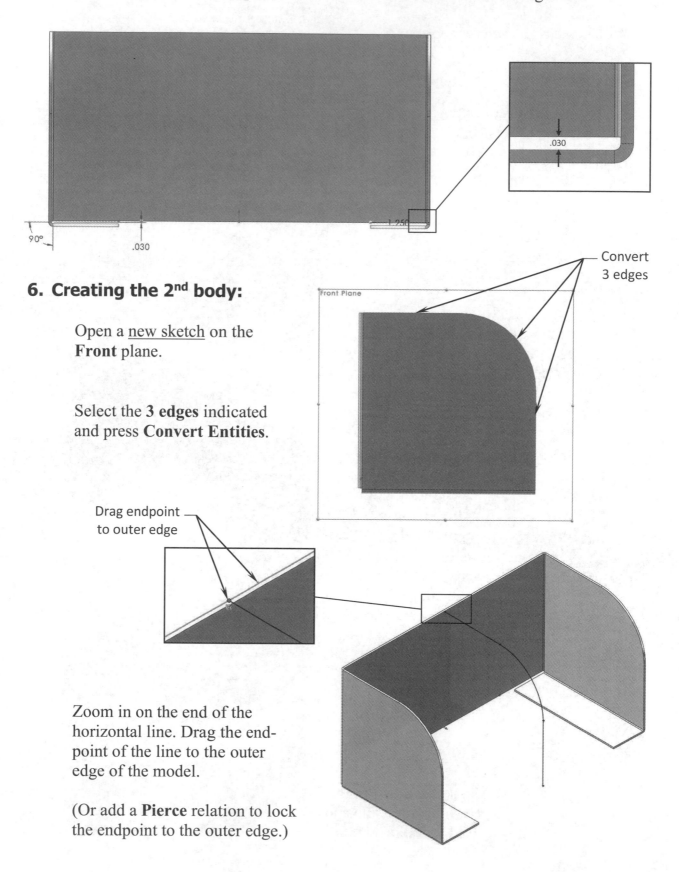

6. Creating the 2nd body:

Open a new sketch on the **Front** plane.

Select the **3 edges** indicated and press **Convert Entities**.

Zoom in on the end of the horizontal line. Drag the end-point of the line to the outer edge of the model.

(Or add a **Pierce** relation to lock the endpoint to the outer edge.)

7. Making the Base Flange:

Select the **Base Flange** command 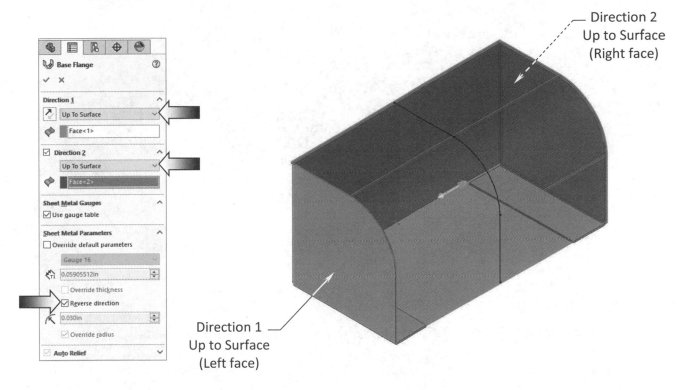.

For Direction 1, select **Up To Surface** from the list and click the **face** on the left side as indicated in the image below.

For Direction 2, select **Up To Surface** again and click the **face** on the right side as noted.

Direction 2
Up to Surface
(Right face)

Direction 1
Up to Surface
(Left face)

Enable the **Reverse Direction** checkbox to add the thickness to the outside.

Leave all other parameters at their default settings.

Click **OK**.

A second solid body is created and listed under the Cut-List (arrow)

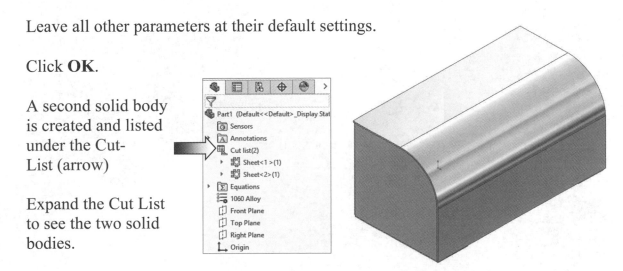

Expand the Cut List to see the two solid bodies.

8. Adding relief cuts:

Select the <u>face</u> as noted and open a **new sketch**.

Sketch a couple of **rectangles** at the 2 corners of the selected face.

Add the **Collinear** relations as indicated to fully define the sketch.

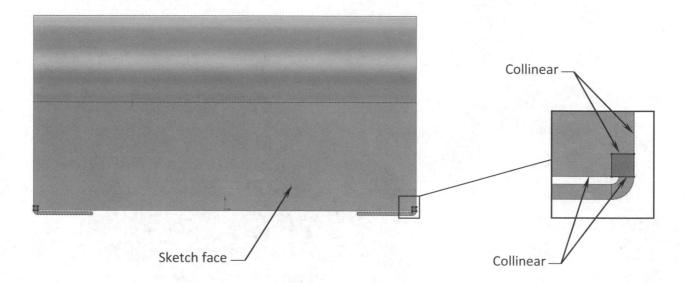

Collinear

Sketch face

Collinear

Switch to the Sheet Metal toolbar and click **Extruded Cut**.

Use the default **Blind** type and enable the **Link To Thickness** checkbox (arrow).

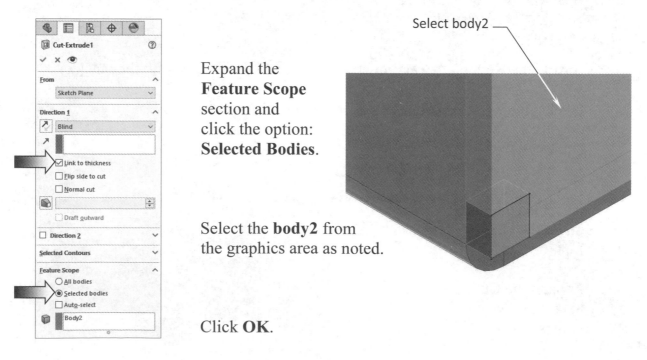

Select body2

Expand the
Feature Scope
section and
click the option:
Selected Bodies.

Select the **body2** from
the graphics area as noted.

Click **OK**.

9. Creating the bottom flange:

Select the <u>outer edge</u> of the **Body2** and click the **Edge Flange** command.

Move the cursor inward and click to define the direction of the flange.

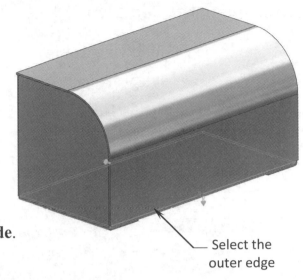

Select the outer edge

For Flange Position, select **Material Inside**.

Enable the Offset checkbox (arrow) and enter: **.030in**.

Click the **Edit-Flange Profile** button (arrow) to edit the sketch.

Add the width dimension **4.00** shown.

4.000

Click **Finish** Finish .

Hide the **Body2**.

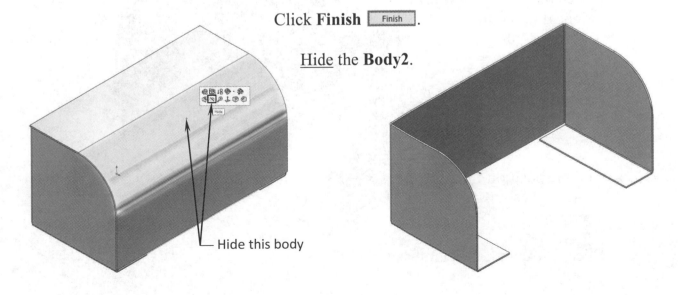

Hide this body

10. Adding the Dimples:

Expand the following: **Design Library, Forming Tools, Embosses**.

Drag and drop the **Dimple** Form Tool onto the <u>upper face</u> of the flange approximately as shown.

Click the **Position** tab (arrow).

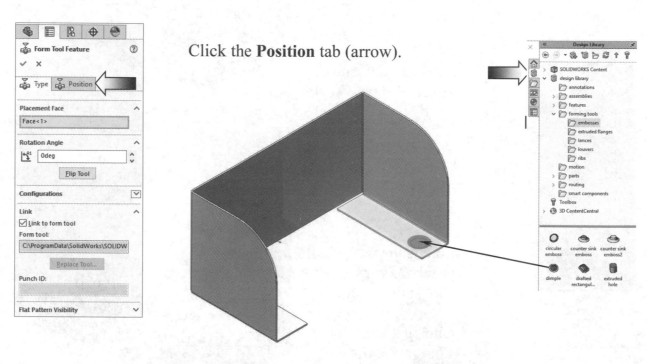

The **Sketch Point** command is selected automatically.

Place another **Dimple** form-tool next to the first one.

Add the **Horizontal** relation between the 2 centers.

Add the position dimensions shown.

Click **OK** to exit the command.

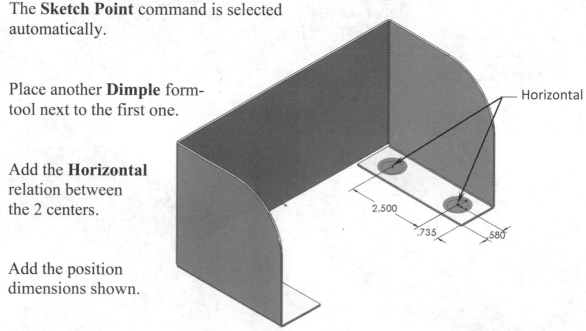

11. Mirroring the formed features:

Switch back to the **Features** toolbar and click the **Mirror** command .

For Mirror Face / Plane, select the **Front** plane from the Feature tree.

For Features to Mirror, select the **Dimple 2017** either from the graphics area or from the Feature tree.

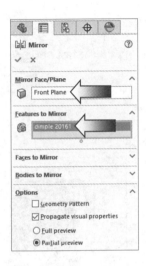

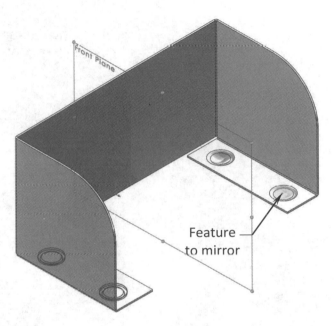

Feature — to mirror

Click **OK**.

12. Adding fillets:

Change to the **Sheet Metal** toolbar and select: **Corners**, **Break-Corner** .

For Break Type, select the **Fillet** option (arrow).

For Radius, enter **.080in**.

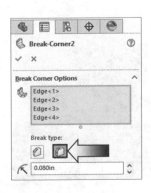

Select the **4 edges** as noted.

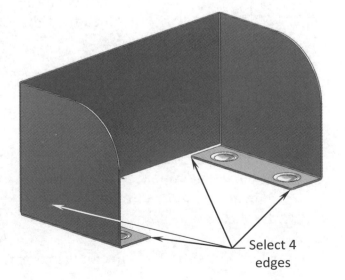

Select 4 edges

Click **OK**.

13. Creating an edge Flange:

Rotate the model to the back side similar to the image shown below.

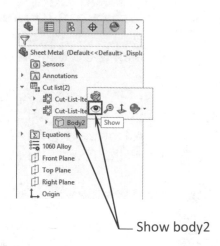

Select the <u>edge</u> indicated and click **Edge Flange**.

Move the cursor downward and click to lock the direction.

For Flange Position, select **Bend Outside**.

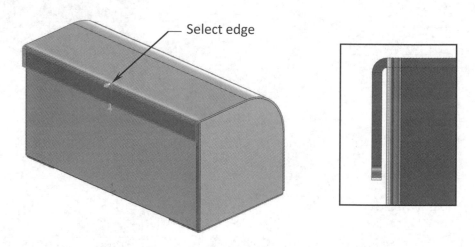

Select edge

Enable the **Offset** checkbox and enter **.030in** for distance.

Click the **Edit Flange Profile** button to edit the sketch.

Drag the 2 vertical lines <u>inward</u> to resize them.

Add a **vertical centerline** and a **circle** to the sketch. Also add the dimensions shown to fully define the sketch.

Click **Finish** [Finish] to exit the command.

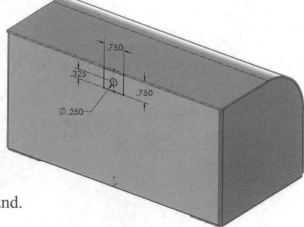

14. Adding more fillets:

Click a face of the Body1 and select **Hide**.

Click **Corners**, **Break-Corner** 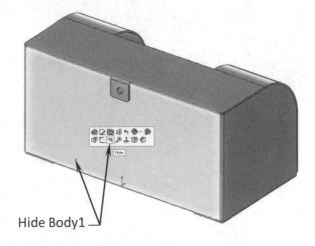.

For Break Type, select the **Fillet** button.

Hide Body1

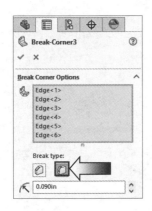

For Radius, enter **.090in**.

Select the **6 edges** as indicated.

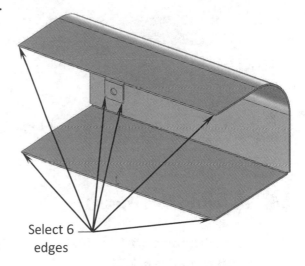

Click **OK**.

Select 6 edges

15. Creating a cutout:

Change to the **Isometric** orientation (Control + 7).

Select the front <u>face</u> as noted and open a **new sketch**.

Sketch a **rectangle** and a **centerline**. Lock one end of the centerline on the origin and the other end on the mid-point of the rectangle.

Add the width dimension **4.00in** to fully define the sketch.

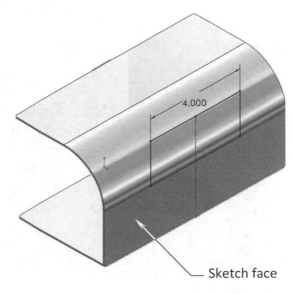

4.000

Sketch face

Click **Extruded Cut**.

For Direction 1, select **Up to Vertex** and select the <u>midpoint</u> of the edge as noted.

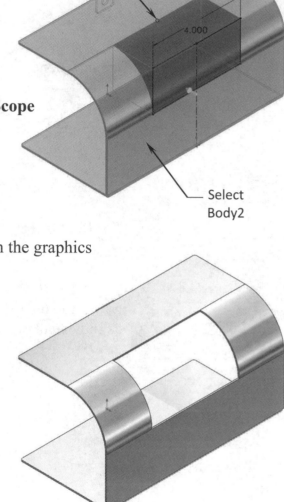

Select vertex

4.000

Select Body2

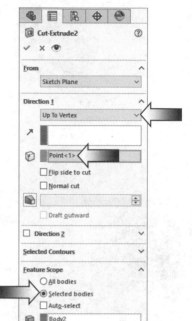

Expand the **Feature Scope** section and click the **Selected Bodies** option (arrow).

Select the **Body2** from the graphics area to cut.

Click **OK**.

16. Showing the Body1:

Expand the **Cut List** feature and the **Cut-List-Item1**.

Note: The Cut List must be updated.

Click the **Body1** and select **Show**.

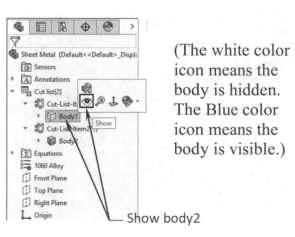

Show body2

(The white color icon means the body is hidden. The Blue color icon means the body is visible.)

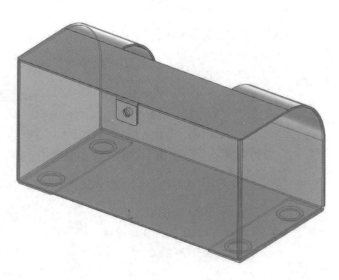

17. Adding the mounting hole:

Open a **new sketch** on the <u>back face</u> of the **Body1**.

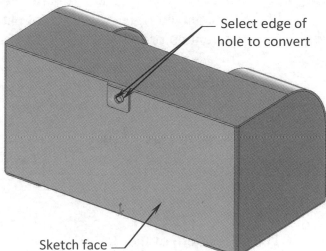

Select edge of hole to convert

Select the <u>circular edge</u> of the hole and click **Convert Entities**.

The selected edge is converted into a circle and projected onto the sketch face.

Sketch face

Change to the **Sheet Metal** toolbar and click **Extruded Cut**.

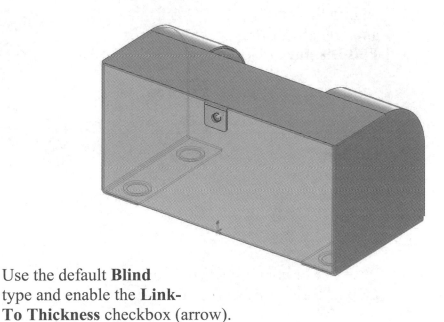

Use the default **Blind** type and enable the **Link-To Thickness** checkbox (arrow).

Expand the **Feature Scope** section and click the **Selected Bodies** option (arrow). Select the **Body1** from the graphics area to cut.

Click **OK**.

18. Toggling the Flat Display:

Any solid body can be flattened in a multi-body sheet metal part. The flat pattern of a body is displayed right over the fold pattern in a different color.

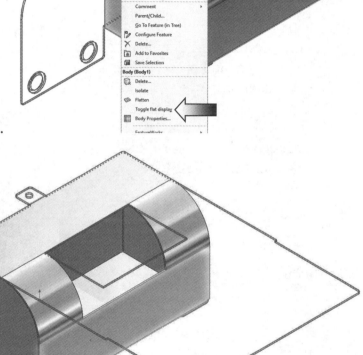

Right-click on **Body1** and select: **Toggle Flat Display** (arrow).

The flat pattern of the Body1 is shown over its folded model.

Right-click on **Body2** and select: **Toggle Flat Display**.

The flat pattern of the Body2 is "superimposed" over the fold pattern.

Push **Esc** or click anywhere in the graphics area to cancel the flat display.

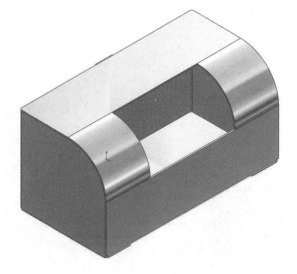

19. Saving your work:

Select **File / Save As**.

Enter: **Multibody Sheet Metal Parts** for the file name and press **Save**.

Using FeatureWorks - Feature Recognition

The FeatureWorks software recognizes features on an imported solid body in a SOLIDWORKS part document.

This exercise will guide you through the basic use of this application to recognize the features of an imported part.

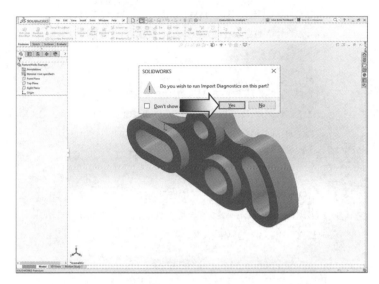

1. Open a document named:
FeatureWorks Example.STEP

2. Click YES to run the Import Diagnostic application.

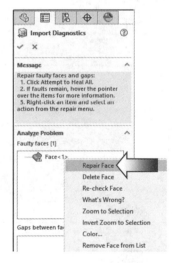

The Import Diagnostic application repairs faulty surfaces, knits repaired surfaces into closed bodies, and makes closed bodies into solids.

The Import Diagnostic reports an error is discovered in one of the faces.

Right-click the **Face<1>** in the Faulty Face section and select **Repair Face**.

The faulty face is corrected and the message (in green color) indicates that there are no faulty faces or gaps remaining in the geometry.

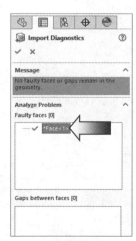

3. Click OK to close the Import-Diagnostic application.

Change the display to **Shaded with Edges**.

4. Setting the options:

Right-click the **Imported1** from the Feature tree and select **FeatureWorks/ Options**.

It is best to "train" the FeatureWorks application on how you want the features to be recognized, and whether to add relations and dimensions automatically.

Go through the 4 options and enable or disable the options as indicated with the arrows below.

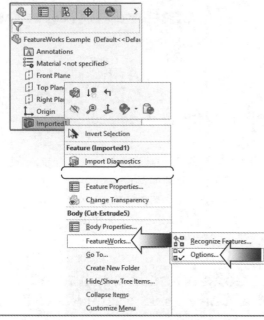

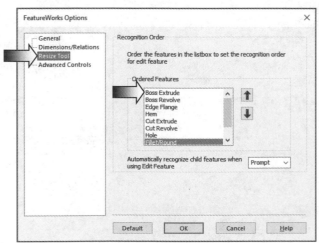

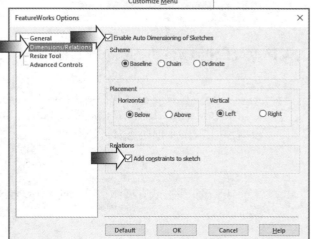

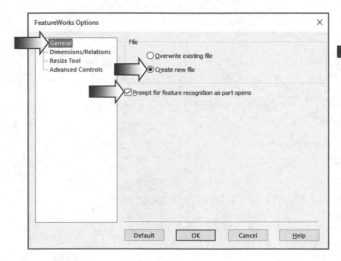

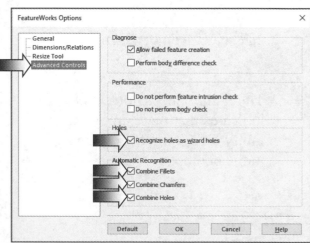

Click **OK** to close the Option dialog box.

5. Recognizing features:

Right-click the feature **Imported1** again and select **Recognize Features** (arrow).

For Recognition Mode, select **Automatic**.

Enable **Extrudes, Holes, and Fillets/Chamfers**.

Clear the other checkboxes and click **Next**.

FeatureWorks recognizes all features except for the main body. It will need to be recognized separately.

Click **OK** to run the Features Recognition. Click OK again to bypass the Intermediate Stage.

The features are recognized and displayed on the Feature tree in the same order that you "trained" the application earlier.

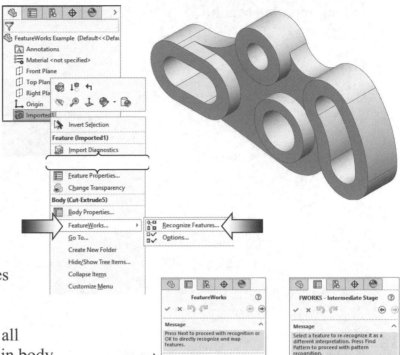

6. Recognizing the base:

Right-click the feature **Imported1** and select **Recognize Features** to re-run the utility.

For Recognition Mode, select **Interactive**.

Leave the Feature Type at **Boss Extrude** and select the **Front** and **Back** faces as noted and click **Recognize**.

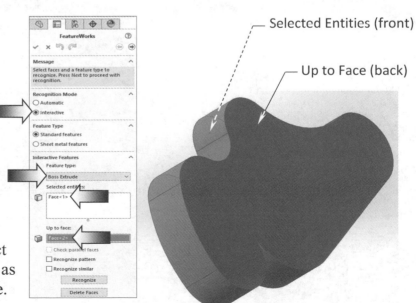

Selected Entities (front)

Up to Face (back)

The Imported1 feature is recognized as an Extruded Boss feature.

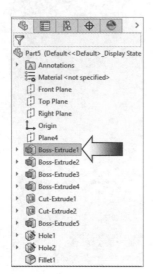 

There are a total of **9 features** that were recognized by the FeatureWorks application.

7. Saving your work:

Save your work as
FeatureWorks Example.sldprt.

Close all documents.

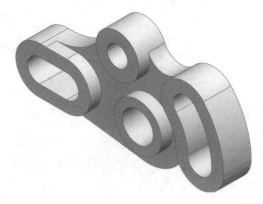

CHAPTER 19

Plastic Parts

Plastic Parts

When designing a plastic part there are a few key points to keep in mind:
First, the wall thickness should be uniform throughout the part and have adequate drafts.

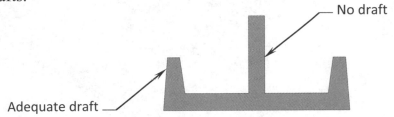

Avoid solid features as thicker areas will shrink and warp more than thin areas.

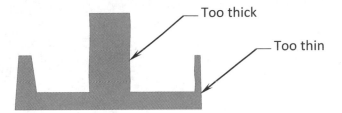

Second, all corners, both internal and external, should be filleted to reduce sink marks. Sharp corners can crack, and they require a longer cooling time.

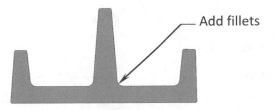

Lastly, add ribs to increase rigidity of the plastic part.

Plastic Parts

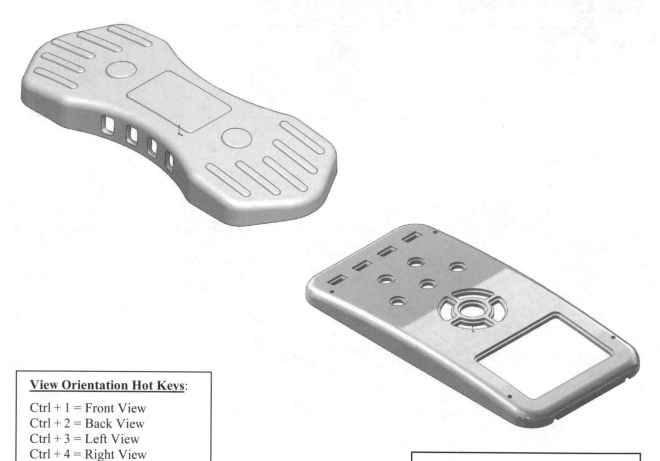

<u>**View Orientation Hot Keys**</u>:

Ctrl + 1 = Front View
Ctrl + 2 = Back View
Ctrl + 3 = Left View
Ctrl + 4 = Right View
Ctrl + 5 = Top View
Ctrl + 6 = Bottom View
Ctrl + 7 = Isometric View
Ctrl + 8 = Normal To Selection

Dimensioning Standards: **ANSI**

Units: **INCHES** – 3 Decimals

Tools Needed:

Centerline	Rectangle	Circle
Offset Entities	Convert Entities	Fillet/Round
Extruded Boss-Base	Extruded Cut	Shell

1. Opening a part document:

Select **File / Open**.

Browse to the Training Folder and open the part document named:
Plastic Part_Design1.sldprt.

This document contains a single sketch which will be used to create the base feature.

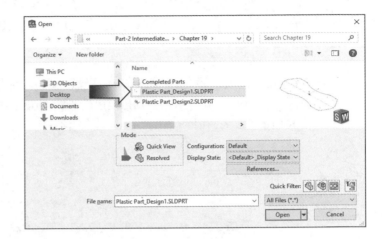

2. Making the base feature:

<u>Edit</u> the **Sketch1** and verify the status of the sketch. It should already be fully defined.

Switch to the **Features** tab and click:
Extruded Boss-Base.

For Direction 1, use the default **Blind** type.

For Depth, enter: **.760in**

For Draft, enter: **2.00deg**

Click **OK**.

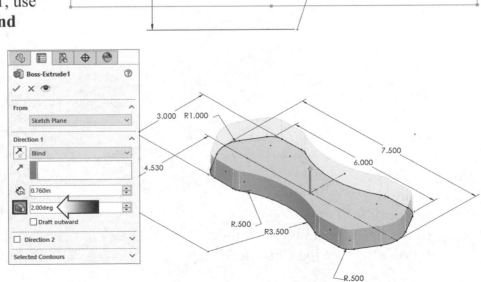

3. Creating a cut feature:

Select the <u>Front</u> plane and open a **new sketch**.

Sketch the profile and add the dimensions/relation as shown below. There are two vertical lines, one horizontal line, and one 3-Point Arc in this sketch.
(Locate the 2 ends of the arc with dimensions, not midpoint relations).

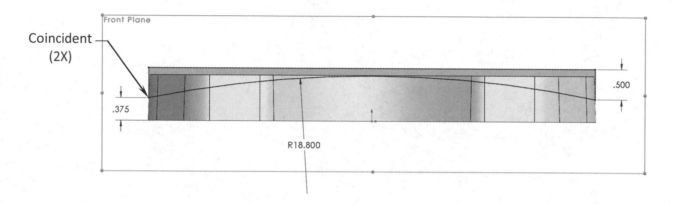

Switch to the **Features** tab and click: **Extruded Cut**.

For Direction 1, select:
Through All-Both.

Click **OK**.

The cut feature creates a curved surface along the top portion of the base. This gives it a little more attractive look than a flat top.

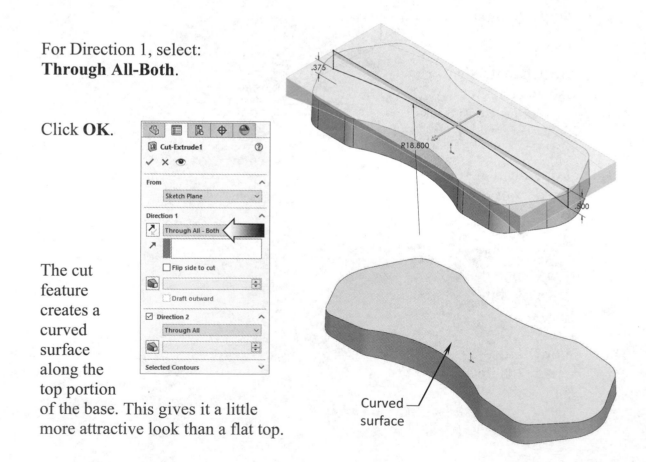

Curved surface

4. Adding the raised features:

Select the Top plane and open a **new sketch**.

Sketch the profile shown below. It is symmetrical about the vertical centerline.

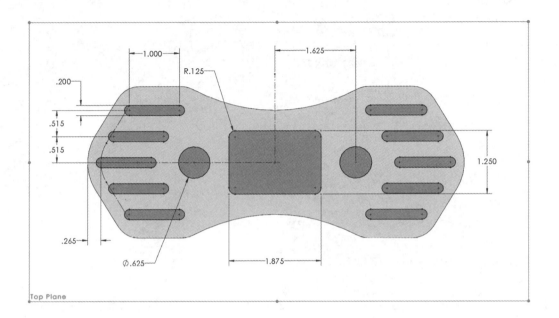

Add the dimensions and relations nccdcd to fully define the sketch.

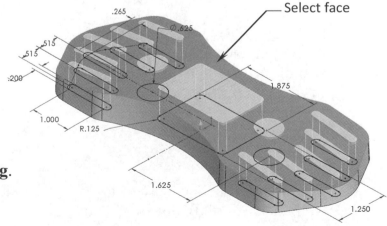

Switch to the **Features** tab and click **Extruded Boss-Base**.

For Direction 1, select **Offset from Surface** and click the face as noted.

Enter **.060in** for Depth and enable the **Reverse Offset** and **Translate Surface** checkboxes.

For Draft angle, enter **1.00deg**.

Click **OK**.

5. Adding a .1875″ fillet:

Click **Fillet**.

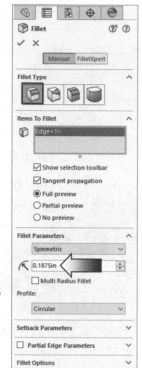

Use the default **Constant Size** option.

Enter **.1875in** for radius size.

Select the <u>edge</u> as indicated. All upper edges are selected automatically.

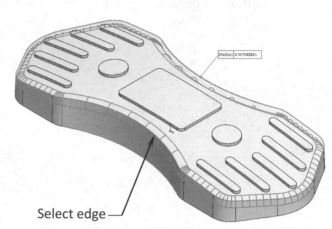

Select edge

Click **OK**.

6. Shelling the model:

Click **Shell**.

For Wall Thickness, enter: **.050in**.

For Faces to Remove, select the <u>bottom face</u> of the model as indicated.

Click **OK**.

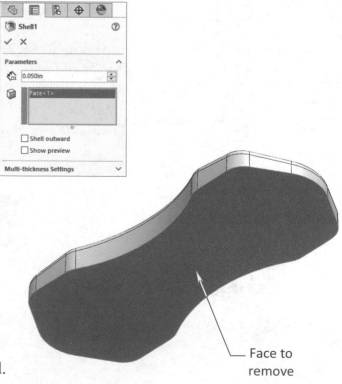

Face to remove

The model is shelled, leaving a wall thickness of .050in all around.

7. Adding the interlock feature:

Select the <u>face</u> as indicated and open a **new sketch**.

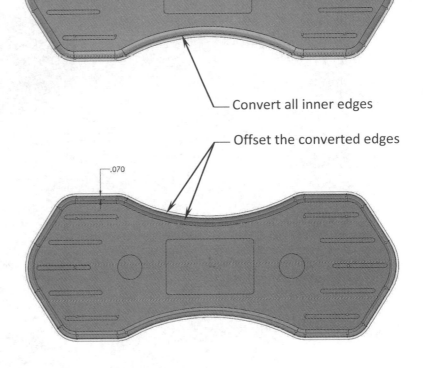

Sketch face

Right-click one of the <u>inside edges</u> and pick: **Select Tangency**.

Click **Convert Entities**.

Convert all inner edges

Offset the converted edges

Right-click one of the converted edges and pick: **Select Chain**.

Click **Offset Entities**.

Enter: **.070in** (inside) for Offset Distance.

.070

Switch to the **Features** tab and click **Extruded Boss-Base**.

For <u>Extrude From</u>, enter: **.070in** and click **Reverse**.

For Direction 1, select: **Up to next**.

For Draft, enter: **2.00deg Outward**.

Click **OK**.

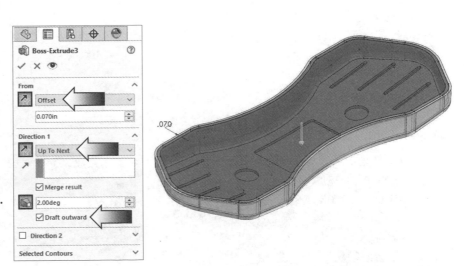

8. Creating a Face Fillet:

Click **Fillet**.

For Fillet Type, select **Face Fillet**.

Select the **Face1** and **Face2** as noted.

Enter **.100in** for radius.

Click **OK**.

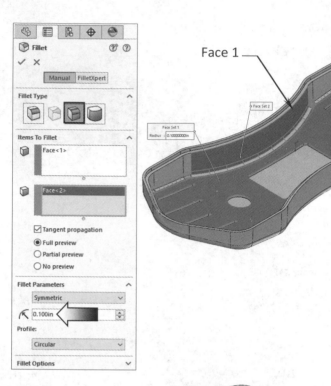

9. Adding a .015" Fillet:

Click **Fillet** again.

For Fillet Type, click the **Constant Size** option.

Enter **.015in** for radius size.

Select the <u>faces</u> and the <u>lower edges</u> of the raised features as shown in the 2 images.

Click **OK**.

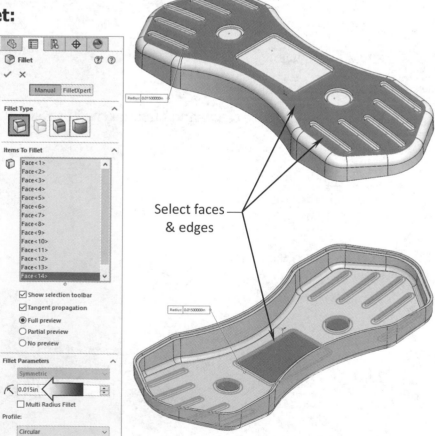

10. Adding another .015" Fillet:

Click **Fillet** again.

Keep all parameters the same as the last fillet.

Change the fillet size to **.040in**.

Select the <u>faces</u> and the <u>upper edges</u> of the raised features as shown in the image.

Click **OK**.

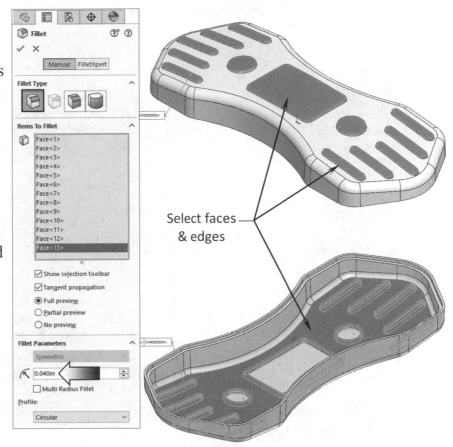

Select faces & edges

Rotate the model to different orientations to inspect the fillets.

Ensure all edges of the raised features are filleted, both inside and outside.

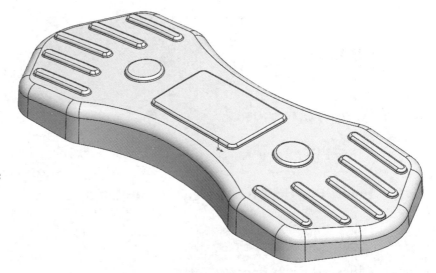

11. Creating the side holes:

Select the <u>Front</u> plane and open a **new sketch**.

Sketch the profile and add the dimensions/ relations to fully define the sketch.

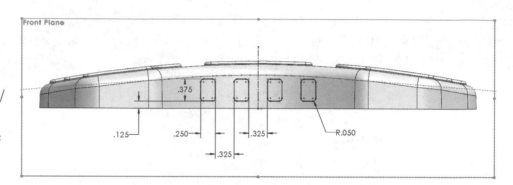

Switch to the **Features** tab and click **Extruded Cut**.

For Direction 1, select **Through All-Both**.

For Draft, enter **1.00deg Outward**.

Also enter **1.00deg Draft Outward** for Direction 2.

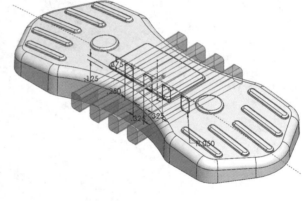

Click **OK**.

12. Saving your work:

Click **File, Save As**.

Enter: **Plastic Part Design1_Completed**.

Click **Save**.

Close all document.

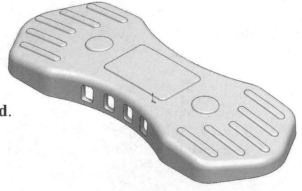

1. Opening a part document:

Select **File / Open**.

Browse to the Training Folder and open the part document named: **Plastic Part_Design2.sldprt**.

This document contains a single sketch which will be used to create the base feature of the part.

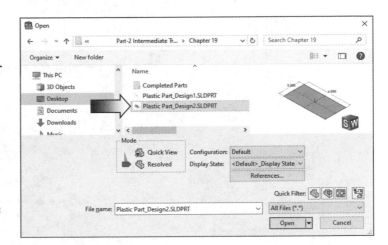

2. Extruding the base feature:

Switch to the **Features** tab.

Click **Extruded Boss-Base**.

For Direction 1, select the **Blind** type.

For Extrude Depth, enter: **.500in**.

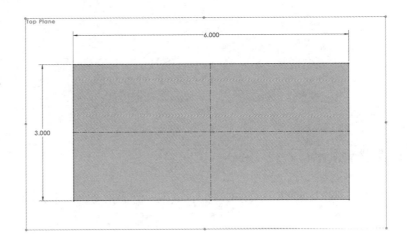

For Draft angle, enter: **1.00deg**.

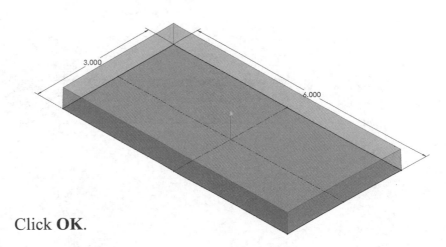

Click **OK**.

3. Adding the corner fillets:

Click **Fillet**.

Use the default **Constant Size** type.

Enter: **.500in** for radius.

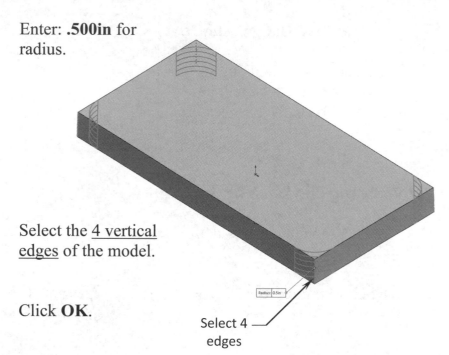

Select the 4 vertical edges of the model.

Click **OK**.

Select 4 edges

4. Making the upper cut:

Select the Front plane and open a **new sketch**.

Sketch the profile and add the dimensions/relations shown below.
(Locate the 2 ends of the arc using a Midpoint relation and a .300 dimension).

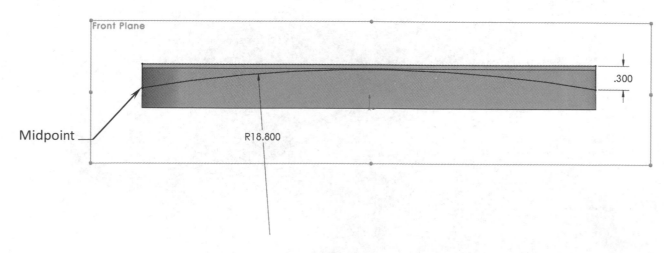

Front Plane

.300

Midpoint

R18.800

Switch to the **Features** tab and click **Extruded Cut**.

For Direction 1, select: **Through All-Both**.

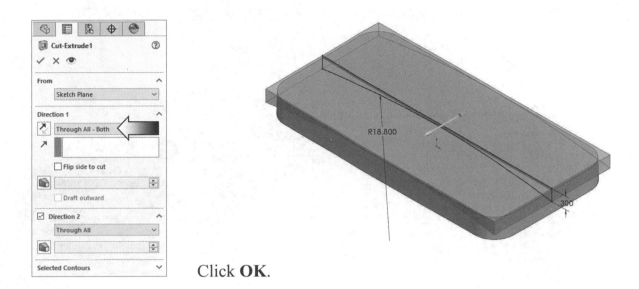

Click **OK**.

5. Adding the .1875in fillets:

Click **Fillet**.

For Fillet Type, use the default **Constant Size Radius** option.

For Radius size, enter: **.1875in**.

For Items to Fillet, select the upper edge as noted.

Select edge

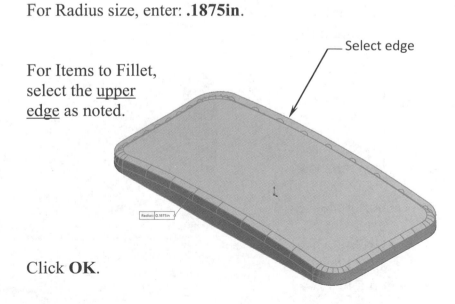

Click **OK**.

6. Creating a new plane:

Click **Reference Geometry, Plane**.

For First Reference, select the <u>Top</u> plane from the FeatureManager tree.

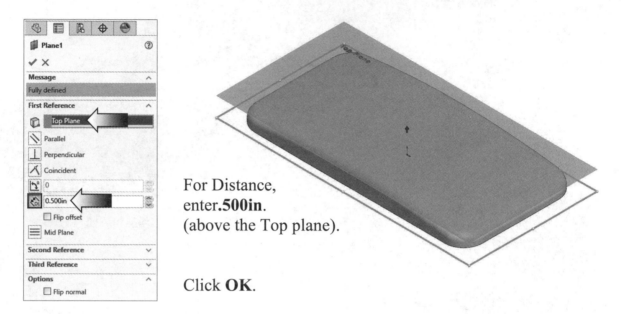

For Distance, enter.**500in**. (above the Top plane).

Click **OK**.

7. Sketching the button holes:

Select the <u>new plane</u> and open a **new sketch**. (If needed, copy & paste the Button Sketch from the Training Folder onto the new plane, position it on the Origin, and continue with step 8.)

Sketch the outline of the buttons as shown in the image.

The sketch is symmetric about the horizontal centerline. Use mirror to speed things up a little.

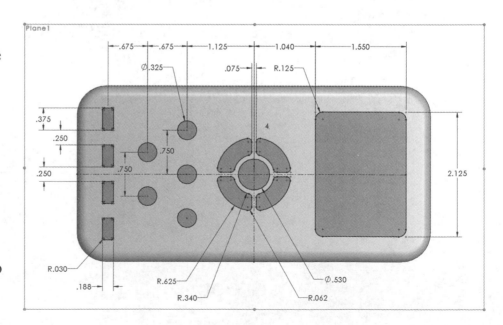

8. Making the cut:

Switch to the **Features** tab.

Click **Extruded Cut**.

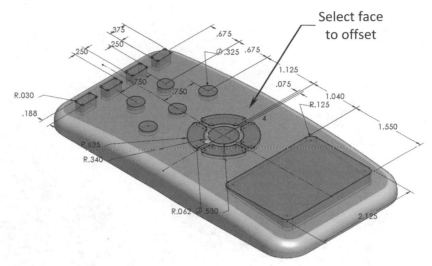

Select face
to offset

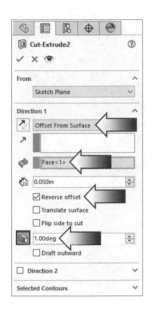

For Direction 1, select:
Offset from Surface.

Select the <u>face</u> as indicated to offset from.

Enable the **Reverse Offset** checkbox.

For Offset Distance, enter: **.050in**.

For Draft Angle, enter: **1.00deg**.

Click **OK**.

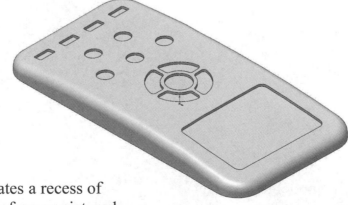

The cut feature creates a recess of
.050in below the surface as pictured.

9. Shelling the model:

Select the **Shell** command from the Features toolbar.

For Wall Thickness, enter: **.070in**.

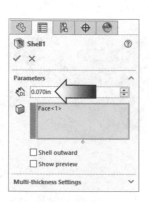

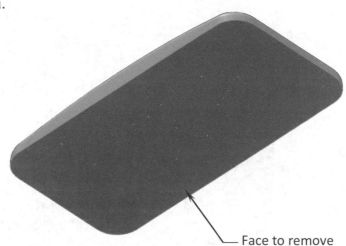

Face to remove

For Faces to Remove, select the <u>bottom face</u> of the model.

Click **OK**.

The model is shelled, leaving a constant wall of .070in all around.

Rotate the model to different orientation to inspect the thickness.

10. Adding a step cut:

Select Plane1 and open a **new sketch**.

Create an **Offset Entities** of all openings. Use a distance of **.040in** for all offsets. All offsets are inside.

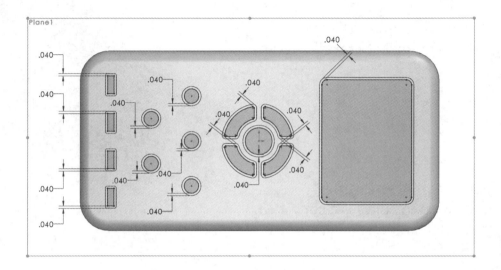

Switch to the **Features** toolbar and click **Extruded Cut**.

For Direction 1, select **Through All**.

For Draft Angle, enter: **1.00deg**.

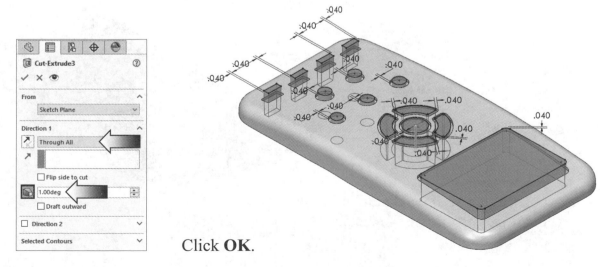

Click **OK**.

11. Making the interlock feature:

Select the <u>face</u> as indicated and open a **new sketch**.

Right-click one of the <u>inside</u> <u>edges</u> and pick: **Select Tangency**.

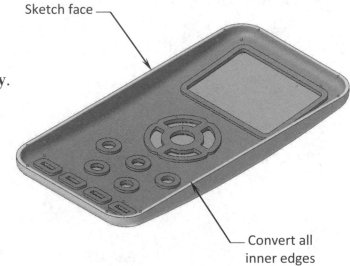

Sketch face

Convert all
inner edges

Click **Convert Entities**.
All selected edges are converted
to sketch entities.

Right-click one of the converted
entities and pick: **Select Chain**.

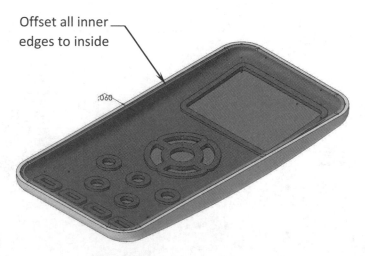

Offset all inner
edges to inside

:.060

Click **Offset Entities**.

Enter: **.060in** for
Offset-Distance.

Click <u>Reverse</u> if needed to place
the offset profile on the <u>inside</u>
of the model.

12. Extruding a boss feature:

Switch to the **Features** toolbar. (From the Isometric View, press Shift+Up Arrow twice to rotate to the bottom isometric view.)

Click **Extruded Boss-Base**.

For Extrude From, select **Offset** and enter **.060in**.

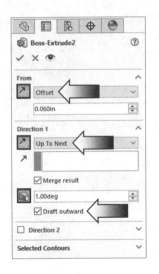

For Direction 1, select: **Up to Next**.

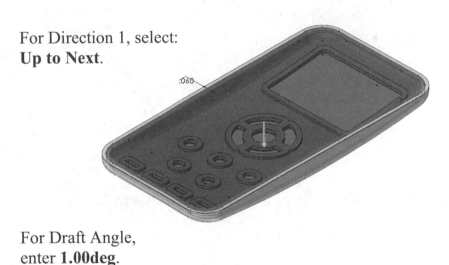

For Draft Angle, enter **1.00deg**.

Enable the **Draft Outward** checkbox.

Click **OK**.

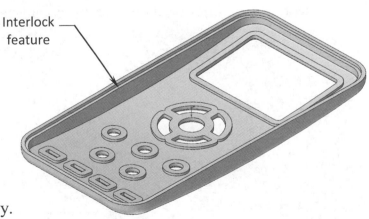

Interlock feature

The interlock feature is added all around the inside of the model. Zoom closer to inspect the feature more closely.

13. Adding a .093" fillet:

Click **Fillet**.

Use the default **Constant Size Radius** type.

Enter **.093in** for radius size.

For Items to Fillet, select the <u>edge</u> as noted.

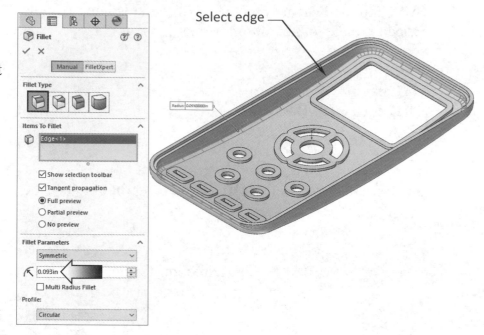

Select edge

Click **OK**.

14. Creating the Flat-Cable outlet cut:

Select the <u>Right</u> plane and open a **new sketch**.

Sketch the profile and add the dimensions shown in the image to fully define the sketch.

(Note: The 2 lines on the left and right have a 1-degree draft on them.)

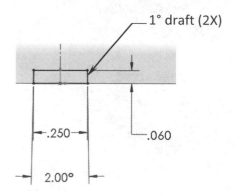

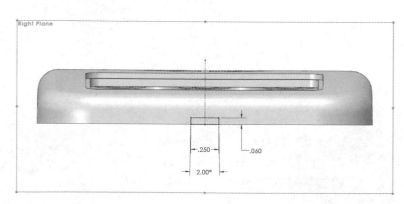

Switch to the **Features** toolbar and click **Extruded Cut**.

For Direction 1, select **Through All**.

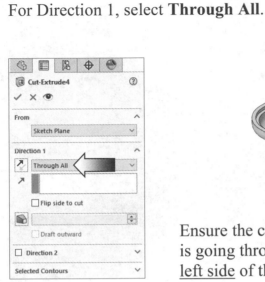

Ensure the cut direction
is going through the
left side of the model.

Click **OK**.

15. Creating the LED Holes:

Select the Right plane and open a **new sketch**.

Sketch a **centerline** that starts from the Origin.

Add **2 circles** that are
symmetric about the
vertical centerline.

Add the dimensions
and relations shown
in the image to fully
define the sketch.

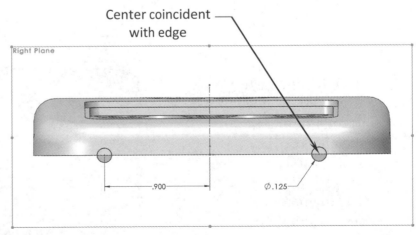

These circles are used
to create the openings for the LED lights in the next step.

Switch to the **Features** tool bar and click **Extruded Cut**.

For Direction 1,
select **Through All**.

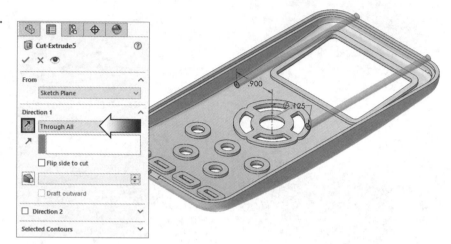

Ensure the cut
direction is going
through the <u>right</u>
<u>side</u> of the model.
Click **Reverse** if
needed.

Click **OK**.

16. Adding the Mounting Bosses:

Select the <u>face</u> as indicated and open a **new sketch**.

Sketch a **centerline** and **4 circles** as shown in the image below.

The circles are symmetric about the
horizontal centerline.

Add the dimensions and
any relations needed
to fully define the
sketch.

The diameter of the
circles should be
Ø.1875 (4 decimals
places).

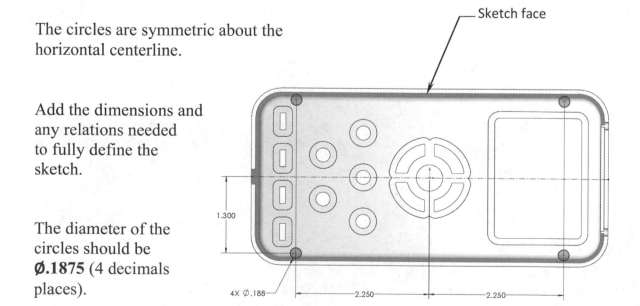

Switch to the **Features** toolbar and click **Extruded Boss-Base**.

For Direction 1, select: **Up to Next**.

For Draft Angle, enter: **3.00deg**.

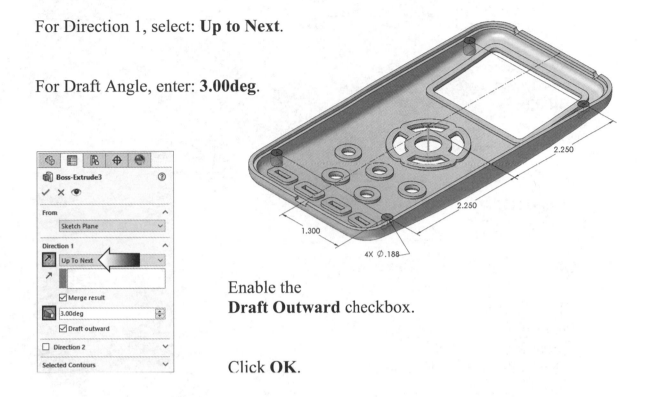

Enable the **Draft Outward** checkbox.

Click **OK**.

17. Adding the Mounting Holes:

Select the <u>face</u> as noted and open a **new sketch**.

Sketch **4 circles** that are **concentric** with the mounting bosses.

Add an **Equal** relation to the 4 circles.

Add the diameter **Ø.090** dimension to fully define the sketch.

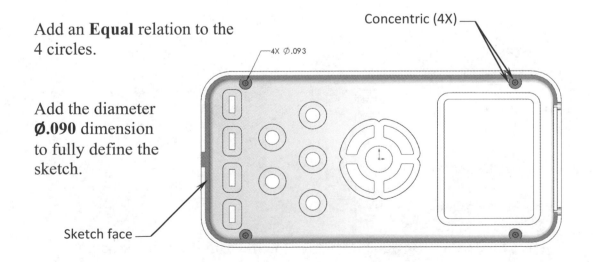

Switch to the **Features** toolbar and click **Extruded Cut**.

For Direction 1, select the **Blind** type.

For Depth, enter **.400in**.

For Draft Angle, enter: **2.00deg**.

Click **OK**.

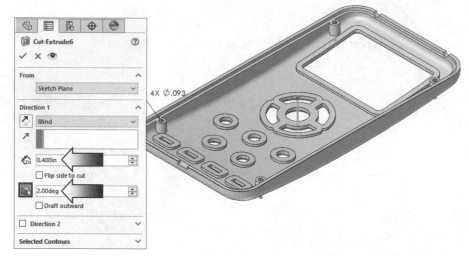

18. Adding the .010" fillet:

This fillet is mainly for breaking all sharp edges around the raised features and the four mounting bosses.

Click **Fillet** and enter **.010in** for radius size.

Select the **faces** (or edges) as shown.

Click **OK**.

Add fillet to all edges of the raised features

19. Saving your work:

Click **File, Save as**.

Enter: **Plastic Part Design2.sldprt** for the file name and click **Save**.

CHAPTER 20

Smart Components

Smart Components
Using Auto Size & Configurator Table

Smart Components are created from components that are used frequently and require the addition of associated components and features to position or mate to one another.

When making a component smart, other components and features can be associated with the Smart Component. However, when the Smart Component is inserted into an assembly, you can choose whether or not to insert the associated components and features.

Auto-sizing capability can be added to cylindrical Smart Components. When the Smart Component is inserted onto a cylindrical component, the size of the Smart Component adjusts to fit the cylindrical component. You can specify, ahead of time, the range of values for which the feature applies by typing a value for Minimum Diameter and Maximum Diameter.

Configurator Table

Specify the configurations for each of the associated features and components for each configuration of the smart component using the drop down list in each cell.

Auto-Size Cap	Minimum Diameter	Maximum Diameter
.375 ID	.371	.375
.425 ID	.421	.425
.500 ID	.496	.500
.550 ID	.546	.550

OK   Cancel   Help

When inserting a Smart Component into an assembly you can position it with mates, exactly as you do for any other component. Then you activate the Smart Feature and choose the associated components and features to add.

You can add, delete, or modify associated components, features, and mates of a Smart Component. You edit the definition of the Smart Component by reconstructing a temporary assembly from the defining data stored in the Smart Component document.

Smart Components
Using Auto Size & Configurator Table

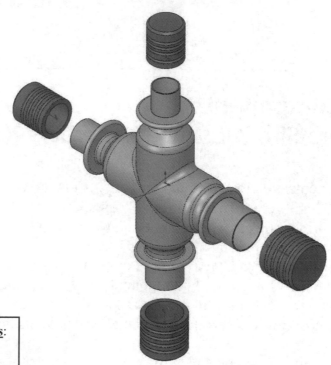

<u>View Orientation Hot Keys</u>:

Ctrl + 1 = Front View
Ctrl + 2 = Back View
Ctrl + 3 = Left View
Ctrl + 4 = Right View
Ctrl + 5 = Top View
Ctrl + 6 = Bottom View
Ctrl + 7 = Isometric View
Ctrl + 8 = Normal To
          Selection

Dimensioning Standards: **ANSI**
Units: **INCHES** – 3 Decimals

Tools Needed:

 Annotations

 Configuration

 Mate References

 Configurator Table

 Auto Size

 Smart Component

1. Opening a part document:

Click **File / Open**.

Browse to the Training
Files folder and open
a part document named:
Auto-Size Cap.sldprt

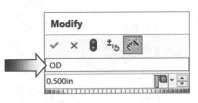

2. Showing the dimensions and their names:

Right-click the Annotation
folder and select **Show Feature Dimensions** (arrow).

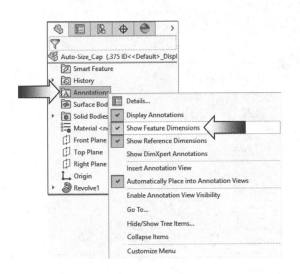

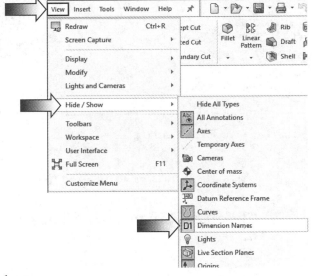

Click the **View** drop-down menu and select
Hide / Show / Dimension Names (arrows).

Double-click the
dimension **Ø.500**
(circled) and
enter **OD** in the
name field.

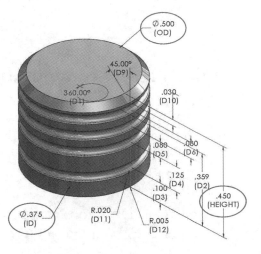

Repeat the same step and rename the other
2 dimensions (**ID** and **Height**).

3. Splitting the FeatureManager tree:

Locate the **Split Handle** on top of the Feature-Manager tree (the round dot) and drag it down about halfway to split it into 2 sections.

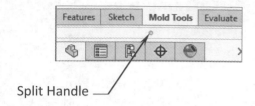

Split Handle —

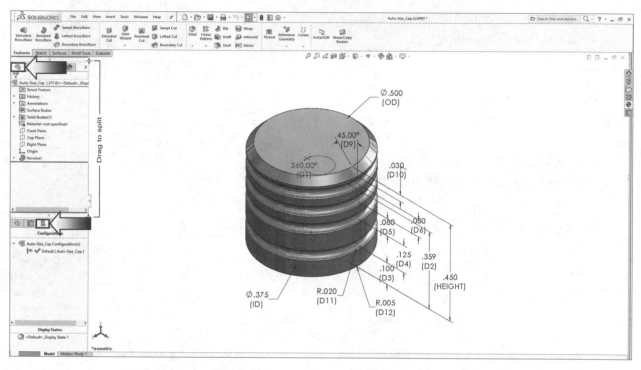

Click the **FeatureManager** tree icon for the upper section.

Click the **ConfigurationManager** for the lower section.

Rename the **Default** configuration to **.375 ID** (arrow).

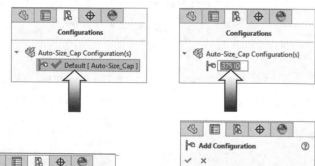

4. Adding a new configuration:

Right-click anywhere inside the Configuration area and select:
Add Configuration (arrow).

For Configuration Name, enter **.425 ID** and click **OK**.

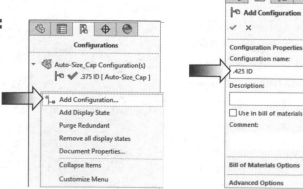

Fast-Double-Click on the dimension **Ø.500 (OD)** to display the **Modify** dialog box.

Click the drop-down arrow at the bottom right of the Modify dialog and select: **This Configuration** option (arrow).

Change the dimension value from **Ø.500** to **Ø.550** and click **Rebuild**.

Change the dimension **Ø.375** to **Ø.425**

Change the dimension **.450** to **.475**

Create a total of **4 configurations** using the dimensions provided below.

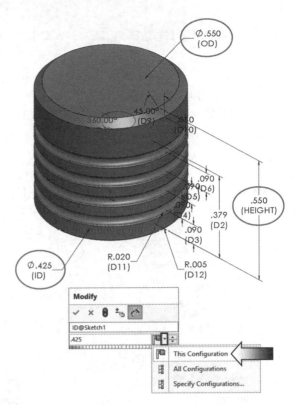

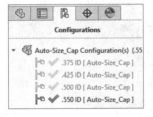

.375 ID	.500 OD	.450 Height
.425 ID	.550 OD	.475 Height
.500 ID	.625 OD	.500 Height
.550 ID	.675 OD	.550 Height

Double-click each configuration to verify the dimension changes.

Right-click the Annotations folder and clear the option **Show Feature Dimensions**.

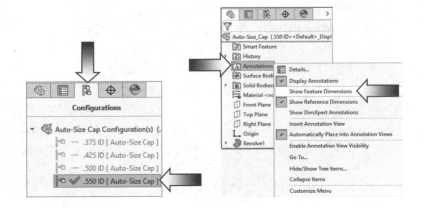

5. Creating mate references:

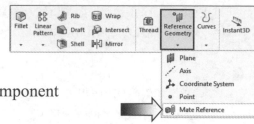

Any components that you typically mate
the same way every time, you can set up mate
references to define the mates used and the component
geometry being mated.

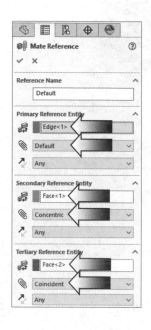

When you drag a component with a mate reference into an
assembly, the software tries to find other combinations of the
same mate reference name and mate type. If the name is the same,
but the type does not match, the software does not add the mate.

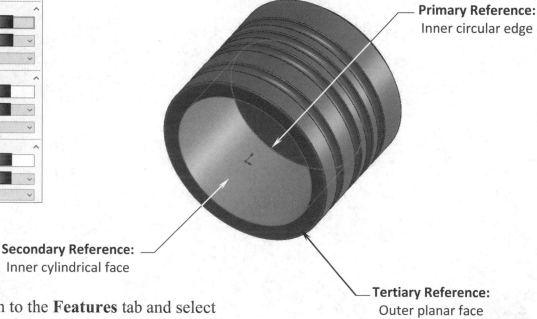

Primary Reference:
Inner circular edge

Secondary Reference:
Inner cylindrical face

Tertiary Reference:
Outer planar face

Switch to the **Features** tab and select
Reference Geometry / Mate References (arrow).

For Primary Reference Entity, select the
inner circular edge as indicated.

For Secondary Reference Entity, select the
inner cylindrical face as shown in the image above.

For Tertiary Reference Entity, select the
outer planar face as noted and click **OK**.

A **MateReferences** folder is added to the FeatureManager
tree (arrow). The references can be edited if necessary.

<u>**Save**</u> the part document.

6. Testing the mate references:

We will test out the mate references to see how they respond in a new assembly document.

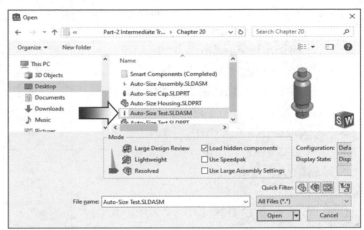

Open an assembly document named:
Auto-Size Test.sldasm

Launch the **Windows Explorer** application
(Hold the Windows button and press E).

Locate the Training Files folder, select the part document named:
Auto-Size Cap.sldprt and <u>drag</u> it into the assembly document.

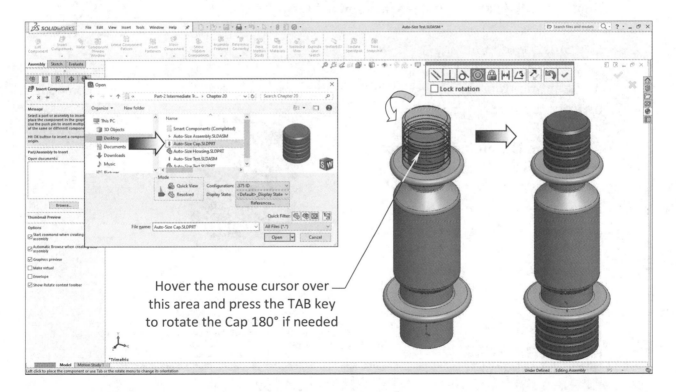

Hover the mouse cursor over this area and press the TAB key to rotate the Cap 180° if needed

<u>Hover</u> the mouse cursor over the upper <u>cylindrical surface</u>. The mate references component is activated and snapped the component onto the cylinder. Press the TAB key to flip the component 180° if needed. Click the Green check to accept.

Drag another instance to the lower cylindrical face. Select the **Ø.550** configuration and click **OK**. <u>Save</u> and close the assembly document.

7. Making a smart component:

The Make Smart Component option is used to place and size a component when its diameter is read. It is used as a trigger for selecting appropriate configuration based on a range of diameter dimensions.

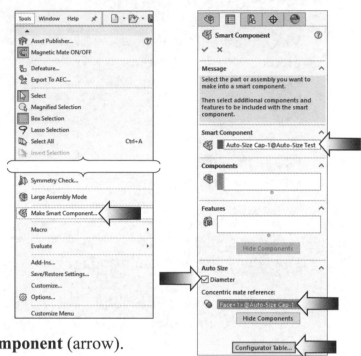

Since it is based on cylindrical references, only cylindrical or turned-type parts can utilize this feature.

Select **Tools / Make Smart Component** (arrow).

For Smart Component, select the **Cap** either in the graphics area or from the tree.

Concentric mate reference — Smart Component

Enable the **Diameter** checkbox in the Auto Size section (arrow).

For Concentric Mate Reference, select the **inside face** of the cap.

Click the **Configurator Table** button (arrow).

The four existing configurations appear in the 1st column.

Enter the range of diameter dimensions in the maximum and minimum columns.

Click **OK**.

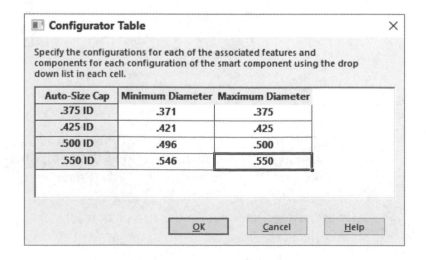

Configurator Table ✕

Specify the configurations for each of the associated features and components for each configuration of the smart component using the drop down list in each cell.

Auto-Size Cap	Minimum Diameter	Maximum Diameter
.375 ID	.371	.375
.425 ID	.421	.425
.500 ID	.496	.500
.550 ID	.546	.550

OK   Cancel   Help

A thunder bolt symbol is added to the icon next to the name of the Cap. A folder called **Smart Features** is also added to the FeatureManager tree.

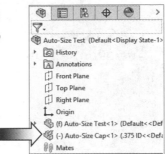

<u>Save</u> and close the part document. We will test it out in another assembly document.

8. Testing the smart component:

Open an assembly document named: **Auto-Size Assembly.sldasm**.

Change to the **Front** view orientation (Control + 1).

Press **Windows + E** to launch the Windows Explorer application.

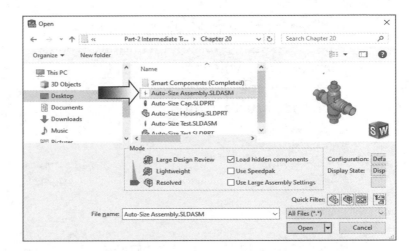

<u>Drag</u> and drop the **Auto-Size Cap** on the <u>cylindrical face</u> shown; press the TAB key to flip 180° if needed. Select the appropriate configuration and press **OK**.

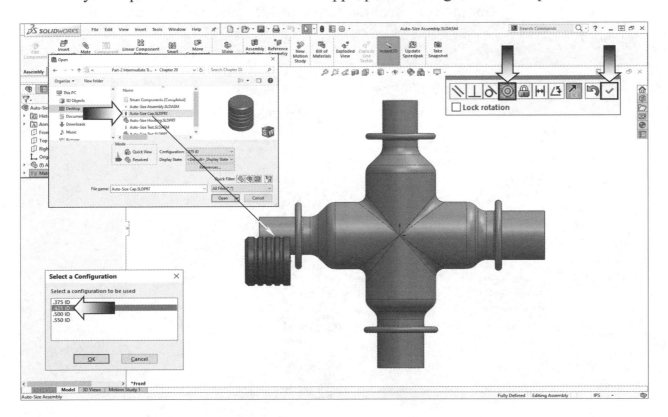

The Concentric mate is selected automatically. Press the green check to accept it and close out of the mate dialog.

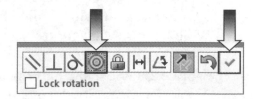

A thunder bolt symbol appears on the component as well as on the Feature Manager tree. This indicates that the Smart Features and References have been added successfully.

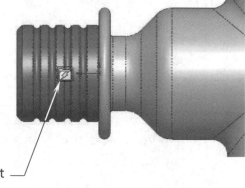

Smart Component symbol

9. Adding other instances of the cap:

Drag and drop 3 more instances of the same cap to the other 3 cylindrical features. Select the appropriate configuration for each side as noted in the image below.

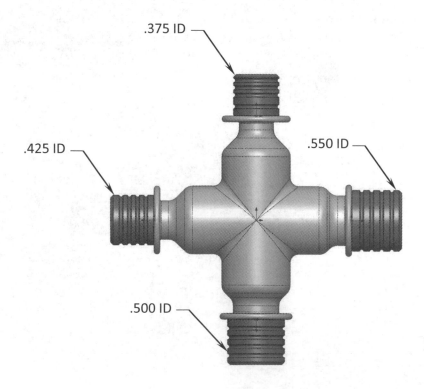

.375 ID

.425 ID

.550 ID

.500 ID

Add the coincident mates between the bottom of the caps and the top of the flanges if the coincident mates did not get added automatically.

10. Optional:

Create a section view to verify the positions of the Caps. The bottom of each cap should be coincident to the top surface of the flanges.

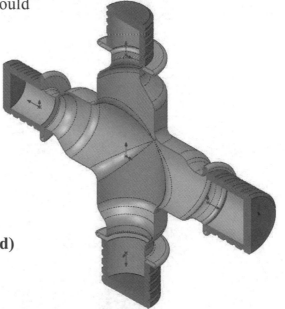

11. Saving your work:

Click **File / Save As**.

Enter **Auto-Size Assembly (Completed)** for the file name.

Press **Save** and close all documents.

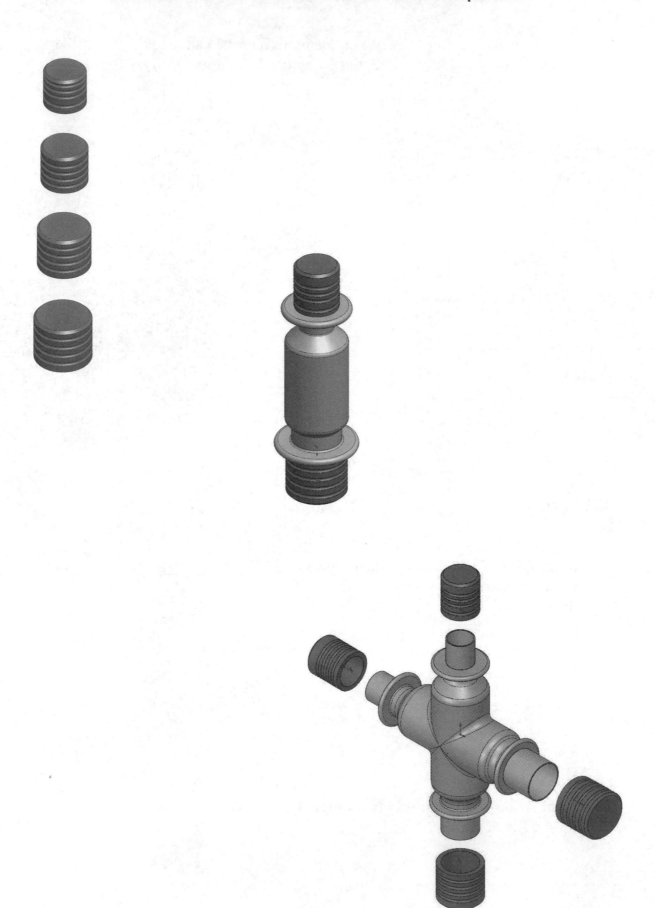

CHAPTER 21

Using Subtract & Intersect Tools

Using Subtract & Intersect Tools
Simple Cavity

The options in the Combine command will only work when they combine bodies contained within one multibody part file.

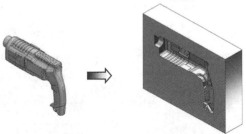

In the Combine PropertyManager, you specify which bodies in a multibody part to add, subtract, or overlap.

* Add: Combines solids of all selected bodies to create a single body.

* Subtract: Removes overlapping material from a selected main body.

* Common: Removes all material except that which overlaps.

The Intersect command can modify existing geometry of solids, surfaces, or planes, or to create new geometry.

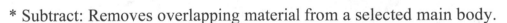

The Intersect command can add open surface geometry to a solid, remove material from a model, or create geometry from an enclosed cavity. It can also merge solids that you define with the Intersect tool, or cap some surfaces to define closed volumes.

Both Combine and Intersect tools are quite useful when it comes to creating cavities in a model because of their simplicity and variety of options to choose from. However, they should only be used to create simple, symmetrical cavities because the Mold Tools were designed to assist you with creating more complex cores and cavities as well as non-planar parting lines.

Using Subtract & Intersect Tools
Simple Cavity

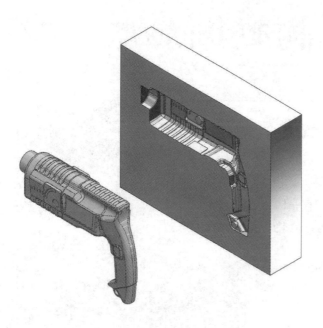

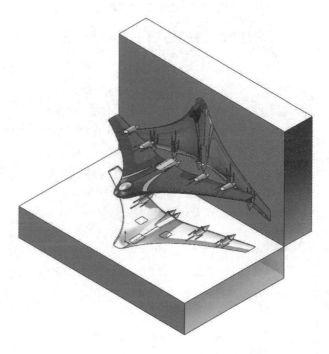

View Orientation Hot Keys:

Ctrl + 1 = Front View
Ctrl + 2 = Back View
Ctrl + 3 = Left View
Ctrl + 4 = Right View
Ctrl + 5 = Top View
Ctrl + 6 = Bottom View
Ctrl + 7 = Isometric View
Ctrl + 8 = Normal To
      Selection

Dimensioning Standards: **ANSI**
Units: **INCHES** – 3 Decimals

Tools Needed:

 Scale

 Move/Copy Bodies

 Combine

 Intersect

 Corner Rectangle

 Extruded Boss-Base

1. Opening a part document:

Select **File / Open**.

Browse to the Training Files folder and open a part document named: **Drill Case.sldprt**

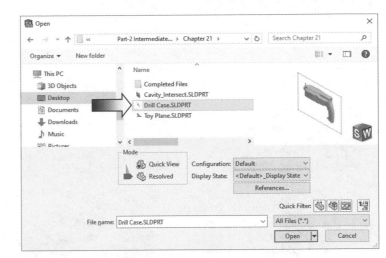

2. Scaling the part:

The Scale feature scales only the geometry of the model for use in data export, cavities, and so on. It does not scale dimensions, sketches, or reference geometry.

The scaled model is used to accommodate the material shrinkage.

Select **Insert / Features / Scale** (arrow).

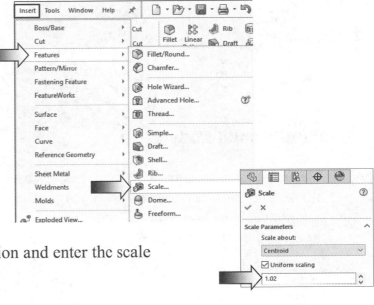

Use the default **Centroid** option and enter the scale of **1.02** (2% larger).

Click **OK**.

3. Extruding the mold block:

Select the **Sketch1** from the FeatureManager tree and click **Extruded Boss-Base**.

Use the **Blind** type and the depth of **3.000in**. Enable **Reverse** Direction.

<u>Clear</u> the **Merge Result** box and click **OK**.

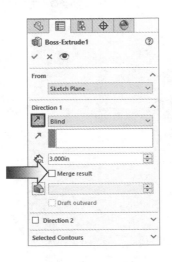

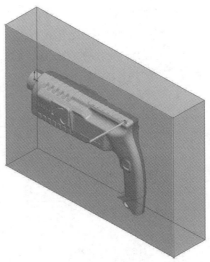

4. Assigning material:

Expand the **Solid Bodies** folder.

Right-click the **Boss-Extruded1** body and select **Material / Plain Carbon Steel**.

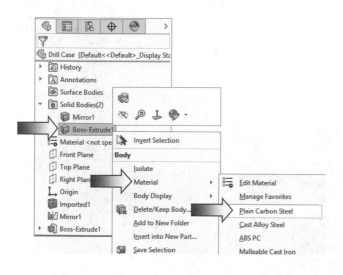
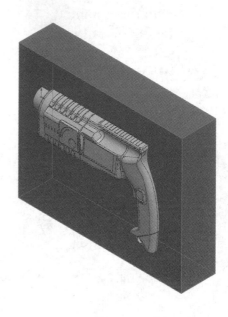

5. Copying a solid body:

We are going to use the Combine-Subtract option to create the cavity in the mold block. The Subtract command consumes the tool body after the subtraction is completed. For clarity, we will make a copy of the Drill Case and use it to create the exploded view later on.

Select **Insert / Features / Move-Copy Bodies**.

For Bodies to Move / Copy, select the **Mirror** body from the Solid Bodies folder.

Enable the **Copy** checkbox.

Leave the X, Y, Z values at **zeros**.

Click **OK**.

Click **OK** again to close the message box.

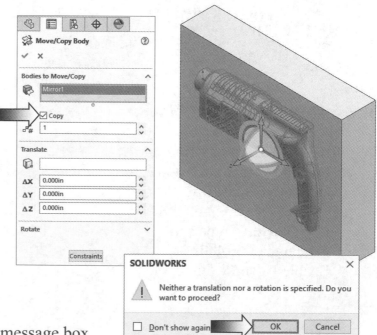

6. Creating the cavity:

Hide the **Body-Move/Copy1** from the Solid Bodies folder (arrow).

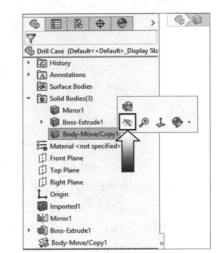

Select **Insert / Features / Combine** (arrow).

For Operation Type, select the **Subtract** option.

For Main Body, select the **Boss-Extrude1** body from the Solid Bodies folder.

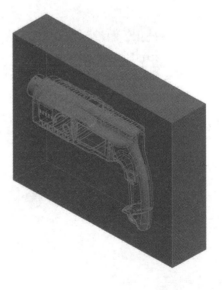

For Bodies to Combine, select the **Mirror1** body also from the Solid Bodies folder.

Click **OK**.

7. Separating the solid bodies:

In the Solid Bodies folder, click **Body-Move/Copy1** and select **show** (arrow).

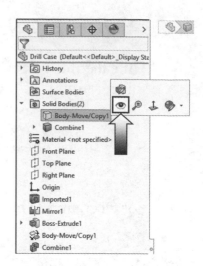

The Exploded View command may be a little bit quicker to separate the solid bodies, but the drawback is that while the solid bodies are exploded, none of the Sketch or Feature tools can be used; they will be grayed out.

We will use the **Move/Copy** command instead.

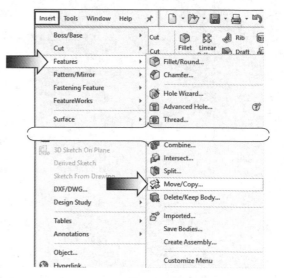

Select **Insert / Features / Move/Copy**.

For Bodies to Move/Copy, select the **Body-Move/Copy1** from the graphics area of the Solid Bodies folder. Clear the **Copy** checkbox.

For Translate Distance, enter **11.000in** in the **Z Direction** box.

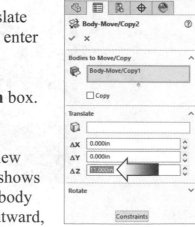

The preview graphics shows the solid body moves outward, along the Z direction.

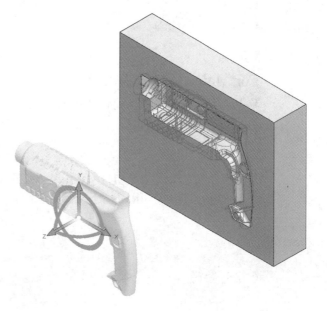

Click **OK**.

8. Saving your work:

Select **File / Save As**.

Enter **Cavity_Combine Subtract.sldprt** for the file name.

Press **Save***.

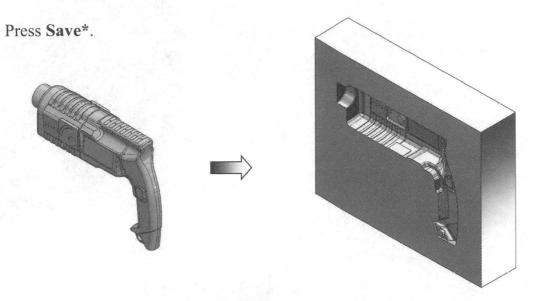

      * Kccp thc part document open to use in the 2ⁿᵈ half of the lesson.

9. Using the Intersect tool:

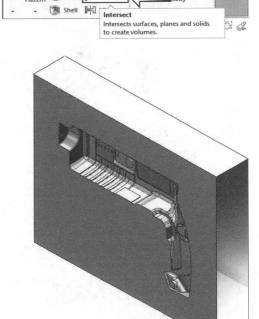

The Intersect tool is used to modify solids, surfaces, and planes. This tool can create a solid from a cavity by merging coincident surface or solid bodies in a part, and then removing the geometry defined by the bodies that enclose the cavity.

To reverse engineer the design we can use the cavity in the mold block and the Front plane to recreate a solid model.

Hide the **Drill Case** body (Body Move/Copy2).

Click the **Intersect** command from the **Features** tool tab (arrow).

For Intersect Selections, select the **mold block** (Combine1) and the **Front** plane.

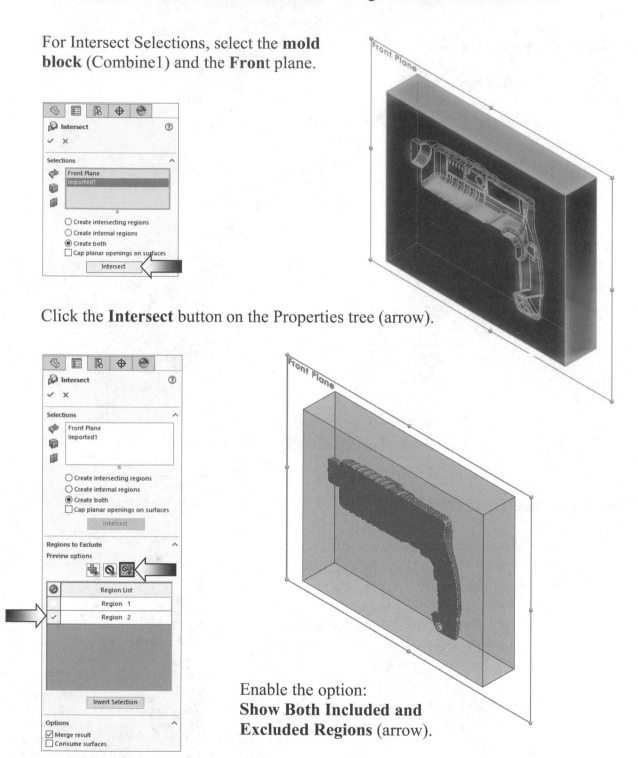

Click the **Intersect** button on the Properties tree (arrow).

Enable the option:
Show Both Included and Excluded Regions (arrow).

For Regions to Exclude, select the **Region 2** (arrow).
The mold block (region 2) will be excluded and the empty area will be filled by the Intersect command.

Click **OK**.

10. Saving your work:

Select **File / Save As**.

Enter **Cavity_to Part.sldprt** for the file name.

Press **Save**.

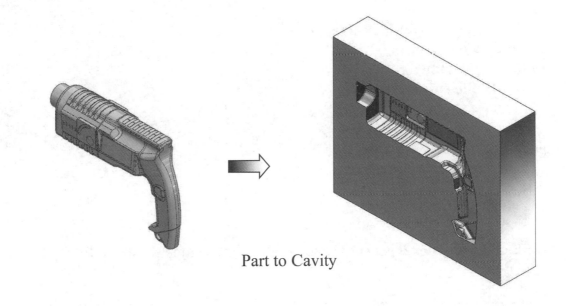

Part to Cavity

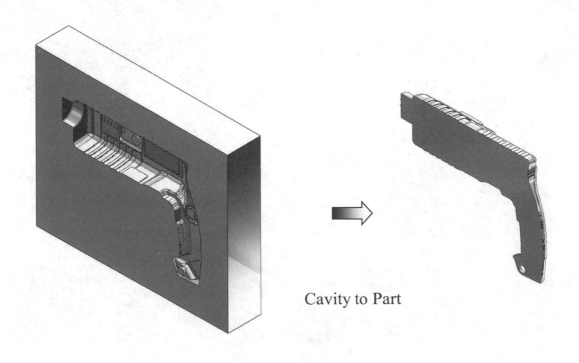

Cavity to Part

Exercise: Creating a Cavity with Combine Subtract

1. Opening a part document:

Select **File / Open**.

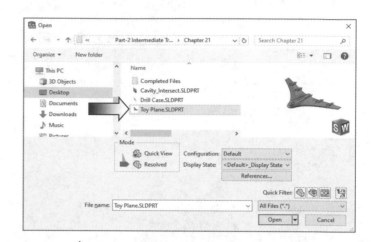

Browse to the Training Files folder and open a part document named: **Toy Plane.sldprt**

2. Creating the first mold block:

Select the **Top** plane and open a <u>new sketch</u>.

Sketch a **Corner Rectangle** and add the dimensions shown to fully define it.

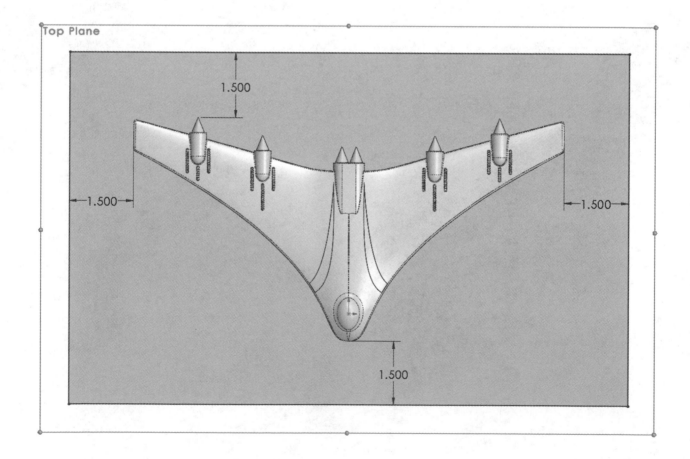

Switch to the **Features** tab and press **Extruded Boss-Base**.

Use the default **Blind** type and a depth of **2.000in**.

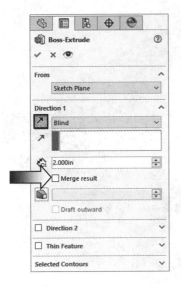

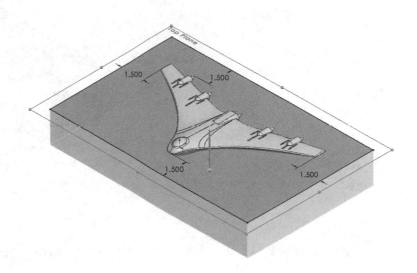

Click the **Reverse Direction** button.

Clear the **Merge Result** checkbox (arrow).

Click **OK**.

3. Creating the first cavity:

Select **Insert / Features / Combine**.

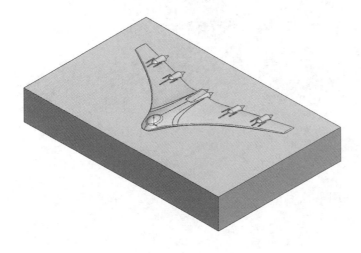

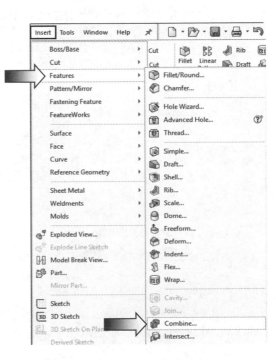

For Operation Type, select the **Subtract** option (arrow).

For Main Body, select the **mold block** body.

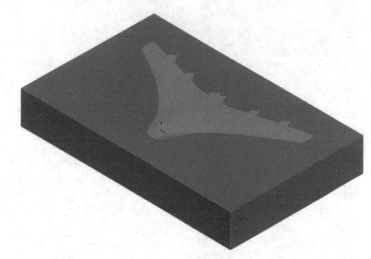

For Bodies to Combine, select the **Toy Plane** body.

Click **OK**.

The Subtract command consumes the tool body after the operation is completed. We need to make a copy of the Toy Plane body and show it in the exploded view.

<u>Drag</u> the Roll-Back line up one step and place it <u>above</u> the Combine1 feature (arrow).

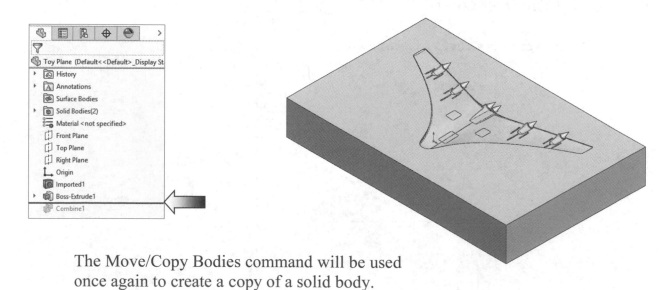

The Move/Copy Bodies command will be used once again to create a copy of a solid body.

4. Copying a solid body:

Select **Insert / Features / Move-Copy**.

For Bodies to Move/Copy, select the **Imported1** body (the Toy Plane).

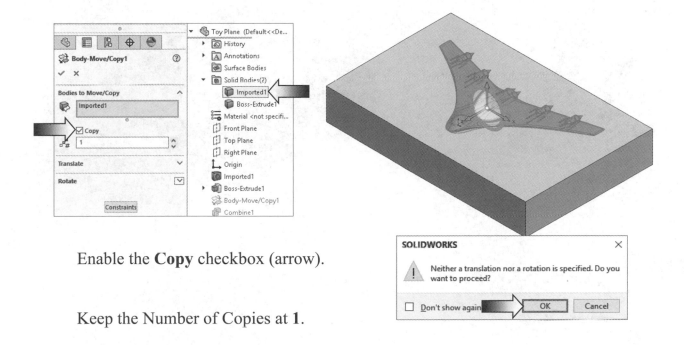

Enable the **Copy** checkbox (arrow).

Keep the Number of Copies at **1**.

Click **OK** and select **OK** again to close the message dialog.

5. Creating the second mold block:

Select the <u>upper face</u> of the block and open a new sketch.

Convert the <u>4 edges</u> on the upper face of the block into <u>4 sketch lines</u> using the Convert Entities command.

(The converted entities have the On-Edge relations embedded in them.)

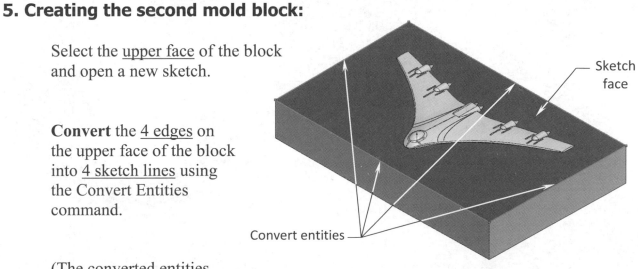

Sketch face

Convert entities

Switch to the **Features** tool tab and click **Extruded Boss-Base**.

Use the **Blind** option and enter a depth of **2.000in**.

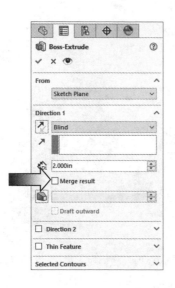 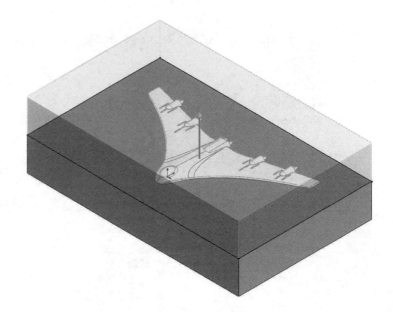

Clear the **Merge Result** checkbox (arrow).

Click **OK**.

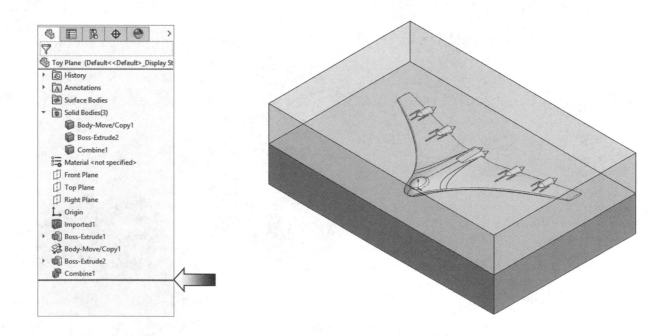

Drag the Roll-Back line to the bottom of the Feature tree (arrow).
There should be a total of 3 Solid Bodies at this point.

6. Creating the second cavity:

Select **Insert / Features / Combine**.

For Operation Type, select the **Subtract** option (arrow).

For Main Body, select the **upper mold block**.

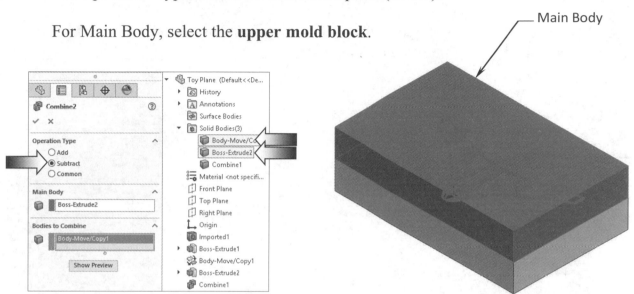

Main Body

For Bodies to Combine, select the **Body-Move/Copy1** body (the Toy Plane).

Click **OK**.

7. Rotating the second mold block:

Select **Insert / Features / Move-Copy**.

For Bodies to Move/Copy, select the **Combine2** body (the second mold block).

Expand the **Rotate** section and select the **horizontal edge** to rotate about.

For Rotate Angle, type **90°** and press **Enter** to see the preview graphic of the mold block.

Click **OK**.

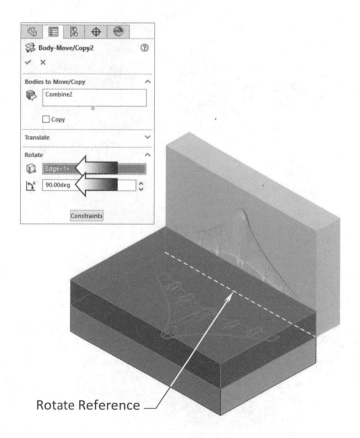

Rotate Reference

8. Assigning materials:

Expand the **Solid Bodies** folder.

Right-click the **Combine1** body and select **Material / Plain Carbon Steel**.

Repeat the same step and assign the same material to the second solid body.

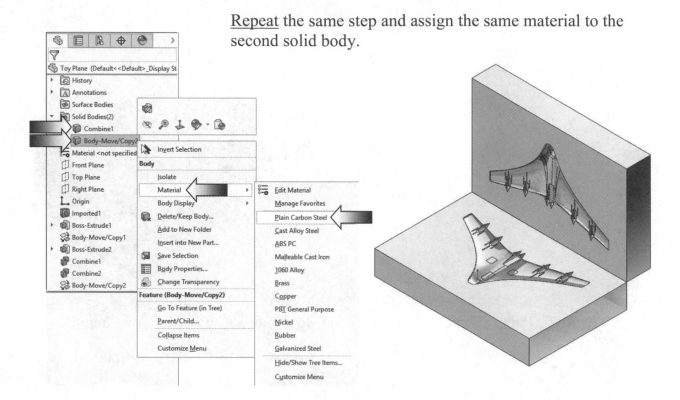

Optional:

Make another copy of the Toy Plane and move it to the position similar to the image below.

9. Saving your work:

Select **File / Save As**.

Enter **Toy Plane_Exe (Completed)** for the file name.

Press **Save**.

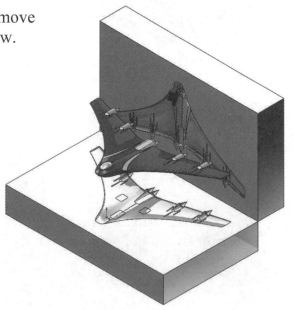

CHAPTER 22

Using Magnetic Mates

Using Magnetic Mates
Deck Assembly

Magnetic mates let you define connection points so that you can quickly place and position components.

Use the Asset Publisher tool to define connection points and a ground plane. Then, as you insert the components and move them within proximity of each other, the components will snap into position based on the predefined connecting points.

While creating the Asset Publisher you can select a model face to define which face of the model to attach to the ground plane when you insert the asset into an assembly. Additionally, you can enter a value to define the offset distance for the selected model face from the ground plane.

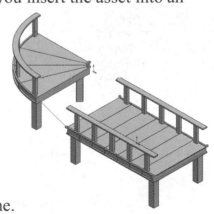

Three basic steps are used when working with Magnetic Mates:

a. **Defining a Ground Plane:** You can define a ground plane in an assembly. When you insert a published asset into the assembly, the asset's ground face snaps to the assembly's ground plane.

b. **Defining Connection Points:** You can define connection points that enable the asset to snap into position relative to other assets in an assembly.

c. **Publishing an Asset:** You can publish a model as an asset. You define connection points that enable the asset to snap into position relative to other assets in an assembly. Optionally, you define a ground face, and you can create a SpeedPak configuration.

Using Magnetic Mates
Deck Assembly

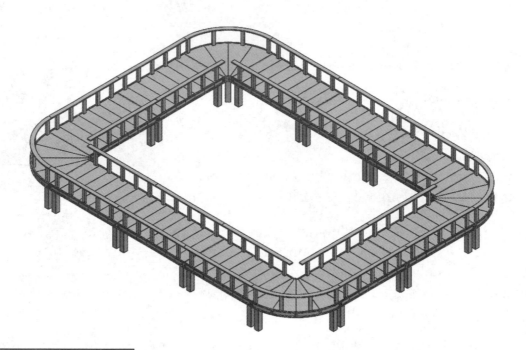

Dimensioning Standards: **ANSI**
Units: **INCHES** – 3 Decimals

Tools Needed:

 Assembly Document

 Asset Publisher

 Magnetic Mates

1. Opening a part document:

Select **File / Open**.

Browse to the **Training Files** folder and open a part document named: **Section1.sldprt**.

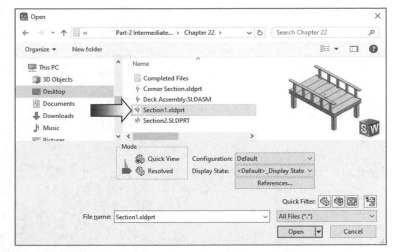

To help focus on the main topic, the part was modeled as a single body and saved as a Parasolid format to reduce the file size. This part will get replicated several times in an assembly document and mated to other components to build a Deck Assembly.

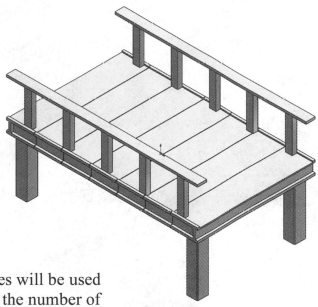

Magnetic Mates will be used to help reduce the number of steps that it usually takes to assemble the components in a large assembly.

The 1st step is to create an Asset Publisher.

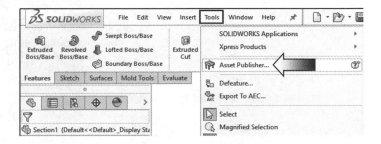

Select **Tools / Asset Publisher** (arrow).

2. Publishing an Asset:

You can publish a model as an asset. You define connection points that enable the asset to snap into position relative to other assets in an assembly. Optionally, you define a ground face, and you can create a SpeedPak configuration.

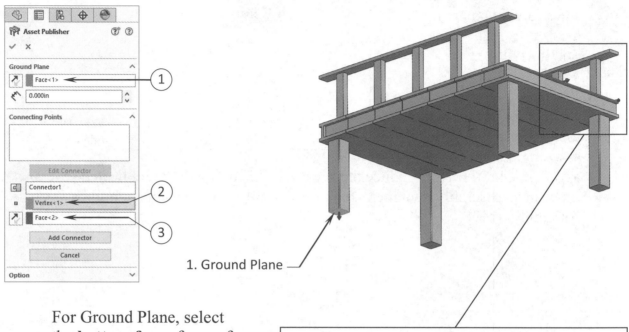

1. Ground Plane

For Ground Plane, select the **bottom face** of one of the legs as indicated in note 1.

For Connector Point, select the **upper vertex** as indicated in note 2.

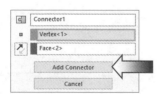

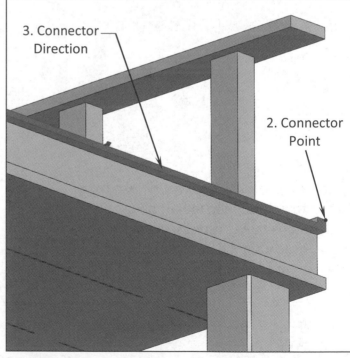

3. Connector Direction

2. Connector Point

For Connector Direction, select the **outer face** as indicated in note 3.

Click **Add Connector** (arrow). The Connector1 is created and stored in the Connector Points section. We will need to create 3 more connectors.

For the 2ⁿᵈ Connector Point, select the **upper vertex** as indicated in note 1.

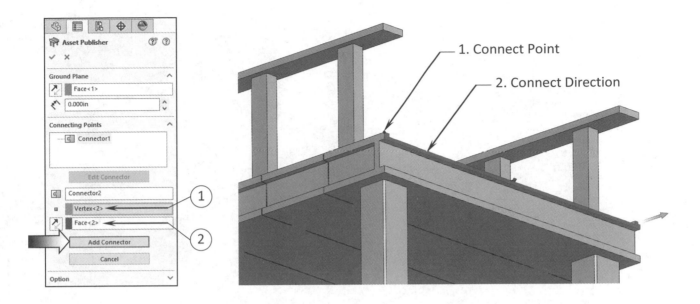

For the 2ⁿᵈ Connector Direction, select the **outer face** as indicated in note 2.

Click the **Add Connector** button (arrow).

Rotate the model and add 2 additional connectors the same way as the other two.

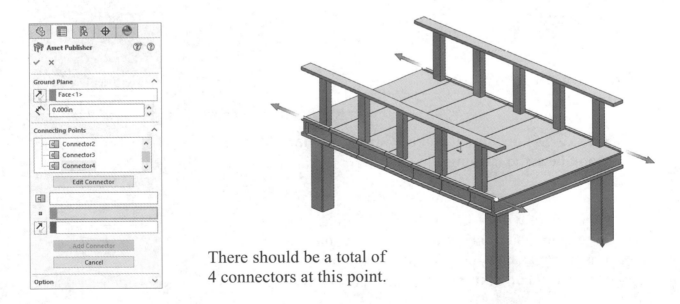

There should be a total of 4 connectors at this point.

Click **OK** to exit the Asset Publisher command.

A new folder called **Published References** is added to the FeatureManager tree. Expand this folder to see the references that were created in the last step (arrow).

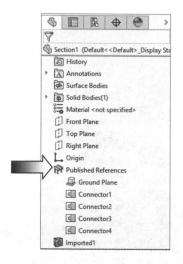

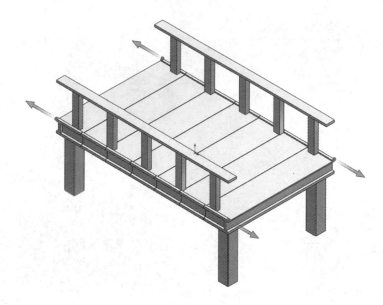

3. Saving the part file:

Select **File / Save** and <u>overwrite</u> the existing document if prompted.

We will open another part document and assign the similar connectors to it, and then assemble both of the parts in an assembly document.

4. Opening another part document:

Open a part document named **Corner Section.sldprt** from the same training file folder.

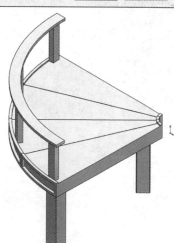

Since there will be no modifications done to any of these parts, and to help us focus on the main topic, this part was also modeled as a single body and saved as a Parasolid format to reduce the file size.

We will need to define four connectors for this part.

5. Defining another Asset Publisher:

Click **Tools / Asset Publisher**.

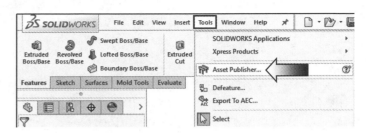

For Ground Plane, select the **bottom face** of one of the legs as indicated in note 1.

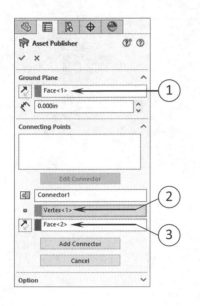

For Connector Point, select the **upper vertex** as indicated in note 2.

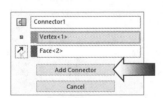

For Connector Direction, select the **outer face** as indicated in note 3.

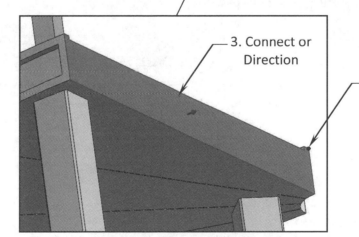

Click Add **Connector** (arrow).

The Connector 1 is created and stored in the Connecting Points section.

Add 3 additional connectors to the other 3 corners of the model.

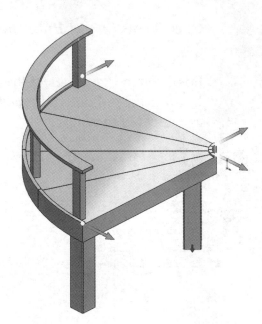

There should be a total of 4 connectors when completed.

Click **OK** to exit the Asset-Publisher command.

6. Saving the part file:

Select **File / Save** and overwrite the existing document if prompted.

We are now ready to try out the magnetic mates when assembling these parts in an assembly document.

7. Opening an assembly document:

Select **File / Open**.

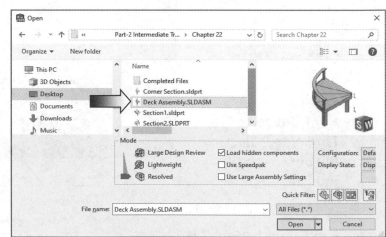

Open an asembly document named: **Deck Assembly.sldasm** from the same training files folder as the last ones.

The Top plane of the assembly has already been assigned to this assembly as the Ground Plane (arrow).

Enable the Magnetic Mate by clicking **Tools / Magnetic Mate ON/OFF** (arrow).

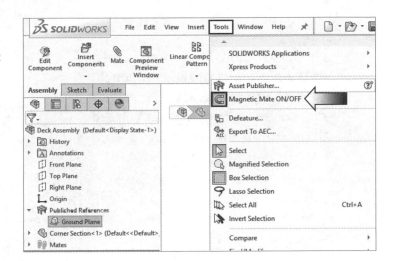

Magnetic mates let you define connection points so that you can quickly place and position components.

8. Adding the Section1 component to the assembly:

We will add the component named **Section1** to the assembly and test out the connecting points that we created and saved earlier.

Click the **Insert Components** command from the **Assembly** tool tab; select the part document named **Section1.sldprt** and click **Open**.

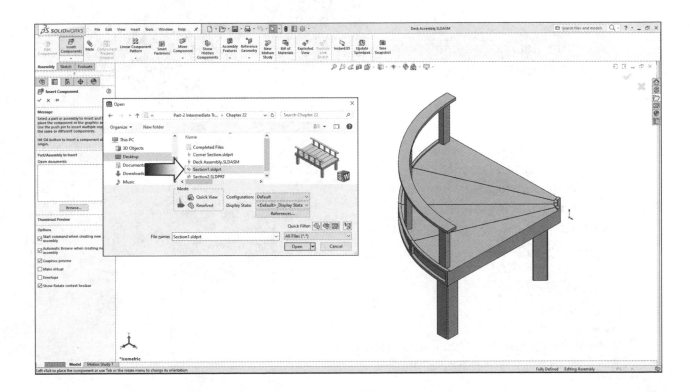

Hover the mouse cursor near one of the 2 corners of the component Corner Section; notice the magnetic mate snap line appears as you drag the Section1 closer to it.

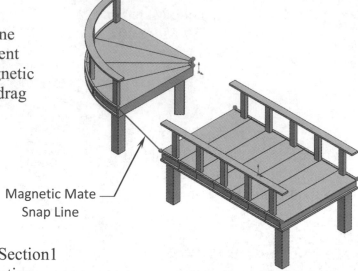

Magnetic Mate
Snap Line

Click the mouse to release the Section1 and let it snap to the Corner Section.

9. Adding another instance of Section1:

Click **Insert Components** once again.

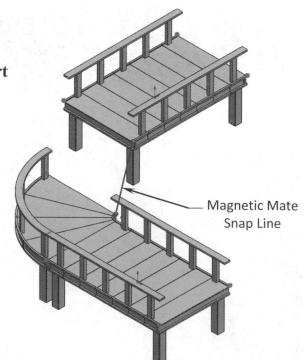

Select the part document **Section1.sldprt** and click **Open**.

Drag the Section1 toward the other side of the Corner Section. Move back and forth near one of the 2 corners until the Magnetic Mate Snap Line appears.

Magnetic Mate
Snap Line

Click the mouse to release the Section1 to let it snap to the Corner Section on its own.

10. Adding more instances:

Click **Insert Components** and add another instance of the Section1 to the assembly.

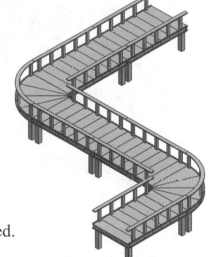

Move the instance close to the connecting point of the other as shown.

Press the **TAB** key to rotate the component **90°**. Click to release when the Snap Line is connecting to the correct corners.

Insert a few more instances of the Section1 and the Corner Section and design your own deck.

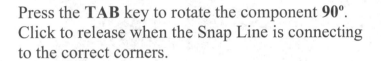

There is a Section2 found in the training files folder, which is a smaller version of Section1. Use it if needed.

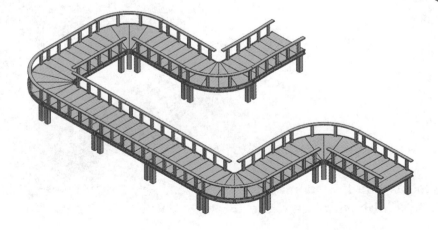

Example 1

Example 2

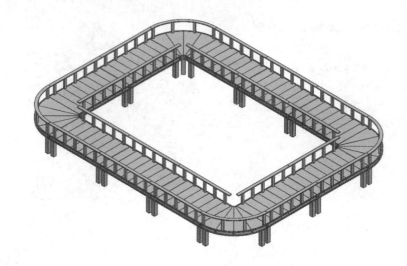

Example 3

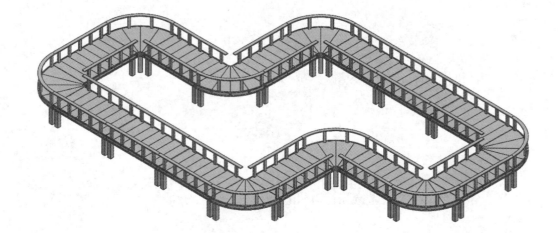

Example 4

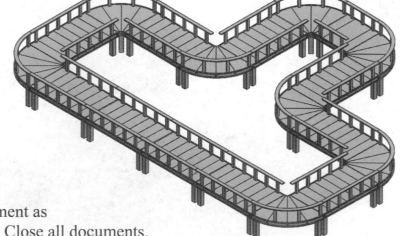

11. Saving your work:

Save the assembly document as
Deck Assembly.sldasm. Close all documents.

CHAPTER 23

Hybrid Modeling

Hybrid Modeling
Remote Control

Hybrid modeling is a method used to create the 3D models by combining the solid and surface tools to develop the geometry. Solid-Surface Hybrid-Modeling is often the best approach. Use surfaces to modify solids or convert a solid model into a surface model to make changes, and then convert it back to a solid with a single or multiple bodies.

A lot of times using only solid tools to create the model can be inefficient and awkward, and using surfaces tools alone can take way too many more steps than needed.

Both solid and surface tools have strengths and weaknesses but used together can greatly boost the ability and flexibility of creating the complex, hard to create types of shapes. In SOLIDWORKS, surfaces are usually an intermediate step to a solid; it is generally best practice to avoid leaving a model as an open surface. A finished model should form a closed volume and a material can be assigned to it for other analysis tasks when needed.

Surfaces are often used in SOLIDWORKS to assist the development of solid geometry; there may be many leftover surfaces when the model is completed. You can choose to delete them to clean up the model; this creates a Body-Delete feature in the FeatureManager tree that can be edited or suppressed.

Because surface models are created one by one and gaps or overlapped geometry are ignored, it is important to make sure that you check your model frequently for errors by using the Check Entity command under the Evaluate tab.

Surface bodies are listed in the Surface Bodies folder at the top of the Feature-Manager tree and solid bodies are listed in their own Solid Bodies folder.

Hybrid Modeling
Remote Control

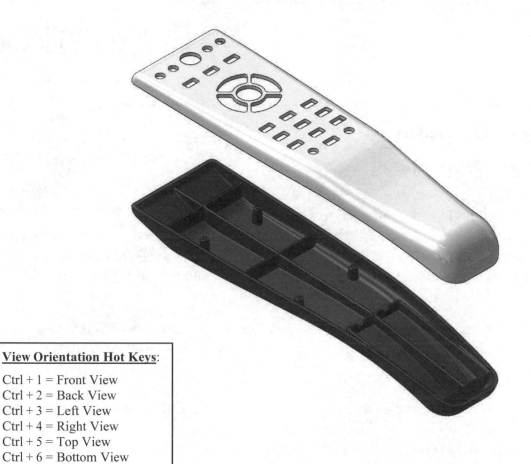

<u>View Orientation Hot Keys</u>:

Ctrl + 1 = Front View
Ctrl + 2 = Back View
Ctrl + 3 = Left View
Ctrl + 4 = Right View
Ctrl + 5 = Top View
Ctrl + 6 = Bottom View
Ctrl + 7 = Isometric View
Ctrl + 8 = Normal To
            Selection

Dimensioning Standards: **ANSI**
Units: **INCHES** – 3 Decimals

Tools Needed

 Extruded Surface   Trim Surface

 Draft   Split Line

 Shell   Thicken

1. Opening a part document:

Select **File / Open**.

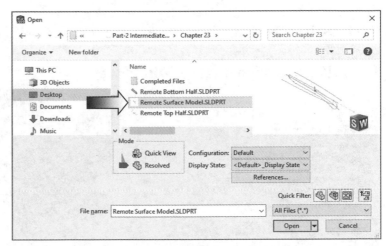

Browse to the Training Files folder and open a part document named: **Remote Surface Model.sldprt**.

There are 4 sketches on the FeatureManager tree; they will be used to create the body of the model.

Right-click on one of the tool tabs and enable the **Surfaces** toolbar.

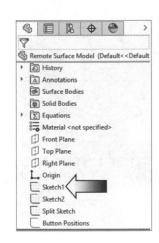

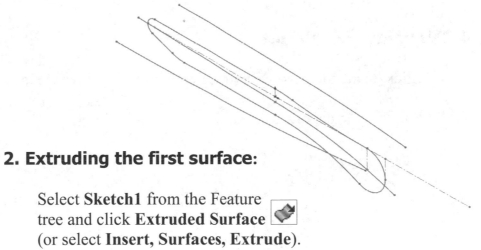

2. Extruding the first surface:

Select **Sketch1** from the Feature tree and click **Extruded Surface** (or select **Insert, Surfaces, Extrude**).

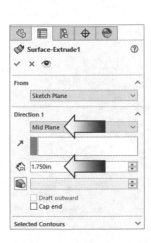

For Direction1, select **Mid Plane**.

For Extrude Depth, enter **1.750in**.

Click **OK**.

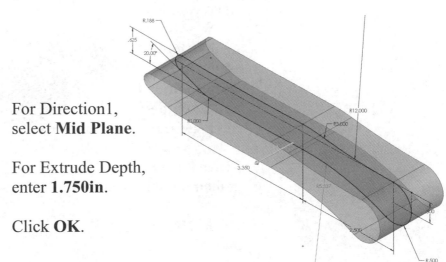

3. Extruding the second surface:

Select **Sketch2** from the Feature tree and press **Extruded Surface** .

For Direction1, select **Mid Plane**.

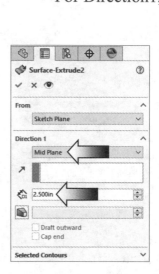

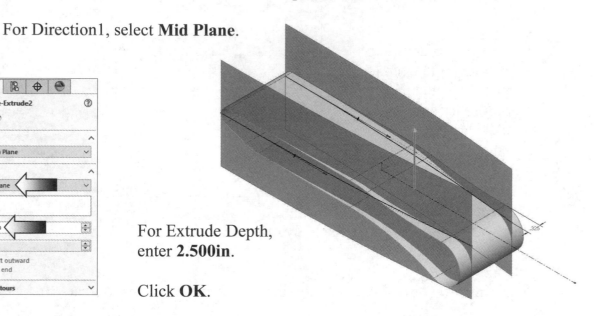

For Extrude Depth, enter **2.500in**.

Click **OK**.

4. Trimming the surfaces:

Click **Trim Surface** (or select **Insert, Surface, Trim**).

For Trim Type, click **Mutual**.

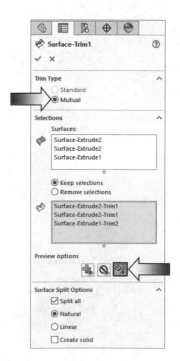

For Selections, select all **3 surfaces** in the graphics area.

Enable the Keep Selections button and select the center portion and the left and right sides as noted.

Enable the button **Show Both Included and Excluded Surfaces** (arrow).

Leave other settings at their default values.

Click **OK**.

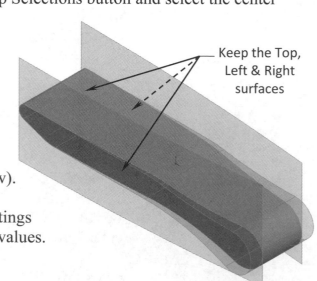

Keep the Top, Left & Right surfaces

5. Thickening the surfaces:

Click **Thicken** 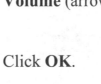 (or select **Insert, Surface, Thicken**).

For Thicken Parameter, select the **surface body** from the graphics area.

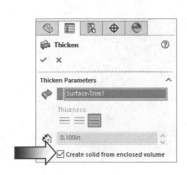

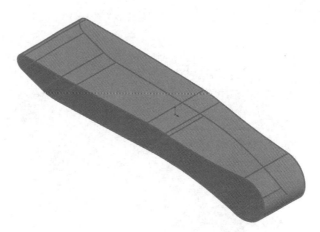

Enable the checkbox: **Create Solid From Enclosed Volume** (arrow).

Click **OK**.

The surface model is converted to a solid model.

6. Creating a split line feature:

Change to the **Features** tool tab.

Select the **Split Line** command 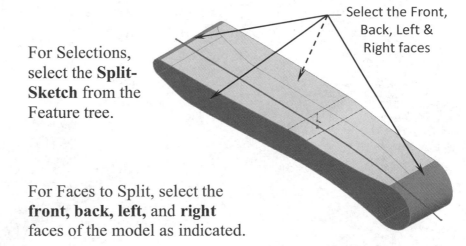 under the **Curves** drop down list.

For Type of Split, select the **Projection** option (arrow).

Select the Front, Back, Left & Right faces

For Selections, select the **Split-Sketch** from the Feature tree.

For Faces to Split, select the **front, back, left,** and **right** faces of the model as indicated.

Click **OK**.

7. Adding draft:

Click the **Draft** command on the **Features** tool tab.

For Type of Draft, select the **Parting Line** option (arrow).

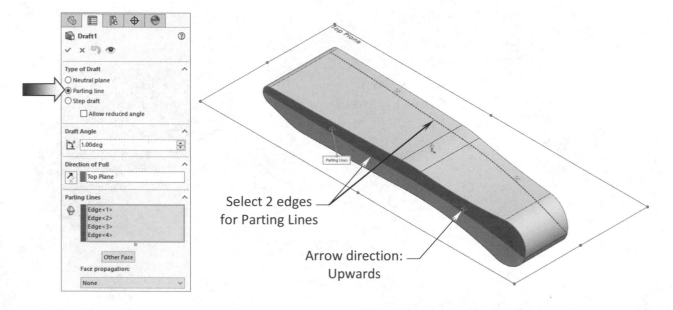

Select 2 edges for Parting Lines

Arrow direction: Upwards

For Draft Angle, enter **1.00deg**.

For Direction of Pull, select the **Top** plane (Direction arrow points upwards).

For parting Lines, select the **2 edges** in the middle of the left and right faces.

Click **OK**.

Repeat step 7 and add the same draft to the bottom half of the model.

Click **Reverse Direction** and make sure the draft direction arrows point downwards as noted.

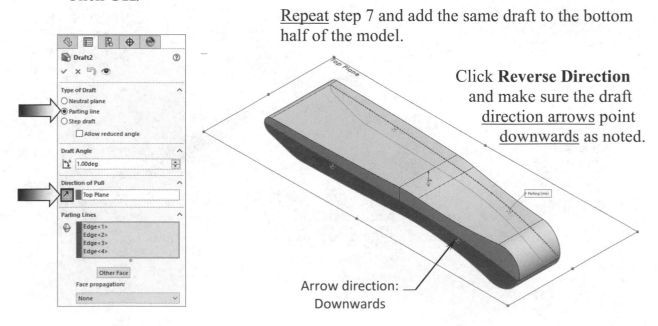

Arrow direction: Downwards

8. Adding fillets:

Click the **Fillet** command 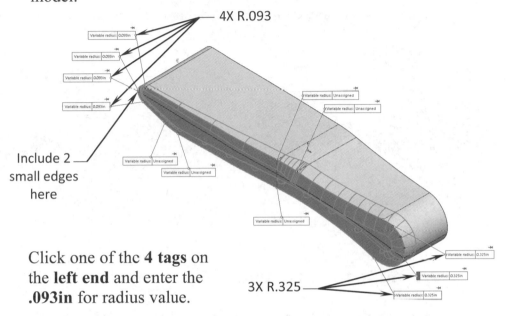 and select the **Variable Size** option (arrow).

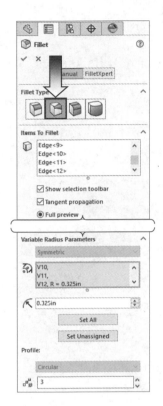

For Items to Fillet, select all the **edges** on the <u>left side</u> of the model.

4X R.093

Include 2 small edges here

3X R.325

Click one of the **4 tags** on the **left end** and enter the **.093in** for radius value.

Enter the radius values as noted for both left and right ends. Skip the 5 edges in the middle.

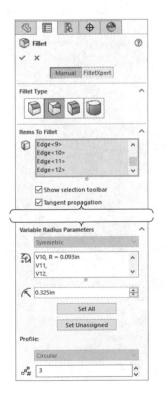

Click **OK**.

4X R.093

3X R.325

Repeat step 8 and add the same fillets on <u>right side</u> of the model.

Click **OK**.

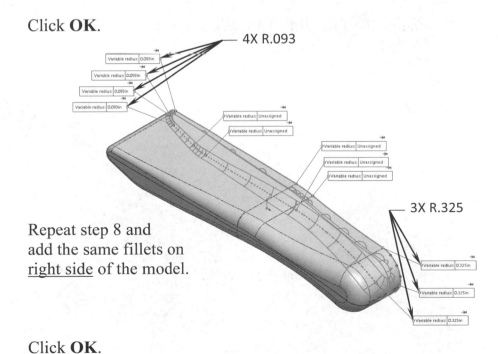

9. Shelling the solid body:

Click the **Shell** command 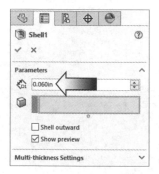 on the **Features** tool tab.

Enter **.060in** for wall thickness.

<u>Do not</u> select any of the faces so that only the <u>inside</u> of the model will be shelled.

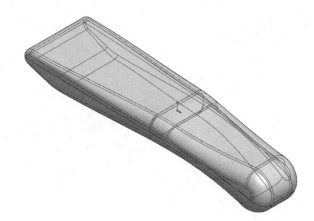

Click **OK**.

10. Cutting the solid body:

Expand the Split Line feature and select the **Split Sketch**. We will use it again to cut the solid body. Click **Extruded Cut**.

For Direction 1, select **Through All – Both**.

Enable the **Flip Side To Cut** checkbox.

The <u>Direction arrow</u> should be pointing <u>downwards</u>.

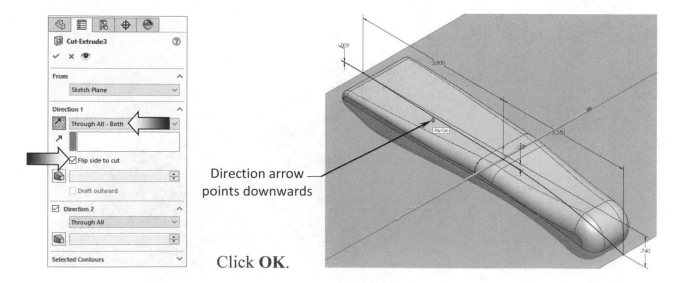

Direction arrow points downwards

Click **OK**.

11. Saving the upper half:

Select **File / Save As**.

For file name, enter **Remote Top Half**.

Browse to the same folder as the original document.

Press **Save**.

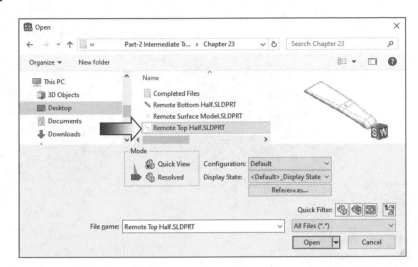

12. Saving the lower half:

Select **File / Save As** once again.

For file name, enter **Remote Bottom Half**.

Select the **Save As Copy and Open** option (arrow).

Click **Save**.

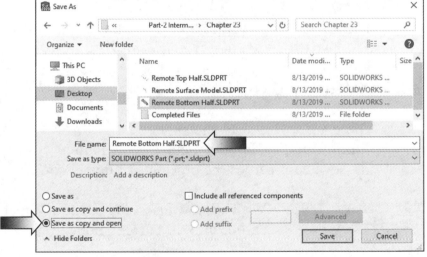

13. Flipping the cut direction:

The **Remote Bottom Half** (the copy) should be the current document at this point.

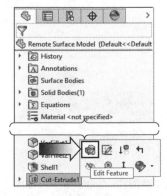

Right-click the **Cut-Extrude1** from the Feature tree and select **Edit Feature**.

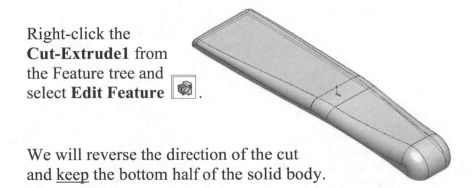

We will reverse the direction of the cut and keep the bottom half of the solid body.

Leave the Direction1 at **Through All – Both**.

Clear the **Flip Side To Cut** checkbox (arrow).

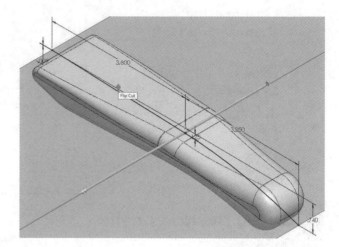

Click **OK**.

The top half of the solid body is removed; the bottom half is now the new body.

14. Inserting the top half:

There are several ways to split a body into multiple bodies, but the method shown in the last few steps serves the objective of this lesson.

We will now insert the upper half that was saved earlier into the current model.

Select **Insert / Part** (arrow).

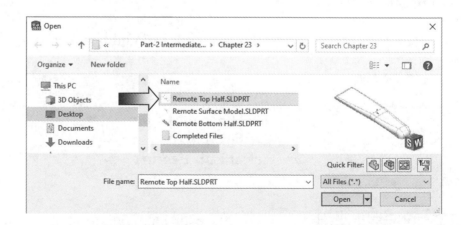

Select the part document **Remote Top Half.sldprt** and click **Open**.

Under the **Transfer** section, select the following:

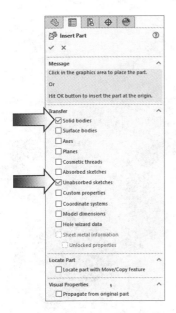

* **Solid Bodies**

* **Unabsorbed**
   Sketches

Click **OK**.

(Depending on the design criteria, other options
such as Surface Bodies, Planes, Model Dimensions, etc.
may need to be selected to help locate or modify the model.)

15. Creating the cut for the keypad:

Expand the Remote Top Half to see the folder that contains
the sketch for the Button Positions (arrow).

Select the sketch **Button Positions** and click
Extruded Cut.

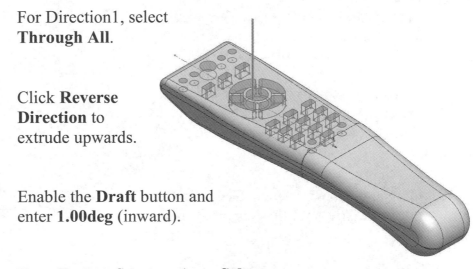

For Direction1, select
Through All.

Click **Reverse
Direction** to
extrude upwards.

Enable the **Draft** button and
enter **1.00deg** (inward).

Keep Feature Scope at **Auto-Select**.

Click **OK**.

16. Separating the two halves:

Although the Exploded View command would probably be quicker to separate the 2 bodies, but when this command is active all other tools will be grayed out. They cannot be used until the exploded view is collapsed.

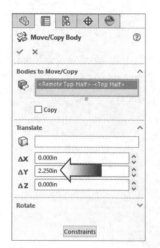

We will use the **Move/Copy** command instead.

Select **Insert / Features / Move-Copy**.

For Bodies to Move/Copy, select the **upper body**.

For Translate, enter **2.250in** in the **Delta Y** box (arrow).

Click **OK**.

17. Changing appearance:

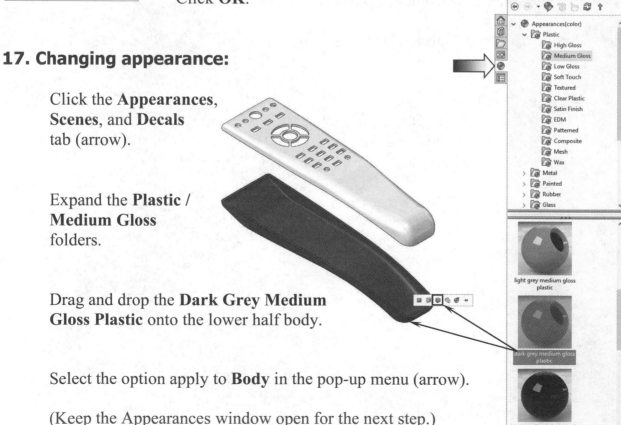

Click the **Appearances**, **Scenes**, and **Decals** tab (arrow).

Expand the **Plastic / Medium Gloss** folders.

Drag and drop the **Dark Grey Medium Gloss Plastic** onto the lower half body.

Select the option apply to **Body** in the pop-up menu (arrow).

(Keep the Appearances window open for the next step.)

Drag and drop the **White Medium Gloss Plastic** onto the upper half body.

Select the option apply to **Body** in the pop-up menu (arrow).

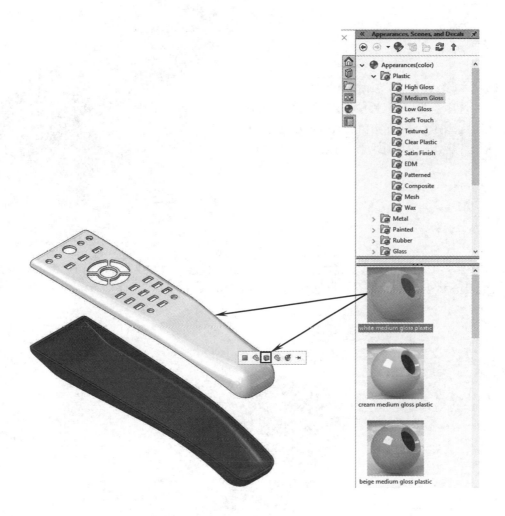

18. Saving your work:

Select **File / Save As**.

Enter: **Remote Hybrid Model (Completed).sldprt** for the file name.

Click **Save**. Close all documents.

Interior details such as support ribs, mounting bosses, etc. can be added to each half of the model either individually or together as multibody part.

Refer to Chapter 3 and chapter 4 in the SOLIDWORKS Part-2 Intermediate textbook for more details on creating and working with multibody parts.

Optional: Insert a few instances of the 2 halves into an assembly document and orient them similar to the image shown below.

Glossary

Alloys:

An Alloy is a mixture of two or more metals (and sometimes a non-metal). The mixture is made by heating and melting the substances together.
Example of alloys are Bronze (Copper and Tin), Brass (Copper and Zinc), and Steel (Iron and Carbon).

Gravity and Mass:

Gravity is the force that pulls everything on earth toward the ground and makes things feel heavy. Gravity makes all falling bodies accelerate at a constant 32ft. per second (9.8 m/s). In the earth's atmosphere, air resistance slows acceleration. Only on airless Moon would a feather and a metal block fall to the ground together.
The mass of an object is the amount of material it contains.
A body with greater mass has more inertia; it needs a greater force to accelerate.
Weight depends on the force of gravity, but mass does not.

When an object spins around another (for example: a satellite orbiting the earth) it is pushed outward. Two forces are at work here: Centrifugal (pushing outward) and Centripetal (pulling inward). If you whirl a ball around you on a string, you pull it inward (Centripetal force). The ball seems to pull outward (Centrifugal force) and if released will fly off in a straight line.

Heat:

Heat is a form of energy and can move from one substance to another in one of three ways: by Convection, by Radiation, and by Conduction.

Convection takes place only in liquids like water (for example: water in a kettle) and gases (for example: air warmed by a heat source such as a fire or radiator).
When liquid or gas is heated, it expands and becomes less dense. Warm air above the radiator rises and cool air moves in to take its place, creating a convection current.

Radiation is movement of heat through the air. Heat forms match set molecules of air moving and rays of heat spread out around the heat source.

Conduction occurs in solids such as metals. The handle of a metal spoon left in boiling liquid warms up as molecules at the heated end moves faster and collide with their neighbors, setting them moving. The heat travels through the metal, which is a good conductor of heat.

Inertia:

A body with a large mass is harder to start and also to stop. A heavy truck traveling at 50mph needs more power breaks to stop its motion than a smaller car traveling at the same speed.
Inertia is the tendency of an object either to stay still or to move steadily in a straight line, unless another force (such as a brick wall stopping the vehicle) makes it behave differently.

Joules:

The Joules is the SI unit of work or energy.
One Joule of work is done when a force of one Newton moves through a distance of one meter. The Joule is named after the English scientist James Joule (1818-1889).

Materials:

Stainless steel is an alloy of steel with chromium or nickel.

Steel is made by the basic oxygen process. The raw material is about three parts melted iron and one-part scrap steel. Blowing oxygen into the melted iron raises the temperature and gets rid of impurities.

All plastics are chemical compounds called polymers.

Glass is made by mixing and heating sand, limestone, and soda ash. When these ingredients melt they turn into glass, which is hardened when it cools.
Glass is in fact not a solid but a "supercooled" liquid, it can be shaped by blowing, pressing, drawing, casting into molds, rolling, and floating across molten tin, to make large sheets.

Ceramic objects, such as pottery and porcelain, electrical insulators, bricks, and roof tiles are all made from clay. The clay is shaped or molded when wet and soft, and heated in a kiln until it hardens.

Machine Tools:

Are powered tools used for shaping metal or other materials, by drilling holes, chiseling, grinding, pressing, or cutting. Often the material (the work piece) is moved while the tool stays still (lathe), or vice versa, the work piece stays while the tool moves (mill).
Most common machine tools are Mill, Lathe, Saw, Broach, Punch press, Grind, Bore and Stamp break.

CNC

Computer Numerical Control is the automation of machine tools that are operated by precisely programmed commands encoded on a storage medium, as opposed to controlled manually via hand wheels or levers, or mechanically automated via cams alone. Most CNC today is computer numerical control in which computers play an integral part of the control.

3D Printing

All methods work by working in layers, adding material, etc. different to other techniques, which are subtractive. Support is needed because almost all methods could support multi material printing, but it is currently only available in certain top tier machines.

A method of turning digital shapes into physical objects. Due to its nature, it allows us to accurately control the shape of the product. The drawback is size restraints and materials are often not durable.

While FDM does not seem like the best method for instrument manufacturing, it is one of the cheapest and most universally available methods.

EDM
Electric Discharge Machining.

FDM
Fused Deposition Modeling.

SLA
Stereo Lithography.

SLS
Selective Laser Sintering.

SLM
Selective Laser Melting.

J-P
Jetted Photopolymer (or Polyjet)

EDM Electric Discharge Machining

The basic EDM process is really quite simple. An electrical spark is created between an electrode and a work piece. The spark is visible evidence of the flow of electricity. This electric spark produces intense heat with temperatures reaching 8000 to 12000 degrees Celsius, melting almost anything.

The spark is very carefully controlled and localized so that it only affects the surface of the material.

The EDM process usually does not affect the heat treat below the surface. With wire EDM the spark always takes place in the dielectric of deionized water. The conductivity of the water is carefully controlled making an excellent environment for the EDM process. The water acts as a coolant and flushes away the eroded metal particles.

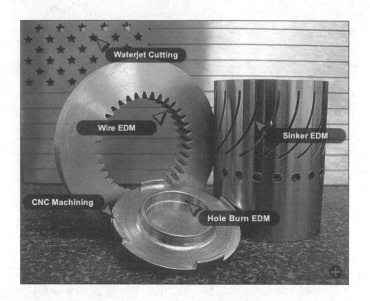

FDM Fused Deposition Modeling

3D printers that run on FDM Technology build parts layer-by-layer by heating thermoplastic material to a semi-liquid state and extruding it according to computer controlled paths.

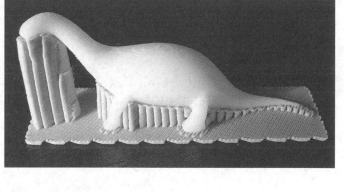

FDM uses two materials to execute a print job: modeling material, which constitutes the finished piece, and support material, which acts as scaffolding. Material filaments are fed from the 3D printer's material bays to the print head, which moves in X and Y coordinates, depositing material to complete each layer before the base moves down the Z axis and the next layer begins.

Once the 3D printer is done building, the user breaks the support material away or dissolves it in detergent and water, and the part is ready to use.

SLA StereoLithograph Apparatus

Stereolithography is an additive fabrication process utilizing a vat of liquid UV-curable photopolymer "resin" and a UV laser to build parts a layer at a time. On each

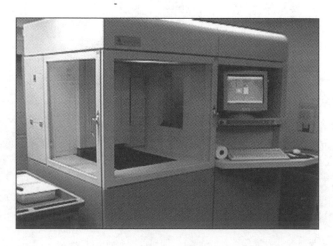

layer, the laser beam traces a part cross-section pattern on the surface of the liquid resin. Exposure to the UV laser light cures, or solidifies, the pattern traced on the resin and adheres it to the layer below.

After a pattern has been traced, the SLA's elevator platform descends by a single layer thickness, typically .0019in to .0059in. Then, a resin-filled blade sweeps across the part cross section, re-coating it with fresh material. On this new liquid surface, the subsequent layer pattern is traced, adhering to the previous layer.

A complete 3-D part is formed by this process. After building, parts are cleaned of excess resin by immersion in a chemical bath and then cured in a UV oven.

SLS Selective Laser Sintering

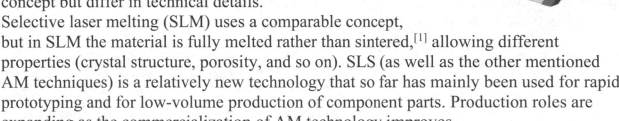

Selective laser sintering (SLS) is an additive manufacturing (AM) technique that uses a laser as the power source to sinter (compacting) powdered material (typically metal), aiming the laser automatically at points in space defined by a 3D model, binding the material together to create a solid structure.

It is similar to direct metal laser sintering (DMLS); the two are instantiations of the same concept but differ in technical details. Selective laser melting (SLM) uses a comparable concept, but in SLM the material is fully melted rather than sintered,[1] allowing different properties (crystal structure, porosity, and so on). SLS (as well as the other mentioned AM techniques) is a relatively new technology that so far has mainly been used for rapid prototyping and for low-volume production of component parts. Production roles are expanding as the commercialization of AM technology improves.

SLM Selective Laser Melting

Selective laser melting is an additive manufacturing process that uses 3D CAD data as a digital information source and energy in the form of a high-power laser beam, to create three-dimensional metal parts by fusing fine metal powders together. Manufacturing applications in aerospace or medical orthopedics are being pioneered.

The process starts by slicing the 3D CAD file data into layers, usually from 20 to 100 micrometres thick (0.00078740157 to 0.00393700787 in) creating a 2D image of each layer; this file format is the industry standard .stl file used on most layer-based 3D printing or stereolithography technologies.

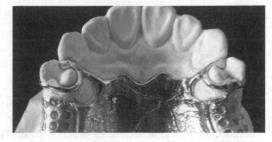

This file is then loaded into a file preparation software package that assigns parameters, values and physical supports that allow the file to be interpreted and built by different types of additive manufacturing machines.

J-P Jetted Photopolymer (or Polyjet)

Photopolymer jetting (or PolyJet) builds prototypes by jetting liquid photopolymer resin from ink-jet style heads. The resin is sprayed from the moving heads, and only the amount of material needed is used.

UV light is simultaneously emitted from the head, which cures each layer of resin immediately after it is applied. The process produces excellent surface finish and feature detail. Photopolymer jetting is used primarily to check form and fit and can handle limited functional tests due to the limited strength of photopolymer resins.

This process offers the unique ability to create prototypes with more than one type of material. For instance, a toothbrush prototype could be composed with a rigid shaft with a rubber-like over-molding for grip.

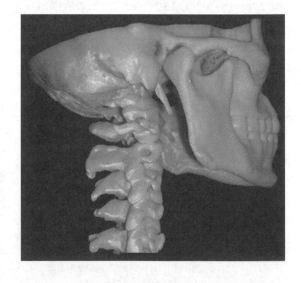

The process works with a variety of proprietary photopolymer resins as opposed to production materials. A tradeoff with this technology is that exposure to ambient heat, humidity, or sunlight can cause dimensional change that can affect tolerance. The process is faster and cleaner than the traditional vat and laser photopolymer processes.

Carbon 3D

The Carbon 3D not only prints composite materials like carbon fiber, but also fiberglass, nylon and PLA. Of course, only one at a time.

The printer employs some pretty nifty advancements too, including a self-leveling printing bed that clicks into position before each print.

The Carbon 3D is groundbreaking 3D printing technology which is 25 to 100 times faster than currently available commercial PolyJet or SLA machines.

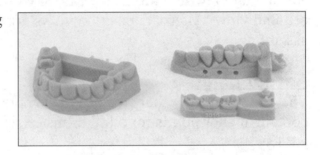

It is a true quantum leap forward for 3D printing speed!

Newton's Law:

1. Every object remains stopped or goes on moving at a steady rate in a straight line unless acted upon by another force. This is the inertia principle.
2. The amount of force needed to make an object change its speed depends on the mass of the object and the amount of the acceleration or deceleration required.
3. To every action there is an equal and opposite reaction. When a body is pushed on way by a force, another force pushes back with equal strength.

Polymers:

A polymer is made of one or more large molecules formed from thousands of smaller molecules. Rubber and Wood are natural polymers. Plastics are synthetic (artificially made) polymers.

Speed and Velocity:

Speed is the rate at which a moving object changes position (how far it moves in a fixed time).
Velocity is speed in a particular direction.
If either speed or direction is changed, velocity also changed.

Absorbed

A feature, sketch, or annotation that is contained in another item (usually a feature) in the FeatureManager design tree. Examples are the profile sketch and profile path in a base-sweep, or a cosmetic thread annotation in a hole.

Align

Tools that assist in lining up annotations and dimensions (left, right, top, bottom, and so on). For aligning parts in an assembly.

Alternate position view

A drawing view in which one or more views are superimposed in phantom lines on the original view. Alternate position views are often used to show range of motion of an assembly.

Anchor point

The end of a leader that attaches to the note, block, or other annotation. Sheet formats contain anchor points for a bill of materials, a hole table, a revision table, and a weldment cut list.

Annotation

A text note or a symbol that adds specific design intent to a part, assembly, or drawing. Specific types of annotations include note, hole callout, surface finish symbol, datum feature symbol, datum target, geometric tolerance symbol, weld symbol, balloon, and stacked balloon. Annotations that apply only to drawings include center mark, annotation centerline, area hatch, and block.

Appearance callouts

Callouts that display the colors and textures of the face, feature, body, and part under the entity selected and are a shortcut to editing colors and textures.

Area hatch

A crosshatch pattern or fill applied to a selected face or to a closed sketch in a drawing.

Assembly

A document in which parts, features, and other assemblies (sub-assemblies) are mated together. The parts and sub-assemblies exist in documents separate from the assembly. For example, in an assembly, a piston can be mated to other parts, such as a connecting rod or cylinder. This new assembly can then be used as a sub-assembly in an assembly of an engine. The extension for a SOLIDWORKS assembly file name is .SLDASM.

Attachment point

The end of a leader that attaches to the model (to an edge, vertex, or face, for example) or to a drawing sheet.

Axis
A straight line that can be used to create model geometry, features, or patterns. An axis can be made in a number of different ways, including using the intersection of two planes.

Balloon
Labels parts in an assembly, typically including item numbers and quantity. In drawings, the item numbers are related to rows in a bill of materials.

Base
The first solid feature of a part.

Baseline dimensions
Sets of dimensions measured from the same edge or vertex in a drawing.

Bend
A feature in a sheet metal part. A bend generated from a filleted corner, cylindrical face, or conical face is a round bend; a bend generated from sketched straight lines is a sharp bend.

Bill of materials
A table inscrtcd into a drawing to keep a record of the parts used in an assembly.

Block
A user-defined annotation that you can use in parts, assemblies, and drawings. A block can contain text, sketch entities (except points), and area hatch, and it can be saved in a file for later use as, for example, a custom callout or a company logo.

Bottom-up assembly
An assembly modeling technique where you create parts and then insert them into an assembly.

Broken-out section
A drawing view that exposes inner details of a drawing view by removing material from a closed profile, usually a spline.

Cavity
The mold half that holds the cavity feature of the design part.

Center mark
A cross that marks the center of a circle or arc.

Centerline
A centerline marks, in phantom font, an axis of symmetry in a sketch or drawing.

Chamfer

Bevels a selected edge or vertex. You can apply chamfers to both sketches and features.

Child

A dependent feature related to a previously built feature. For example, a chamfer on the edge of a hole is a child of the parent hole.

Click-release

As you sketch, if you click and then release the pointer, you are in click-release mode. Move the pointer and click again to define the next point in the sketch sequence.

Click-drag

As you sketch, if you click and drag the pointer, you are in click-drag mode. When you release the pointer, the sketch entity is complete.

Closed profile

Also called a closed contour, it is a sketch or sketch entity with no exposed endpoints, for example, a circle or polygon.

Collapse

The opposite of explode. The collapse action returns an exploded assembly's parts to their normal positions.

Collision Detection

An assembly function that detects collisions between components when components move or rotate. A collision occurs when an entity on one component coincides with any entity on another component.

Component

Any part or sub-assembly within an assembly

Configuration

A variation of a part or assembly within a single document. Variations can include different dimensions, features, and properties. For example, a single part such as a bolt can contain different configurations that vary the diameter and length.

ConfigurationManager

Located on the left side of the SOLIDWORKS window, it is a means to create, select, and view the configurations of parts and assemblies.

Constraint

The relations between sketch entities, or between sketch entities and planes, axes, edges, or vertices.

Construction geometry
The characteristic of a sketch entity that the entity is used in creating other geometry but is not itself used in creating features.

Coordinate system
A system of planes used to assign Cartesian coordinates to features, parts, and assemblies. Part and assembly documents contain default coordinate systems; other coordinate systems can be defined with reference geometry. Coordinate systems can be used with measurement tools and for exporting documents to other file formats.

Cosmetic thread
An annotation that represents threads.

Crosshatch
A pattern (or fill) applied to drawing views such as section views and broken-out sections.

Curvature
Curvature is equal to the inverse of the radius of the curve. The curvature can be displayed in different colors according to the local radius (usually of a surface).

Cut
A feature that removes material from a part by such actions as extrude, revolve, loft, sweep, thicken, cavity, and so on.

Dangling
A dimension, relation, or drawing section view that is unresolved. For example, if a piece of geometry is dimensioned, and that geometry is later deleted, the dimension becomes dangling.

Degrees of freedom
Geometry that is not defined by dimensions or relations is free to move. In 2D sketches, there are three degrees of freedom: movement along the X and Y axes, and rotation about the Z axis (the axis normal to the sketch plane). In 3D sketches and in assemblies, there are six degrees of freedom: movement along the X, Y, and Z axes, and rotation about the X, Y, and Z axes.

Derived part
A derived part is a new base, mirror, or component part created directly from an existing part and linked to the original part such that changes to the original part are reflected in the derived part.

Derived sketch

A copy of a sketch, in either the same part or the same assembly that is connected to the original sketch. Changes in the original sketch are reflected in the derived sketch.

Design Library

Located in the Task Pane, the Design Library provides a central location for reusable elements such as parts, assemblies, and so on.

Design table

An Excel spreadsheet that is used to create multiple configurations in a part or assembly document.

Detached drawing

A drawing format that allows opening and working in a drawing without loading the corresponding models into memory. The models are loaded on an as-needed basis.

Detail view

A portion of a larger view, usually at a larger scale than the original view.

Dimension line

A linear dimension line references the dimension text to extension lines indicating the entity being measured. An angular dimension line references the dimension text directly to the measured object.

DimXpertManager

Located on the left side of the SOLIDWORKS window, it is a means to manage dimensions and tolerances created using DimXpert for parts according to the requirements of the ASME Y.14.41-2003 standard.

DisplayManager

The DisplayManager lists the appearances, decals, lights, scene, and cameras applied to the current model. From the DisplayManager, you can view applied content, and add, edit, or delete items. When PhotoView 360 is added in, the DisplayManager also provides access to PhotoView options.

Document

A file containing a part, assembly, or drawing.

Draft

The degree of taper or angle of a face usually applied to molds or castings.

Drawing

A 2D representation of a 3D part or assembly. The extension for a SOLIDWORKS drawing file name is .SLDDRW.

Drawing sheet
A page in a drawing document.

Driven dimension
Measurements of the model, but they do not drive the model and their values cannot be changed.

Driving dimension
Also referred to as a model dimension, it sets the value for a sketch entity. It can also control distance, thickness, and feature parameters.

Edge
A single outside boundary of a feature.

Edge flange
A sheet metal feature that combines a bend and a tab in a single operation.

Equation
Creates a mathematical relation between sketch dimensions, using dimension names as variables, or between feature parameters, such as the depth of an extruded feature or the instance count in a pattern.

Exploded view
Shows an assembly with its components separated from one another, usually to show how to assemble the mechanism.

Export
Save a SOLIDWORKS document in another format for use in other CAD/CAM, rapid prototyping, web, or graphics software applications.

Extension line
The line extending from the model indicating the point from which a dimension is measured.

Extrude
A feature that linearly projects a sketch to either add material to a part (in a base or boss) or remove material from a part (in a cut or hole).

Face
A selectable area (planar or otherwise) of a model or surface with boundaries that help define the shape of the model or surface. For example, a rectangular solid has six faces.

Fasteners

A SOLIDWORKS Toolbox library that adds fasteners automatically to holes in an assembly.

Feature

An individual shape that, combined with other features, makes up a part or assembly. Some features, such as bosses and cuts, originate as sketches. Other features, such as shells and fillets, modify a feature's geometry. However, not all features have associated geometry. Features are always listed in the FeatureManager design tree.

FeatureManager design tree

Located on the left side of the SOLIDWORKS window, it provides an outline view of the active part, assembly, or drawing.

Fill

A solid area hatch or crosshatch. Fill also applies to patches on surfaces.

Fillet

An internal rounding of a corner or edge in a sketch, or an edge on a surface or solid.

Forming tool

Dies that bend, stretch, or otherwise form sheet metal to create such form features as louvers, lances, flanges, and ribs.

Fully defined

A sketch where all lines and curves in the sketch, and their positions, are described by dimensions or relations, or both, and cannot be moved. Fully defined sketch entities are shown in black.

Geometric tolerance

A set of standard symbols that specify the geometric characteristics and dimensional requirements of a feature.

Graphics area

The area in the SOLIDWORKS window where the part, assembly, or drawing appears.

Guide curve

A 2D or 3D curve used to guide a sweep or loft.

Handle

An arrow, square, or circle that you can drag to adjust the size or position of an entity (a feature, dimension, or sketch entity, for example).

Helix

A curve defined by pitch, revolutions, and height. A helix can be used, for example, as a path for a swept feature cutting threads in a bolt.

Hem

A sheet metal feature that folds back at the edge of a part. A hem can be open, closed, double, or teardrop.

HLR

(Hidden lines removed) a view mode in which all edges of the model that are not visible from the current view angle are removed from the display.

HLV

(Hidden lines visible) A view mode in which all edges of the model that are not visible from the current view angle are shown gray or dashed.

Import

Open files from other CAD software applications into a SOLIDWORKS document.

In-context feature

A feature with an external reference to the geometry of another component; the in-context feature changes automatically if the geometry of the referenced model or feature changes.

Inference

The system automatically creates (infers) relations between dragged entities (sketched entities, annotations, and components) and other entities and geometry. This is useful when positioning entities relative to one another.

Instance

An item in a pattern or a component in an assembly that occurs more than once. Blocks are inserted into drawings as instances of block definitions.

Interference detection

A tool that displays any interference between selected components in an assembly.

Jog

A sheet metal feature that adds material to a part by creating two bends from a sketched line.

Knit

A tool that combines two or more faces or surfaces into one. The edges of the surfaces must be adjacent and not overlapping, but they cannot ever be planar. There is no difference in the appearance of the face or the surface after knitting.

Layout sketch

A sketch that contains important sketch entities, dimensions, and relations. You reference the entities in the layout sketch when creating new sketches, building new geometry, or positioning components in an assembly. This allows for easier updating of your model because changes you make to the layout sketch propagate to the entire model.

Leader

A solid line from an annotation (note, dimension, and so on) to the referenced feature.

Library feature

A frequently used feature, or combination of features, that is created once and then saved for future use.

Lightweight

A part in an assembly or a drawing has only a subset of its model data loaded into memory. The remaining model data is loaded on an as-needed basis. This improves performance of large and complex assemblies.

Line

A straight sketch entity with two endpoints. A line can be created by projecting an external entity such as an edge, plane, axis, or sketch curve into the sketch.

Loft

A base, boss, cut, or surface feature created by transitions between profiles.

Lofted bend

A sheet metal feature that produces a roll form or a transitional shape from two open profile sketches. Lofted bends often create funnels and chutes.

Mass properties

A tool that evaluates the characteristics of a part or an assembly such as volume, surface area, centroid, and so on.

Mate

A geometric relationship, such as coincident, perpendicular, tangent, and so on, between parts in an assembly.

Mate reference

Specifies one or more entities of a component to use for automatic mating. When you drag a component with a mate reference into an assembly, the software tries to find other combinations of the same mate reference name and mate type.

Mates folder
A collection of mates that are solved together. The order in which the mates appear within the Mates folder does not matter.

Mirror
(a) A mirror feature is a copy of a selected feature, mirrored about a plane or planar face.
(b) A mirror sketch entity is a copy of a selected sketch entity that is mirrored about a centerline.

Miter flange
A sheet metal feature that joins multiple edge flanges together and miters the corner.

Model
3D solid geometry in a part or assembly document. If a part or assembly document contains multiple configurations, each configuration is a separate model.

Model dimension
A dimension specified in a sketch or a feature in a part or assembly document that defines some entity in a 3D model.

Model item
A characteristic or dimension of feature geometry that can be used in detailing drawings.

Model view
A drawing view of a part or assembly.

Mold
A set of manufacturing tooling used to shape molten plastic or other material into a designed part. You design the mold using a sequence of integrated tools that result in cavity and core blocks that are derived parts of the part to be molded.

Motion Study
Motion Studies are graphical simulations of motion and visual properties with assembly models. Analogous to a configuration, they do not actually change the original assembly model or its properties. They display the model as it changes based on simulation elements you add.

Multibody part
A part with separate solid bodies within the same part document. Unlike the components in an assembly, multibody parts are not dynamic.

Native format
DXF and DWG files remain in their original format (are not converted into SOLIDWORKS format) when viewed in SOLIDWORKS drawing sheets (view only).

Open profile
Also called an open contour, it is a sketch or sketch entity with endpoints exposed. For example, a U-shaped profile is open.

Ordinate dimensions
A chain of dimensions measured from a zero ordinate in a drawing or sketch.

Origin
The model origin appears as three gray arrows and represents the (0,0,0) coordinate of the model. When a sketch is active, a sketch origin appears in red and represents the (0,0,0) coordinate of the sketch. Dimensions and relations can be added to the model origin, but not to a sketch origin.

Out-of-context feature
A feature with an external reference to the geometry of another component that is not open.

Over defined
A sketch is over defined when dimensions or relations are either in conflict or redundant.

Parameter
A value used to define a sketch or feature (often a dimension).

Parent
An existing feature upon which other features depend. For example, in a block with a hole, the block is the parent to the child hole feature.

Part
A single 3D object made up of features. A part can become a component in an assembly, and it can be represented in 2D in a drawing. Examples of parts are bolt, pin, plate, and so on. The extension for a SOLIDWORKS part file name is .SLDPRT.

Path
A sketch, edge, or curve used in creating a sweep or loft.

Pattern
A pattern repeats selected sketch entities, features, or components in an array, which can be linear, circular, or sketch driven. If the seed entity is changed, the other instances in the pattern update.

Physical Dynamics

An assembly tool that displays the motion of assembly components in a realistic way. When you drag a component, the component applies a force to other components it touches. Components move only within their degrees of freedom.

Pierce relation

Makes a sketch point coincident to the location at which an axis, edge, line, or spline pierces the sketch plane.

Planar

Entities that can lie on one plane. For example, a circle is planar, but a helix is not.

Plane

Flat construction geometry. Planes can be used for a 2D sketch, section view of a model, a neutral plane in a draft feature, and others.

Point

A singular location in a sketch, or a projection into a sketch at a single location of an external entity (origin, vertex, axis, or point in an external sketch).

Predefined view

A drawing view in which the view position, orientation, and so on can be specified before a model is inserted. You can save drawing documents with predefined views as templates.

Profile

A sketch entity used to create a feature (such as a loft) or a drawing view (such as a detail view). A profile can be open (such as a U shape or open spline) or closed (such as a circle or closed spline).

Projected dimension

If you dimension entities in an isometric view, projected dimensions are the flat dimensions in 2D.

Projected view

A drawing view projected orthogonally from an existing view.

PropertyManager

Located on the left side of the SOLIDWORKS window, it is used for dynamic editing of sketch entities and most features.

RealView graphics

A hardware (graphics card) support of advanced shading in real time; the rendering applies to the model and is retained as you move or rotate a part.

Rebuild

Tool that updates (or regenerates) the document with any changes made since the last time the model was rebuilt. Rebuild is typically used after changing a model dimension.

Reference dimension

A dimension in a drawing that shows the measurement of an item but cannot drive the model and its value cannot be modified. When model dimensions change, reference dimensions update.

Reference geometry

Includes planes, axes, coordinate systems, and 3D curves. Reference geometry is used to assist in creating features such as lofts, sweeps, drafts, chamfers, and patterns.

Relation

A geometric constraint between sketch entities or between a sketch entity and a plane, axis, edge, or vertex. Relations can be added automatically or manually.

Relative view

A relative (or relative to model) drawing view is created relative to planar surfaces in a part or assembly.

Reload

Refreshes shared documents. For example, if you open a part file for read-only access while another user makes changes to the same part, you can reload the new version, including the changes.

Reorder

Reordering (changing the order of) items is possible in the FeatureManager design tree. In parts, you can change the order in which features are solved. In assemblies, you can control the order in which components appear in a bill of materials.

Replace

Substitutes one or more open instances of a component in an assembly with a different component.

Resolved

A state of an assembly component (in an assembly or drawing document) in which it is fully loaded in memory. All the component's model data is available, so its entities can be selected, referenced, edited, and used in mates, and so on.

Revolve

A feature that creates a base or boss, a revolved cut, or revolved surface by revolving one or more sketched profiles around a centerline.

Rip

A sheet metal feature that removes material at an edge to allow a bend.

Rollback

Suppresses all items below the rollback bar.

Section

Another term for profile in sweeps.

Section line

A line or centerline sketched in a drawing view to create a section view.

Section scope

Specifies the components to be left uncut when you create an assembly drawing section view.

Section view

A section view (or section cut) is (1) a part or assembly view cut by a plane, or (2) a drawing view created by cutting another drawing view with a section line.

Seed

A sketch or an entity (a feature, face, or body) that is the basis for a pattern. If you edit the seed, the other entities in the pattern are updated.

Shaded

Displays a model as a colored solid.

Shared values

Also called linked values, these are named variables that you assign to set the value of two or more dimensions to be equal.

Sheet format

Includes page size and orientation, standard text, borders, title blocks, and so on. Sheet formats can be customized and saved for future use. Each sheet of a drawing document can have a different format.

Shell

A feature that hollows out a part, leaving open the selected faces and thin walls on the remaining faces. A hollow part is created when no faces are selected to be open.

Sketch

A collection of lines and other 2D objects on a plane or face that forms the basis for a feature such as a base or a boss. A 3D sketch is non-planar and can be used to guide a sweep or loft, for example.

Smart Fasteners

Automatically adds fasteners (bolts and screws) to an assembly using the SOLIDWORKS Toolbox library of fasteners.

SmartMates

An assembly mating relation that is created automatically.

Solid sweep

A cut sweep created by moving a tool body along a path to cut out 3D material from a model.

Spiral

A flat or 2D helix, defined by a circle, pitch, and number of revolutions.

Spline

A sketched 2D or 3D curve defined by a set of control points.

Split line

Projects a sketched curve onto a selected model face, dividing the face into multiple faces so that each can be selected individually. A split line can be used to create draft features, to create face blend fillets, and to radiate surfaces to cut molds.

Stacked balloon

A set of balloons with only one leader. The balloons can be stacked vertically (up or down) or horizontally (left or right).

Standard 3 views

The three orthographic views (front, right, and top) that are often the basis of a drawing.

StereoLithography

The process of creating rapid prototype parts using a faceted mesh representation in STL files.

Sub-assembly

An assembly document that is part of a larger assembly. For example, the steering mechanism of a car is a sub-assembly of the car.

Suppress

Removes an entity from the display and from any calculations in which it is involved. You can suppress features, assembly components, and so on. Suppressing an entity does not delete the entity; you can un-suppress the entity to restore it.

Surface

A zero-thickness planar or 3D entity with edge boundaries. Surfaces are often used to create solid features. Reference surfaces can be used to modify solid features.

Sweep

Creates a base, boss, cut, or surface feature by moving a profile (section) along a path. For cut sweeps, you can create solid sweeps by moving a tool body along a path.

Tangent arc

An arc that is tangent to another entity, such as a line.

Tangent edge

The transition edge between rounded or filleted faces in hidden lines visible or hidden lines removed modes in drawings.

Task Pane

Located on the right-side of the SOLIDWORKS window, the Task Pane contains SOLIDWORKS Resources, the Design Library, and the File Explorer.

Template

A document (part, assembly, or drawing) that forms the basis of a new document. It can include user-defined parameters, annotations, predefined views, geometry, and so on.

Temporary axis

An axis created implicitly for every conical or cylindrical face in a model.

Thin feature

An extruded or revolved feature with constant wall thickness. Sheet metal parts are typically created from thin features.

TolAnalyst

A tolerance analysis application that determines the effects that dimensions and tolerances have on parts and assemblies.

Top-down design

An assembly modeling technique where you create parts in the context of an assembly by referencing the geometry of other components. Changes to the referenced components propagate to the parts that you create in context.

Triad

Three axes with arrows defining the X, Y, and Z directions. A reference triad appears in part and assembly documents to assist in orienting the viewing of models. Triads also assist when moving or rotating components in assemblies.

Under defined

A sketch is under defined when there are not enough dimensions and relations to prevent entities from moving or changing size.

Vertex

A point at which two or more lines or edges intersect. Vertices can be selected for sketching, dimensioning, and many other operations.

Viewports

Windows that display views of models. You can specify one, two, or four viewports. Viewports with orthogonal views can be linked, which links orientation and rotation.

Virtual sharp

A sketch point at the intersection of two entities after the intersection itself has been removed by a feature such as a fillet or chamfer. Dimensions and relations to the virtual sharp are retained even though the actual intersection no longer exists.

Weldment

A multibody part with structural members.

Weldment cut list

A table that tabulates the bodies in a weldment along with descriptions and lengths.

Wireframe

A view mode in which all edges of the part or assembly are displayed.

Zebra stripes

Simulate the reflection of long strips of light on a very shiny surface. They allow you to see small changes in a surface that may be hard to see with a standard display.

Zoom

To simulate movement toward or away from a part or an assembly.

Index

SOLIDWORKS Quick Guide
Quick Reference Guide to SOLIDWORKS Command Icons & Toolbars

STANDARD Toolbar

 Creates a new document.

 Opens an existing document.

 Saves an active document.

 Make Drawing from Part/Assembly.

 Make Assembly from Part/Assembly.

 Prints the active document.

 Print preview.

 Cuts the selection & puts it on the clipboard.

 Copies the selection & puts it on the clipboard.

 Inserts the clipboard contents.

 Deletes the selection.

 Reverses the last action.

 Rebuilds the part / assembly / drawing.

 Redo the last action that was undone.

 Saves all documents.

 Edits material.

 Closes an existing document.

 Shows or hides the Selection Filter toolbar.

 Shows or hides the Web toolbar.

 Properties.

File properties.

 Loads or unloads the 3D instant website add-in.

 Select tool.

 Select the entire document.

 Checks read-only files.

 Options.

 Help.

 Full screen view.

 OK.

 Cancel.

 Magnified selection.

SKETCH TOOLS Toolbar

 Select.

 Sketch.

 3D Sketch.

 Sketches a rectangle from the center.

 Sketches a CenterPoint arc slot.

 Sketches a 3-point arc slot.

 Sketches a straight slot.

 Sketches a CenterPoint straight slot.

 Sketches a 3-point arc.

 Creates sketched ellipses.

Quick Reference Guide to SOLIDWORKS Command Icons & Toolbars

SKETCH TOOLS Toolbar

3D sketch on plane.	Partial ellipses.
Sets up Grid parameters.	Adds a Parabola.
Creates a sketch on a selected plane or face.	Adds a spline.
Equation driven curve.	Sketches a polygon.
Modifies a sketch.	Sketches a corner rectangle.
Copies sketch entities.	Sketches a parallelogram.
Scales sketch entities.	Creates points.
Rotates sketch entities.	Creates sketched centerlines.
Sketches 3-point rectangle from the center.	Adds text to sketch.
Sketches 3-point corner rectangle.	Converts selected model edges or sketch entities to sketch segments.
Sketches a line.	Creates a sketch along the intersection of multiple bodies.
Creates a center point arc: center, start, end.	Converts face curves on the selected face into 3D sketch entities.
Creates an arc tangent to a line.	Mirrors selected segments about a centerline.
Sketches splines on a surface or face.	Fillets the corner of two lines.
Sketches a circle.	Creates a chamfer between two sketch entities.
Sketches a circle by its perimeter.	Creates a sketch curve by offsetting model edges or sketch entities at a specified distance.
Makes a path of sketch entities.	Trims a sketch segment.
Mirrors entities dynamically about a centerline.	Extends a sketch segment.
Insert a plane into the 3D sketch.	Splits a sketch segment.
Instant 2D.	Construction Geometry.
Sketch numeric input.	Creates linear steps and repeat of sketch entities.
Detaches segment on drag.	Creates circular steps and repeat of sketch entities.
Sketch picture.	

Quick Reference Guide to SOLIDWORKS Command Icons & Toolbars

SHEET METAL Toolbar

 Add a bend from a selected sketch in a Sheet Metal part.

 Shows flat pattern for this sheet metal part.

 Shows part without inserting any bends.

 Inserts a rip feature to a sheet metal part.

 Create a Sheet Metal part or add material to existing Sheet Metal part.

 Inserts a Sheet Metal Miter Flange feature.

 Folds selected bends.

 Unfolds selected bends.

 Inserts bends using a sketch line.

 Inserts a flange by pulling an edge.

 Inserts a sheet metal corner feature.

 Inserts a Hem feature by selecting edges.

 Breaks a corner by filleting/chamfering it.

 Inserts a Jog feature using a sketch line.

 Inserts a lofted bend feature using 2 sketches.

 Creates inverse dent on a sheet metal part.

 Trims out material from a corner, in a sheet metal part.

 Inserts a fillet weld bead.

 Converts a solid/surface into a sheet metal part.

 Adds a Cross Break feature into a selected face.

 Sweeps an open profile along an open/closed path.

 Adds a gusset/rib across a bend.

 Corner relief.

 Welds the selected corner.

SURFACES Toolbar

 Creates mid surfaces between offset face pairs.

 Patches surface holes and external edges.

 Creates an extruded surface.

 Creates a revolved surface.

 Creates a swept surface.

 Creates a lofted surface.

 Creates an offset surface.

 Radiates a surface originating from a curve, parallel to a plane.

 Knits surfaces together.

 Creates a planar surface from a sketch or a set of edges.

 Creates a surface by importing data from a file.

 Extends a surface.

 Trims a surface.

 Surface flattens.

 Deletes Face(s).

 Replaces Face with Surface.

 Patches surface holes and external edges by extending the surfaces.

 Creates parting surfaces between core & cavity surfaces.

 Inserts ruled surfaces from edges.

WELDMENTS Toolbar

 Creates a weldment feature.

 Creates a structure member feature.

 Adds a gusset feature between 2 planar adjoining faces.

 Creates an end cap feature.

 Adds a fillet weld bead feature.

 Trims or extends structure members.

Weld bead.

Quick Reference Guide to SOLIDWORKS Command Icons & Toolbars

DIMENSIONS/RELATIONS Toolbar

 Inserts dimension between two lines.

 Creates a horizontal dimension between selected entities.

 Creates a vertical dimension between selected entities.

 Creates a reference dimension between selected entities.

 Creates a set of ordinate dimensions.

 Creates a set of Horizontal ordinate

 Creates a set of Vertical ordinate dimensions.

 Creates a chamfer dimension.

 Adds a geometric relation.

 Automatically Adds Dimensions to the current sketch.

 Displays and deletes geometric relations.

 Fully defines a sketch.

 Scans a sketch for elements of equal length or radius.

 Angular Running dimension.

 Display / Delete dimension.

 Isolate changed dimension.

 Path length dimension.

BLOCK Toolbar

 Makes a new block.

 Edits the selected block.

 Inserts a new block to a sketch or drawing.

Adds/Removes sketch entities to/from blocks.

 Updates parent sketches affected by this block.

 Saves the block to a file.

 Explodes the selected block.

 Inserts a belt.

STANDARD VIEWS Toolbar

 Front view.

 Back view.

 Left view.

 Right view.

 Top view.

 Bottom view.

 Isometric view.

 Trimetric view.

 Dimetric view.

 Normal to view.

 Links all views in the viewport together.

 Displays viewport with front & right

 Displays a 4-view viewport with 1st or 3rd angle of projection.

 Displays viewport with front & top.

 Displays viewport with a single view.

 View selector.

 New view.

FEATURES Toolbar

 Creates a boss feature by extruding a sketched profile.

 Creates a revolved feature based on profile and angle parameter.

 Creates a cut feature by extruding a sketched profile.

 Creates a cut feature by revolving a sketched profile.

 Thread.

 Creates a cut by sweeping a closed profile along an open or closed path.

 Loft cut.

 Creates a cut by thickening one or more adjacent surfaces.

 Adds a deformed surface by push or pull on points.

 Creates a lofted feature between two or more profiles.

 Creates a solid feature by thickening one or more adjacent surfaces.

 Creates a filled feature.

 Chamfers an edge or a chain of tangent edges.

 Inserts a rib feature.

 Combine.

 Creates a shell feature.

 Applies draft to a selected surface.

 Creates a cylindrical hole.

 Inserts a hole with a pre-defined cross section.

 Puts a dome surface on a face.

 Model break view.

 Applies global deformation to solid or surface bodies.

 Wraps closed sketch contour(s) onto a face.

 Curve Driven pattern.

 Suppresses the selected feature or component.

 Un-suppresses the selected feature or component.

 Flexes solid and surface bodies.

 Intersect.

 Variable Patterns.

 Live Section Plane.

 Mirrors.

 Scale.

 Creates a Sketch Driven pattern.

 Creates a Table-Driven Pattern.

 Inserts a split Feature.

 Hole series.

 Joins bodies from one or more parts into a single part in the context of an assembly.

 Deletes a solid or a surface.

 Instant 3D.

 Inserts a part from file into the active part document.

 Moves/Copies solid and surface bodies or moves graphics bodies.

 Merges short edges on faces.

 Pushes solid / surface model by another solid / surface model.

 Moves face(s) of a solid.

 FeatureWorks Options.

 Linear Pattern.

 Fill Pattern.

 Cuts a solid model with a

 Boundary Boss/Base.

 Boundary Cut.

 Circular Pattern.

 Recognize Features.

 Grid System.

MOLD TOOLS Toolbar

 Extracts core(s) from existing tooling split.

 Constructs a surface patch.

 Moves face(s) of a solid.

 Creates offset surfaces.

 Inserts cavity into a base part.

 Scales a model by a specified factor.

 Applies draft to a selected surface.

 Inserts a split line feature.

 Creates parting lines to separate core & cavity surfaces.

 Finds & creates mold shut-off surfaces.

 Creates a planar surface from a sketch or a set of edges.

 Knits surfaces together.

 Inserts ruled surfaces from edges.

 Creates parting surfaces between core & cavity surfaces.

 Creates multiple bodies from a single body.

 Inserts a tooling split feature.

 Creates parting surfaces between the core & cavity.

 Inserts surface body folders for mold operation.

SELECTION FILTERS Toolbar

 Turns selection filters on and off.

 Clears all filters.

 Selects all filters.

 Inverts current selection.

 Allows selection of edges only.

 Allows selection filter for vertices only.

 Allows selection of faces only.

 Adds filter for Surface Bodies.

 Adds filter for Solid Bodies.

 Adds filter for Axes.

 Adds filter for Planes.

 Adds filter for Sketch Points.

 Allows selection for sketch only.

 Adds filter for Sketch Segments.

 Adds filter for Midpoints.

 Adds filter for Center Marks.

 Adds filter for Centerline.

 Adds filter for Dimensions and Hole Callouts.

 Adds filter for Surface Finish Symbols.

 Adds filter for Geometric Tolerances.

 Adds filter for Notes / Balloons.

 Adds filter for Weld Symbols.

 Adds filter for Weld beads.

 Adds filter for Datum Targets.

 Adds filter for Datum feature only.

 Adds filter for blocks.

 Adds filter for Cosmetic Threads.

Adds filter for Dowel pin symbols.

 Adds filter for connection points.

 Adds filter for routing points.

SOLIDWORKS Add-Ins Toolbar

 Loads/unloads CircuitWorks add-in.

 Loads/unloads the Design Checker add-in.

 Loads/unloads the PhotoView 360 add-in.

 Loads/unloads the Scan-to-3D add-in.

 Loads/unloads the SOLIDWORKS Motions add-in.

 Loads/unloads the SOLIDWORKS Routing add-in.

 Loads/unloads the SOLIDWORKS Simulation add-in.

 Loads/unloads the SOLIDWORKS Toolbox add-in.

 Loads/unloads the SOLIDWORKS TolAnalysis add-in.

 Loads/unloads the SOLIDWORKS Flow Simulation add-in.

 Loads/unloads the SOLIDWORKS Plastics add-in.

 Loads/unloads the SOLIDWORKS MBD SNL license.

FASTENING FEATURES Toolbar

 Creates a parameterized mounting boss.

 Creates a parameterized snap hook.

 Creates a groove to mate with a hook feature.

 Uses sketch elements to create a vent for air flow.

 Creates a lip/groove feature.

SCREEN CAPTURE Toolbar

 Copies the current graphics window to the clipboard.

 Records the current graphics window to an AVI file.

 Stops recording the current graphics window to an AVI file.

EXPLODE LINE SKETCH Toolbar

 Adds a route line that connect entities.

 Adds a jog to the route lines.

LINE FORMAT Toolbar

 Changes layer properties.

 Changes the current document layer.

 Changes line color.

 Changes line thickness.

 Changes line style.

 Hides / Shows a hidden edge.

 Changes line display mode.

Did you know??

* Ctrl+Q will force a rebuild on all features of a part.

* Ctrl+B will rebuild the feature being worked on and all of its dependents.

2D-To-3D Toolbar

 Makes a Front sketch from the selected entities.

 Makes a Top sketch from the selected entities.

 Makes a Right sketch from the selected entities.

Makes a Left sketch from the selected entities.

Makes a Bottom sketch from the selected entities.

Makes a Back sketch from the selected entities.

Makes an Auxiliary sketch from the selected entities.

Creates a new sketch from the selected entities.

Repairs the selected sketch.

Aligns a sketch to the selected point.

Creates an extrusion from the selected sketch segments, starting at the selected sketch point.

Creates a cut from the selected sketch segments, optionally starting at the selected sketch point.

ALIGN Toolbar

Aligns the left side of the selected annotations with the leftmost annotation.

Aligns the right side of the selected annotations with the rightmost annotation.

Aligns the top side of the selected annotations with the topmost annotation.

Aligns the bottom side of the selected annotations with the lowermost annotation.

Evenly spaces the selected annotations horizontally.

Evenly spaces the selected annotations vertically.

Centrally aligns the selected annotations horizontally.

Centrally aligns the selected annotations vertically.

Compacts the selected annotations horizontally.

Compacts the selected annotations vertically.

Creates a group from the selected items.

Deletes the grouping between these items.

Aligns & groups selected dimensions along a line or an arc.

Aligns & groups dimensions at uniform distances.

Evenly spaces selected dimensions.

Aligns collinear selected dimensions.

Aligns stagger selected dimensions.

SOLIDWORKS MBD Toolbar

Captures 3D view.

Manages 3D PDF templates.

Creates shareable 3D PDF presentations.

Toggles dynamic annotation views.

MACRO Toolbar

Runs a Macro.

Stops Macro recorder.

Records (or pauses recording of) actions to create a Macro.

Launches the Macro Editor and begins editing a new macro.

Opens a Macro file for editing.

Creates a custom macro.

SMARTMATES icons

Concentric & Coincident 2 circular edges.

Concentric 2 cylindrical faces.

Coincident 2 linear edges.

Coincident 2 planar faces.

Coincident 2 vertices.

Coincident 2 origins or coordinate systems.

TABLE Toolbar

 Adds a hole table of selected holes from a specified origin datum.

 Adds a Bill of Materials.

 Adds a revision table.

 Displays a Design table in a drawing.

 Adds a weldments cuts list table.

 Adds a Excel based of Bill of Materials

 Adds a weldment cut list table.

REFERENCE GEOMETRY Toolbar

 Adds a reference plane.

 Creates an axis.

 Creates a coordinate system.

 Adds the center of mass.

 Specifies entities to use as references using SmartMates.

SPLINE TOOLS Toolbar

 Inserts a point to a spline.

 Displays all points where the concavity of selected spline changes.

 Displays minimum radius of selected spline.

 Displays curvature combs of selected spline.

 Reduces numbers of points in a selected spline.

 Adds a tangency control.

 Adds a curvature control.

 Adds a spline based on selected sketch entities & edges.

 Displays the spline control polygon.

ANNOTATIONS Toolbar

 Inserts a note.

 Inserts a surface finish symbol.

 Inserts a new geometric tolerancing symbol.

 Attaches a balloon to the selected edge or face.

 Adds balloons for all components in selected view.

 Inserts a stacked balloon.

 Attaches a datum feature symbol to a selected edge / detail.

 Inserts a weld symbol on the selected edge / face / vertex.

 Inserts a datum target symbol and / or point attached to a selected edge / line.

 Selects and inserts block.

 Inserts annotations & reference geometry from the part / assembly into the selected.

 Adds center marks to circles on model.

 Inserts a Centerline.

 Inserts a hole callout.

 Adds a cosmetic thread to the selected cylindrical feature.

 Inserts a Multi-Jog leader.

 Selects a circular edge or and arc for Dowel pin symbol insertion.

 Adds a view location symbol.

 Inserts latest version symbol.

 Adds a cross hatch patterns or solid fill.

 Adds a weld bead caterpillar on an edge.

 Adds a weld symbol on a selected entity.

 Inserts a revision cloud.

 Inserts a magnetic line.

 Hides/shows annotation.

DRAWINGS Toolbar

 Updates the selected view to the model's current stage.

 Creates a detail view.

 Creates a section view.

 Inserts an Alternate Position view.

 Unfolds a new view from an existing view.

 Generates a standard 3-view drawing (1st or 3rd angle).

 Inserts an auxiliary view of an inclined surface.

 Adds an Orthogonal or Named view based on an existing part or assembly.

 Adds a Relative view by two orthogonal faces or planes.

 Adds a Predefined orthogonal projected or Named view with a model.

 Adds an empty view.

 Adds vertical break lines to selected view.

 Crops a view.

 Creates a Broken-out section.

QUICK SNAP Toolbar

 Snap to points.

 Snap to center points.

 Snap to midpoints.

 Snap to quadrant points.

 Snap to intersection of 2 curves.

 Snap to nearest curve.

 Snap tangent to curve.

 Snap perpendicular to curve.

 Snap parallel to line.

 Snap horizontally / vertically to points.

 Snap horizontally / vertically.

 Snap to discrete line lengths.

 Snap to angle.

LAYOUT Toolbar

 Creates the assembly layout sketch.

 Sketches a line.

 Sketches a corner rectangle.

 Sketches a circle.

 Sketches a 3-point arc.

 Rounds a corner.

 Trims or extends a sketch.

 Adds sketch entities by offsetting faces, edges curves.

 Mirrors selected entities about a centerline.

 Adds a relation.

 Creates a dimension.

 Displays / Deletes geometric relations.

 Makes a new block.

 Edits the selected block.

 Inserts a new block to the sketch or drawing.

 Adds / Removes sketch entities to / from a block.

 Saves the block to a file.

 Explodes the selected block.

 Creates a new part from a layout sketch block.

 Positions 2 components relative to one another.

CURVES Toolbar

 Projects sketch onto selected surface.

 Inserts a split line feature.

 Creates a composite curve from selected edges, curves and sketches.

 Creates a curve through free points.

 Creates a 3D curve through reference points.

 Helical curve defined by a base sketch and shape parameters.

VIEW Toolbar

 Displays a view in the selected orientation.

 Reverts to previous view.

 Redraws the current window.

 Zooms out to see entire model.

 Zooms in by dragging a bounding box.

 Zooms in or out by dragging up or down.

 Zooms to fit all selected entities.

 Dynamic view rotation.

 Scrolls view by dragging.

 Displays image in wireframe mode.

 Displays hidden edges in gray.

 Displays image with hidden lines removed.

 Controls the visibility of planes.

 Controls the visibility of axis.

 Controls the visibility of parting lines.

 Controls the visibility of temporary axis.

 Controls the visibility of origins.

 Controls the visibility of coordinate systems.

 Controls the visibility of reference curves.

 Controls the visibility of sketches.

 Controls the visibility of 3D sketch planes.

 Controls the visibility of 3D sketch.

 Controls the visibility of all annotations.

 Controls the visibility of reference points.

 Controls the visibility of routing points.

 Controls the visibility of lights.

 Controls the visibility of cameras.

 Controls the visibility of sketch relations.

 Changes the display state for the current configuration.

 Rolls the model view.

 Turns the orientation of the model view.

 Dynamically manipulate the model view in 3D to make selection.

 Changes the display style for the active view.

 Displays a shade view of the model with its edges.

 Displays a shade view of the model.

 Toggles between draft quality & high quality HLV.

 Cycles through or applies a specific scene.

 Views the models through one of the model's cameras.

 Displays a part or assembly w/different colors according to the local radius of curvature.

 Displays zebra stripes.

 Displays a model with hardware accelerated shades.

 Applies a cartoon affect to model edges & faces.

 Views simulations symbols.

TOOLS Toolbar

 Calculates the distance between selected items.

 Adds or edits equation.

 Calculates the mass properties of the model.

 Checks the model for geometry errors.

 Inserts or edits a Design Table.

 Evaluates section properties for faces and sketches that lie in parallel planes.

 Reports Statistics for this Part/Assembly.

 Deviation Analysis.

 Runs the SimulationXpress analysis wizard powered by SOLIDWORKS Simulation.

 Checks the spelling.

 Import diagnostics.

 Runs the DFMXpress analysis wizard.

 Runs the SOLIDWORKS FloXpress analysis wizard.

ASSEMBLY Toolbar

 Creates a new part & inserts it into the assembly.

 Adds an existing part or sub-assembly to the assembly.

 Creates a new assembly & inserts it into the assembly.

 Turns on/off large assembly mode for this document.

Hides / shows model(s) associated with the selected model(s).

Toggles the transparency of components.

Changes the selected components to suppressed or resolved.

Inserts a belt.

Toggles between editing part and assembly.

 Smart Fasteners.

 Positions two components relative to one another.

 External references will not be created.

 Moves a component.

 Rotates an un-mated component around its center point.

 Replaces selected components.

 Replaces mate entities of mates of the selected components on the selected Mates group.

 Creates a New Exploded view.

 Creates or edits explode line sketch.

 Interference detection.

 Shows or Hides the Simulation toolbar.

 Patterns components in one or two linear directions.

 Patterns components around an axis.

 Sets the transparency of the components other than the one being edited.

 Sketch driven component pattern.

 Pattern driven component pattern.

 Curve driven component pattern.

 Chain driven component pattern.

 SmartMates by dragging & dropping components.

 Checks assembly hole alignments.

Mirrors subassemblies and parts.

SOLIDWORKS Quick-Guide©
STANDARD Keyboard Shortcuts

Rotate the model

* Horizontally or Vertically: _____ Arrow keys

* Horizontally or Vertically 90°: _____ Shift + Arrow keys

* Clockwise or Counterclockwise: _____ Alt + left or right Arrow

* Pan the model: _____ Ctrl + Arrow keys

* Zoom in: _____ Z (shift + Z or capital Z)

* Zoom out: _____ z (lower case z)

* Zoom to fit: _____ F

* Previous view: _____ Ctrl+Shift+Z

View Orientation

* View Orientation Menu: _____ Space bar

* Front: _____ Ctrl+1

* Back: _____ Ctrl+2

* Left: _____ Ctrl+3

* Right: _____ Ctrl+4

* Top: _____ Ctrl+5

* Bottom: _____ Ctrl+6

* Isometric: _____ Ctrl+7

Selection Filter & Misc.

* Filter Edges: _____ e

* Filter Vertices: _____ v

* Filter Faces: _____ x

* Toggle Selection filter toolbar: _____ F5

* Toggle Selection Filter toolbar (on/off): _____ F6

* New SOLIDWORKS document: _____ F1

* Open Document: _____ Ctrl+O

* Open from Web folder: _____ Ctrl+W

* Save: _____ Ctrl+S

* Print: _____ Ctrl+P

* Magnifying Glass Zoom _____ g

* Switch between the SOLIDWORKS documents _____ Ctrl + Tab

Function Keys

F1 _____ SW-Help

F2 _____ 2D Sketch

F3 _____ 3D Sketch

F4 _____ Modify

F5 _____ Selection Filters

F6 _____ Move (2D Sketch)

F7 _____ Rotate (2D Sketch)

F8 _____ Measure

F9 _____ Extrude

F10 _____ Revolve

F11 _____ Sweep

F12 _____ Loft

Sketch

C _____ Circle

P _____ Polygon

E _____ Ellipse

O _____ Offset Entities

Alt + C _____ Convert Entities

M _____ Mirror

Alt + M _____ Dynamic Mirror

Alt + F _____ Sketch Fillet

T _____ Trim

Alt + X _____ Extend

D _____ Smart Dimension

Alt + R _____ Add Relation

Alt + P _____ Plane

Control + F _____ Fully Define Sketch

Control + Q _____ Exit Sketch